T0140709

Ulf Martin Engel

On Quantum Chaos, Stochastic Webs and Localization in a Quantum Mechanical Kick System

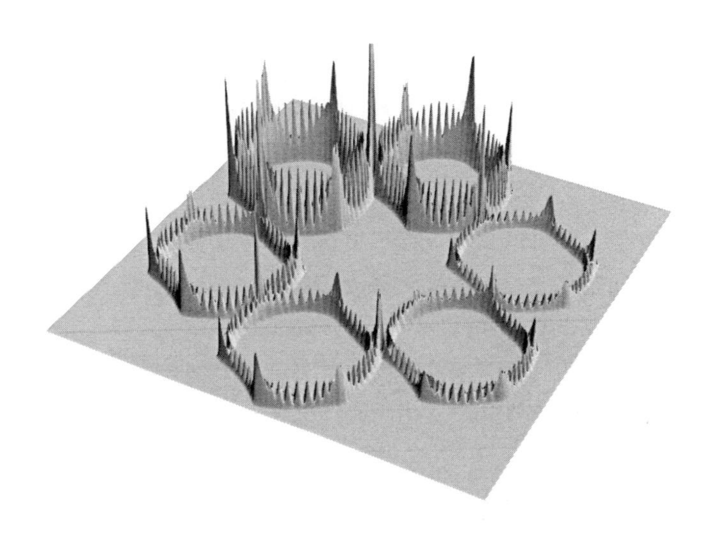

λογος

Reihe Nichtlineare und Stochastische Physik
herausgegeben von:
Prof. Dr. Lutz Schimansky-Geier
Institut für Physik
Humboldt-Universität zu Berlin
Newtonstr. 15
D-12489 Berlin

email: alsg@physik.hu-berlin.de

Bibliografische Information der Deutschen Nationalbibliothek

Die Deutsche Nationalbibliothek verzeichnet diese Publikation in der Deutschen Nationalbibliografie; detaillierte bibliografische Daten sind im Internet über http://dnb.d-nb.de abrufbar.

ISBN 978-3-8325-1653-6
ISSN 1435-7151

Logos Verlag Berlin
Comeniushof, Gubener Str. 47,
10243 Berlin
Tel.: +49 030 42 85 10 90
Fax: +49 030 42 85 10 92
INTERNET: http://www.logos-verlag.de

Für Helene Engel

Contents

Introduction

Quantum chaos — is there any?

JOSEPH FORD
Directions in Chaos Vol. 2 [For88]

On Quantum Chaos

Quantum chaos is a controversial subject [SZ94]. On the basis of the correspondence principle one might be tempted to take quantum chaos for granted, because classical mechanics is supposed to be a special limiting case of quantum mechanics, and clearly there are classical dynamical systems that are known to exhibit chaos — cf. for example [AFH95, KH95a] and the comprehensive lists of references therein.

But this approach is problematic: while stating that in the limit $\hbar = 0$ the quantum dynamics transforms into the classical dynamics of the system, the correspondence principle does not say *how* this quantum-classical transition comes about. There is no well-defined algorithm that maps the two dynamical theories onto each other; the semiclassical limit $\hbar \to 0$ of quantum mechanics is mathematically singular [Ber01]. With such an algorithm, given a classically chaotic system, it would be possible to uniquely identify those properties of the corresponding quantum system that mark its quantum chaoticity, based on the well-established concepts of classical chaos theory: LIAPUNOV exponents, (fractal) dimensions, entropies, etc. Conversely, since there is no such algorithm and the concepts of chaotic dynamics originate in the theory of classical mechanics, until today nobody can define in a straightforward way what makes a quantum system chaotic and thus what quantum chaos really is. With such a definition it would be possible to identify a quantum system as chaotic without having to discuss its classical mechanics beforehand.

Another aspect of the same problem is the fact that quantum mechanical phase space is quantized into cells of finite size, given by $\hbar$. As a consequence, the concept of sensitive dependence on initial conditions — an essential ingredient of the theory of classical chaos — cannot be applied to quantum mechanics in a straightforward way. What is more, if the (region of) phase space under consideration is of finite volume $\mathcal{A}$ then the corresponding quantum dynamics effectively evolves on $\mathcal{A}/\hbar$ different phase space cells only, which makes the HILBERT space effectively finite-dimensional, brings about quasiperiodic dynamics and thus excludes chaos in the classical sense [KW96]. This is to be contrasted with the continuum of phase space points in any nontrivial classical phase space and illustrates again the fundamental differences between the two dynamical theories.

Note that the above reasoning leading to the absence of classical chaos in quantum mechanics is also based on the *linearity* of the SCHRÖDINGER equation; the canonical equations of motion of classical mechanics, on the other hand, are in general *nonlinear*. BERRY considers this distinction to be "the ultimate reason for the absence of quantum chaos" [Ber89].

To give an example, since the days of POINCARÉ the three-body problem [Bru94] is a paradigm of classically chaotic nonlinear dynamics — see [Klu97, Hip97] for some recent studies of classical molecular three-body systems and [Con02] for applications in astronomy. However, its quantum counterparts, for example the H_2^+ ion and the He atom, can be analyzed with relative ease using conventional quantum techniques [Nol02], their eigenvalues and eigenfunctions are obtained accurately, and from this point of view it is difficult to see why these apparently well-behaved quantum three-body problems should be examples of quantum chaos.

Nevertheless, motivated by random matrix theory, one also has the BOHIGAS-GIANNONI-SCHMIT conjecture [BGS84] which claims that *there are* manifestations of classical chaos in quantum energy level spacing distributions: classical integrability is associated with POISSON distributions and level clustering, whereas for (hard) classical chaos WIGNER distributions and level repulsion are expected. The conjecture is yet unproven, but there is much evidence for its validity [Stö99]. — Similarly, there is the general conjecture that classically chaotic systems with diffusive dynamics are characterized by quantum suppression of diffusion [CCIF79, Ber89]: another quantum fingerprint of classical chaos.

A pragmatic way out of this dilemma is to adopt a working definition of quantum chaos that avoids the correspondence principle altogether, and to define *quantum chaos* as the study of quantum systems whose classical counterparts exhibit classical chaos [CGS93]. This definition — often also termed *quantum chaology* [Ber89], trying to avoid the impression that the quantum dynamics might be chaotic per se — appears to be accepted by more and more researchers in the field (see for example [GVZJ91, CC95, Stö99, Haa01]). Note that this definition makes no ad hoc statement about the actual properties of the quantum systems, other than their classical chaoticity.

With the above definition, quantum chaos integrates smoothly into the development of the theory of nonlinear dynamics, or chaos theory, as it has taken place since the middle of the last century. Using both the increasing theoretical understanding of the dynamics and computers which constantly became more powerful, a lot of knowledge about classical dynamical systems has been accumulated. The natural next step is to look for the quantum manifestations of classical chaos. With the present study I want to make a contribution to this development by investigating the quantum mechanics of a system the chaotic classical mechanics of which has been studied in some detail in the past: the *kicked harmonic oscillator.*

Among the classically chaotic systems, those that exhibit *hard chaos* take an exceptional role. Hard chaotic systems are characterized by complete absence of regular regions in classical phase space: all orbits, including the periodic orbits, are unstable. Once all periodic orbits of such a system are known, the quantum mechanics of the system can be studied in the semiclassical approximation $\hbar \approx 0$ using, for example, GUTZWILLER's trace formulas [Gut90, Gut91]. This theory is especially successful for bound dynamics [BB97], but has been extended to scattering dynamics as well [Wir99, CAM$^+$03].

Billiards represent a frequently studied class of model systems for semiclassical quantization. Inside a billiard, the dynamics is free, and therefore easy to solve classically and quantum mechanically. The billiard boundaries then account for the nontrivial part of the dynamics; in particular, defocusing boundaries can be chosen such that hard chaotic classical dynamics with a complete symbolic dynamics is obtained, and the methods of semiclassical quantization can be applied [Hor93, Jun97, Hau00]. Two-dimensional quantum billiard systems have the striking additional advantage of being accessible to direct experimental verification of the numerical and analytical results: these systems can be modelled by microwave resonators, also known as microwave billiards, which allow to measure the wave function of the system using a macroscopic experimental setup; this method is based on the fact that for two-dimensional systems, both the SCHRÖDINGER equation for the quantum billiard and the MAXWELL equations for the microwave resonator give rise to the same HELMHOLTZ equation with essentially identical boundary conditions [Stö99].

The generic situation in nonlinear Hamiltonian dynamical systems, on the other hand, is *soft* or *weak chaos*. Here, the phase portrait is mixed and consists of regions of regular and irregular or chaotic dynamics at the same time. Often such systems can be analyzed from a perturbation point of view, starting from a completely integrable system to which an increasingly strong perturbation is added, rendering the system nonintegrable and leading to growing phase space regions of chaotic dynamics.

The kicked harmonic oscillator is a typical representative of this class of systems, the integrable component being the well-known harmonic oscillator to which an impulsive time-periodic forcing is added. The kicked harmonic oscillator is *un*typical, however, in that respect that it can generate a very special kind of classical phase portraits, namely *stochastic webs*. These extended, periodic structures in phase space contain both bounded cells of regular dynamics and infinitely long channels of chaotic, or stochastic, motion. Due to the degeneracy of the harmonic oscillator, its classical dynamics cannot be analyzed in terms of the elegant standard theory of perturbed integrable systems, KAM theory [LL92, JS98]. Rather, a theory of its own had to be developed in order to understand the stochastic webs that are generated by this system [ZSUC91].

In contrast to classically hard chaotic systems, the quantum mechanics of systems exhibiting weak chaos is much less understood from a general point of view. The methods of semiclassical quantization via periodic orbit theory, as mentioned above, cannot be applied in a straightforward way. Therefore, a significant part of the quantum investigations of weakly chaotic systems have been performed using quantum mechanical methods without $\hbar$-related approximations, allowing to consider both the full quantum regime and the semiclassical case [Zas85, Hei92, CC95, BR97] (and references therein). The same approach is followed in the present study.

Beyond the lack of a suitable semiclassical theory, studying the quantum dynamics of the kicked harmonic oscillator is further complicated by the quantum fingerprints of its diffusive dynamics in the channels of the classical webs: in cases of resonance between the harmonic oscillator's eigenfrequency and the frequency of the perturbation, the evolving quantum states can extend quite rapidly into large areas in phase space, which greatly increases the numerical effort needed to study the system. It is interesting to note how this numerical complexity of the system has been assessed by some prominent researchers in this field:

> "The numerical investigation of this model [the quantum kicked harmonic oscillator] is much more difficult than in the kicked HARPER case, since the dynamics leads to spreading in the whole (p, x)-plane... This seems to be the reason why there were practically no attempts to investigate the quantum dynamics of [this] model" [SS92].

While this remark is slightly outdated by now, it is still true that the kicked harmonic oscillator is a seldom studied object with respect to quantum chaos. To put that into perspective, the quantum dynamics of the paradigmatic example of nonlinear Hamiltonian kick dynamics, the *kicked rotor* which gives rise to the *standard map*, has been studied in hundreds of publications, while for the kicked harmonic oscillator there are still only a few in comparison. Among other — more theoretical — reasons, this is mainly due to the fact that it is *much* easier to numerically model the kicked rotor than the kicked harmonic oscillator; this might become clearer in chapter 5, where both systems are studied along parallel lines.

Outline of this Study

Chapter 1 is the exposition of the principal model system used in the present study, the kicked harmonic oscillator. The Hamiltonian of the model system is introduced and some sample applications to physical systems are discussed. After scaling, the POINCARÉ map describing the classical dynamics over one period of the excitation is derived. In cases of resonance, this *web map* is then used to numerically generate classical stochastic webs in phase space and to explain some of their most important properties, such as the topology and the symmetries of the webs, and the diffusive dynamics in the channels. In cases of nonresonance, the discrete map does not give rise to web-like structures and the dynamics is typically diffusive. — While this chapter does not contain much original material, it serves two important purposes. First, in comparison with the existing literature on the subject it gives a more readable account of — some important aspects of — the theory of stochastic webs. Second, it provides the basic classical results with which the quantum results of the following chapters are to be compared.

In chapter 2 I describe the quantum formulation of the problem. FLOQUET theory is employed to derive the *quantum map* which is the quantum analogue of the classical POINCARÉ map. The similarities and the fundamental differences of these two mappings are discussed.

Studying the quantum analogue of stochastic webs requires the iteration of the quantum map for a very large number of times. This cannot be done analytically; it can only be accomplished using numerical means. Since the numerical effort to be spent for a single iteration of the quantum map is *much* larger than for one iteration of the classical POINCARÉ map, it is important to select the most efficient algorithm that is available. In chapter 3 I present and compare three numerical methods that can be used to implement the quantum map on a computer. It turns out that representing the FLOQUET operator in the eigenbasis of the (unkicked) harmonic oscillator is better suited for the present study than using conventional finite differences methods.

Chapters 4 and 5 contain the core results of this study: for several parameter combinations, I iterate the quantum map very often, compare the resulting sequences of quantum states with the corresponding classical dynamics, and give analytical explanations for the observations. For this comparison, a technique is needed that is reviewed in appendix A: the theory of *quantum phase space distribution functions* can be used to define a quantum analogue of classical phase space; the quantum states are then described equivalently in terms of distribution functions in this quantum phase space that take the role of the classical LIOUVILLE distribution. In this way a direct comparison of the classical and quantum results becomes possible.

In chapter 4, the quantum dynamics in the resonance cases is studied numerically, with the result that there exist *quantum stochastic webs* if and only if there are classical stochastic webs. The quantum webs resemble their classical counterparts as closely as allowed by the value of $\hbar$, and generically the dynamics in the channels of the webs is diffusive, as in the classical case. In other words, in the quantum webs the dynamics is as classical as can be expected from a quantum wave packet. These numerical findings are then explained analytically using an argument that relies on exploiting the symmetries of the FLOQUET operator and on constructing groups of mutually commuting translation operators in the phase plane that also commute with the FLOQUET operator [BR95].

The complementary case of nonresonance is dealt with in chapter 5. It turns out that in this case the dynamics is similar — in a well-defined way — to the dynamics of the quantum kicked rotor, which is known for some time already to exhibit quantum suppression of diffusion, or *quantum localization* [CCIF79, FGP82]. In chapter 5 I show both numerically and analytically that the same is true for the model system considered here: the nonresonant quantum kicked harmonic oscillator is ANDERSON-localized.

Finally, appendix B contains some technical material needed for the proof of localization in chapter 5, and appendix C is a collection of sample quantum phase portraits of the dynamics of the kicked harmonic oscillator, both in cases of resonance and nonresonance.

Chapter 1

Stochastic Webs in Classical Mechanics

Oh what a tangled web we weave ...

Marmion
SIR WALTER SCOTT

A *stochastic web* is a web-like structure in the phase space of a classical dynamical system (normally with a single degree of freedom) — in the present case: of the kicked harmonic oscillator. This structure completely covers the phase space and possesses certain (approximate) symmetry properties: rotational and, in some cases, translational invariance. The dynamics on the web can be given, for example, by ZASLAVSKY's *web map*. There are two distinct types of dynamics in the web: regular motion in the meshes of the web and irregular, stochastic motion in the channels; the second provides the motivation for characterizing the web as "stochastic".

Stochastic webs can be obtained by considering suitable discrete mappings derived from the classical dynamics of the kicked harmonic oscillator. I therefore begin the exposition of the subject in section 1.1 by introducing and briefly discussing this model system. Then, in section 1.2, I make use of the kicked harmonic oscillator to generate the stochastic webs that are in the focus of interest here and provide the classical background for the entire present study. Finally, in section 1.3, I address the complementary case characterized by nonexistence of stochastic webs.

1.1 The Kicked Harmonic Oscillator

There are several different ways to motivate the discussion of the kicked harmonic oscillator. In each case it describes — more or less approximately — certain physical systems. In this section I discuss the most common of these model systems that finds its application for example in plasma physics [AS83]. ZASLAVSKY and co-workers discuss this approach to the problem in some detail as well [ZSUC91].

Consider the dynamics of a charged point particle in a homogeneous stationary magnetic field $\vec{B}$ and a time-dependent electric field $\vec{E}$ orthogonal to $\vec{B}$:

$$\vec{B} = B_0 \, \vec{e}_z \tag{1.1a}$$

$$\vec{E}(x,t) = E_0 \sum_{n=-\infty}^{\infty} \sin(kx - n\omega t) \, \vec{e}_x \tag{1.1b}$$

with real-valued constants B_0, E_0, k, ω. The electric field can be interpreted as a wave packet that is periodic both in space and time, and that consists of FOURIER-like components $\sin(kx - n\omega t)$ each of which propagates in x-direction and contributes to the complete packet with equal weight.

Field configurations of this type can also be used, for instance, to describe the beams of charged particles in storage rings under the influence of beam-beam-interactions [Hel83, Ten83]. The applicability in cases like this becomes more obvious when the series in equation (1.1b) is reformulated in a certain way. Taking into account that

$$\sum_{n=-\infty}^{\infty} \sin(kx - n\omega t) = \sin kx \sum_{n=-\infty}^{\infty} \cos n\omega t = T \sin kx \sum_{n=-\infty}^{\infty} \delta(t - nT), \tag{1.2}$$

with the period $T = 2\pi/\omega$, one obtains

$$\vec{E}(x,t) = E_0 T \sin kx \sum_{n=-\infty}^{\infty} \delta(t - nT) \, \vec{e}_x. \tag{1.3}$$

This demonstrates that the electric field, although it is written as a *propagating* wave packet in equation (1.1b), can in fact be interpreted as a *standing* harmonic wave that is switched on stroboscopically at discrete points of time given by nT. At these times the charged particle is submitted to an impulsive force, similar to the situation one encounters in a particle accelerator or storage ring, where the particle beams intersect with each other at certain discrete points, namely in the interaction regions.

1.1.1 Newtonian Equations of Motion

Let the particle be of mass m_0 and charge q_0. Then the dynamics of the particle is governed by the Newtonian equation of motion

$$m_0\ddot{\vec{r}} = q_0\left(\vec{E} + \dot{\vec{r}} \times \vec{B}\right) \tag{1.4}$$

with the vector $\vec{r} = (x, y, z)^t$ in position space. Using the electromagnetic field (1.1) this yields for the components of $\vec{r}$:

$$m_0\ddot{x} = q_0B_0\dot{y} + q_0E_0T\sin kx \sum_{n=-\infty}^{\infty} \delta(t - nT) \tag{1.5a}$$

$$m_0\ddot{y} = -q_0B_0\dot{x} \tag{1.5b}$$

$$m_0\ddot{z} = 0. \tag{1.5c}$$

The dynamics in z-direction is trivial (rectilinear and uniform) and can thus be neglected. From equation (1.5b) one obtains

$$m_0\dot{y} = -q_0B_0x + \text{const.}, \tag{1.6}$$

where the constant can be set to zero without loss of generality (by an appropriate choice of the origin of the x-axis). Substitution into equation (1.5a) finally yields

$$\ddot{x} + \omega_0^2 x = \frac{q_0E_0T}{m_0}\sin kx \sum_{n=-\infty}^{\infty} \delta(t - nT), \tag{1.7}$$

the equation of motion of a *kicked harmonic oscillator* with eigenfrequency equal to the cyclotron frequency,

$$\omega_0 := \frac{q_0B_0}{m_0}, \tag{1.8}$$

which is essentially given by the magnetic field. The right hand side of equation (1.7) describes the impulsive force that is driving the oscillator; the strength of this driving is essentially determined by the amplitude of the electric field. The functional dependence on x of the driving force is specified by the *kick function* which in equation (1.7) is proportional to $\sin kx$. In subsection 1.1.4 I briefly return to the issue of choosing the kick function in a variety of other cases.

At the times nT the impulse — the *kick* — changes the momentum of the particle instantaneously, whereas its position remains unchanged. For all other times the dynamics is just that of an unperturbed (i.e. free) harmonic oscillator.

1.1.2 Canonical Formulation

In principle, the Newtonian equation of motion (1.7) allows a complete analysis of the classical dynamics of the kicked harmonic oscillator. Nevertheless, it is advantageous to use the Hamiltonian formulation of the problem instead, because many classical results can be derived in this formulation with much more ease (cf. section 1.2). What is more, for the discussion of the corresponding quantum problem the Hamiltonian operator is indispensable anyway.

The dynamics of a charged particle in an electromagnetic field can be described by the Hamiltonian [Gol80]

$$H(\vec{r},\vec{p},t) = \frac{1}{2m_0}\left\{\vec{p}-q_0\vec{A}(\vec{r},t)\right\}^2 + q_0\phi(\vec{r},t), \tag{1.9}$$

with the momentum $\vec{p}=(p_x,p_y,p_z)^t$ of the particle and the vector potential $\vec{A}$ and the scalar potential ϕ for the electromagnetic field. The potentials have to be chosen in such a way that the magnetic and electric fields are obtained via $\vec{B}=\vec{\nabla}\times\vec{A}$ and $\vec{E}=-\vec{\nabla}\phi-\partial/\partial t\,\vec{A}$. For the fields of equation (1.1), this is achieved, for example, by choosing

$$\vec{A}(\vec{r}) = B_0 x\,\vec{e}_y \tag{1.10a}$$

$$\phi(\vec{r},t) = \frac{E_0T}{k}\cos kx \sum_{n=-\infty}^{\infty}\delta(t-nT), \tag{1.10b}$$

such that the corresponding Hamiltonian becomes

$$H(\vec{r},\vec{p},t) = \frac{1}{2m_0}\left\{p_x^2+(p_y-q_0B_0x)^2+p_z^2\right\} + \frac{q_0E_0T}{k}\cos kx \sum_{n=-\infty}^{\infty}\delta(t-nT). \tag{1.11}$$

H is cyclic in y and z. As in the previous subsection it is clear that the dynamics in z-direction is that of a free particle and separates from the rest of the dynamics; therefore in the following I drop the z-dynamics altogether. Cyclicity of H in y means that p_y is a constant of motion. Since this constant enters the Hamiltonian via the term $p_y - q_0B_0x$ only, changing the value of p_y just results in a shift of the origin of the x-axis (cf. with the constant in equation (1.6)). Hence p_y can be set to zero without loss of generality. The remaining "essential" part of the Hamiltonian is

$$H(x,p_x,t) = \frac{1}{2m_0}p_x^2+\frac{1}{2}m_0\omega_0^2x^2+V_0T\cos kx\sum_{n=-\infty}^{\infty}\delta(t-nT), \tag{1.12}$$

with the parameter

$$V_0 \;:=\; \frac{q_0 E_0}{k} \tag{1.13}$$

which controls the amplitude of the kick; V_0 has the dimension of an energy. As from here on only the momentum p_x, conjugate to x, is of importance and no confusion with other momenta can arise, I now drop the index x.

Naturally, the two canonical equations that follow from equation (1.12),

$$\begin{aligned} \dot{x} &= \frac{\partial H}{\partial p} = \frac{1}{m_0}p \\ \dot{p} &= -\frac{\partial H}{\partial x} = -m_0\omega_0^2 x + kV_0 T \sin kx \sum_{n=-\infty}^{\infty} \delta(t-nT), \end{aligned} \tag{1.14}$$

can be combined to obtain once again the Newtonian equation of motion (1.7).

In order to minimize the number of parameters of the system I introduce dimensionless variables by the scaling transformation

$$k\,x \;\longmapsto\; x \tag{1.15a}$$

$$\frac{k}{m_0\omega_0}\,p \;\longmapsto\; p \tag{1.15b}$$

$$\omega_0\,t \;\longmapsto\; t \tag{1.15c}$$

$$\frac{k^2}{m_0\omega_0^2}\,H \;\longmapsto\; H. \tag{1.15d}$$

Similarly, the parameters V_0, T are replaced with dimensionless versions:

$$\begin{aligned} \frac{k^2 T}{m_0\omega_0}\,V_0 &\longmapsto V_0 \\ \omega_0\,T &\longmapsto T. \end{aligned} \tag{1.16}$$

The Hamiltonian under investigation then reads

$$\boxed{H(x,p,t) \;=\; \frac{1}{2}p^2 + \frac{1}{2}x^2 + V_0 \cos x \sum_{n=-\infty}^{\infty} \delta(t-nT).} \tag{1.17}$$

For simplicity, the function

$$V(x) \;=\; V_0 \cos x \tag{1.18}$$

is often referred to as the *kick potential*.[1] It is one of the advantages of the scaling used here that the two remaining parameters of the system, the (scaled) period T of the perturbation and its (scaled) amplitude V_0, concern only the kick part of the Hamiltonian. The unperturbed part of H after scaling is just a harmonic oscillator with — formally — unit mass and unit frequency.

Note that, while equation (1.12) reduces to the Hamiltonian of the CHIRIKOV-TAYLOR map [Chi79] in the limit $\omega_0 \to 0$, no analogous statement holds for the scaled version (1.17), since $\omega_0 \neq 0$ is a necessary condition for the scaling (1.15). The CHIRIKOV-TAYLOR map is an important standard example of classical kicked dynamical systems. I return to this map in subsection 5.1.1, where its bifurcation scenario and diffusive dynamics are described.

1.1.3 Discrete Dynamics — The Web Map

The dynamics that is generated by the Hamiltonian (1.17) can obviously be split into two parts, namely the trivial dynamics of a free harmonic oscillator between two successive kicks, and the kick dynamics itself. Because of the stroboscopic nature of the kick it makes sense to consider the dynamic variables x and p at those times only when kicks occur. Let

$$x_n := \lim_{t \nearrow nT} x(t) \tag{1.19a}$$

$$p_n := \lim_{t \nearrow nT} p(t) \tag{1.19b}$$

be the values of x and p immediately before the n-th kick. Then the discrete dynamics generated by (1.17) is given by the POINCARÉ map[2]

$$\begin{pmatrix} x_{n+1} \\ p_{n+1} \end{pmatrix} = M \begin{pmatrix} x_n \\ p_n \end{pmatrix} = M_{\text{free}}\, M_{\text{kick}} \begin{pmatrix} x_n \\ p_n \end{pmatrix} \tag{1.20a}$$

with

$$M_{\text{kick}} = M_{\text{kick}}(V_0) = \begin{pmatrix} 1 & 0 \\ V_0 \sin & 1 \end{pmatrix} \tag{1.20b}$$

$$M_{\text{free}} = M_{\text{free}}(T) = \begin{pmatrix} \cos T & \sin T \\ -\sin T & \cos T \end{pmatrix}, \tag{1.20c}$$

[1]The full driving term $V_0 \cos x \sum_{n=-\infty}^{\infty} \delta(t - nT)$ in equation (1.17) is correctly referred to as a potential only after scaling, i.e. using dimensionless time, as in the present case. Before scaling, this term has the dimension of energy over time, rather than energy.

[2]Alternatively, x and p immediately *after* the kicks could be considered. The resulting POINCARÉ map would be topologically conjugate to M as given by equations (1.20).

or

$$\boxed{\begin{aligned} x_{n+1} &= x_n \cos T + (p_n + V_0 \sin x_n) \sin T \\ p_{n+1} &= -x_n \sin T + (p_n + V_0 \sin x_n) \cos T. \end{aligned}} \tag{1.21}$$

Starting at the phase space point $(x_n, p_n)^t$, first the momentum is shifted by $V_0 \sin x_n$ due to the kick; then the new point $(x_n, p_n + V_0 \sin x_n)^t$ is submitted to harmonic rotation in phase space for a period of time of length T. Figure 1.1 gives a graphical account of this dynamics of the map.

In the following I pay particular attention to the *resonance cases*

$$T_{\text{res}} = \frac{P}{Q}\pi \quad \text{with} \quad P, Q \in \mathbb{N}. \tag{1.22}$$

The most important resonances are those for which there are exactly q kicks per period 2π of the unforced oscillator:

$$T_{\text{res}} = \frac{2\pi}{q} \quad \text{with} \quad q \in \mathbb{N}. \tag{1.23}$$

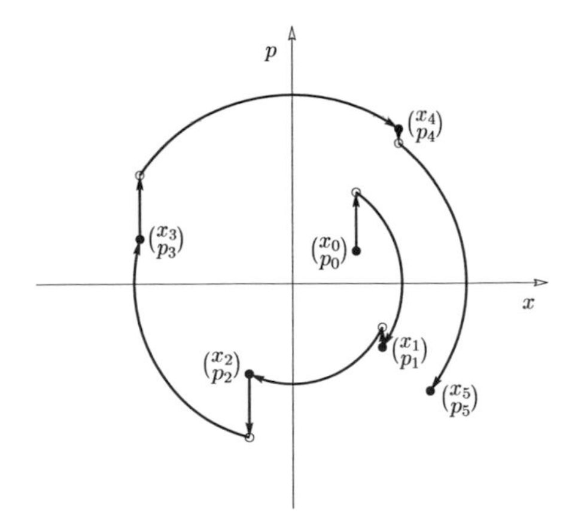

Figure 1.1: Schematic sketch of the dynamics of the map (1.21). The first five iterates of $(x_0, p_0)^t = (2, 1)^t$ are shown for $T = \pi/2$ and $V_0 = 2$. • und ∘ correspond to the situation immediately before and after the kicks, respectively.

It is just in these resonance cases that a stochastic web develops, as will be seen in section 1.2. In fact, webs with both translational and rotational symmetry can only occur for just a few special values of q; this is discussed in subsection 1.2.1. Due to this connection between stochastic webs and the map (1.21) in these particular cases of resonance, maps of this type are commonly called *web maps*. (See for example [ZSUC91], but note that there a different scaling is used.) In the formulation used here the web map is given by

$$M_q : \begin{cases} x_{n+1} = x_n \cos\frac{2\pi}{q} + (p_n + V_0 \sin x_n)\sin\frac{2\pi}{q} \\ p_{n+1} = -x_n \sin\frac{2\pi}{q} + (p_n + V_0 \sin x_n)\cos\frac{2\pi}{q}; \end{cases} \tag{1.24}$$

it depends on the two parameters $q \in \mathbb{N}$ and $V_0 \in \mathbb{R}_+$.

I now briefly discuss the web map for some specific values of q. The case $q = 1$ (i.e. $T = 2\pi$) corresponds to *cyclotron resonance*, where the frequency of the kick coincides with the frequency of the harmonic oscillator:

$$M_1 : \begin{cases} x_{n+1} = x_n \\ p_{n+1} = p_n + V_0 \sin x_n \end{cases} \tag{1.25}$$

This iteration is trivially solved by

$$\begin{aligned} x_n &= x_0 = \text{const.} \\ p_n &= p_0 + nV_0 \sin x_0 \,. \end{aligned} \tag{1.26}$$

In the context of the original model system of a particle moving within an electromagnetic field (subsection 1.1.1), the x-component of the particle position at times of kick is subject to uniform acceleration, as shown in figure 1.2.

For $q = 2$ ($T = \pi$) the dynamics is similar to that of equation (1.26): the map is given by

$$M_2 : \begin{cases} x_{n+1} = -x_n \\ p_{n+1} = -p_n - V_0 \sin x_n, \end{cases} \tag{1.27}$$

with the explicit solution

$$\begin{aligned} x_n &= (-1)^n x_0 \\ p_n &= (-1)^n (p_0 + nV_0 \sin x_0)\,. \end{aligned} \tag{1.28}$$

In this case of half-integer cyclotron resonance the particle's momentum at times of kick increases in quite the same way as for $q = 1$, but separately for odd and even n; the solution (1.28) can be obtained from equation (1.26) by successively reflecting the orbit points about the origin of phase space.

Because of their evident simplicity — in particular, the dynamics is confined to one-dimensional lines in phase space and no two-dimensional web structures can arise — I do not discuss the maps M_1 or M_2 any further.

For $q \geq 3$ the dynamics becomes much more interesting, and complicated structures in phase space can develop. The most important example is the web map for $q = 4$ ($T = \pi/2$), the dynamics of which has been sketched in figure 1.1:

$$M_4 : \quad \begin{cases} x_{n+1} &= p_n + V_0 \sin x_n \\ p_{n+1} &= -x_n. \end{cases} \tag{1.29}$$

Each kick is followed by a rotation in phase space through a quarter circle. The web maps for $q = 3$ and $q = 6$ ($T = 2\pi/3$ and $T = \pi/3$, respectively) are also of importance but not given here explicitly, because the corresponding formulae cannot be further simplified much beyond the form of equation (1.24).

The structures that evolve in phase space when M_q is iterated for a large number of times are discussed in section 1.2, with emphasis laid on the cases $q = 3, 4, 6$.

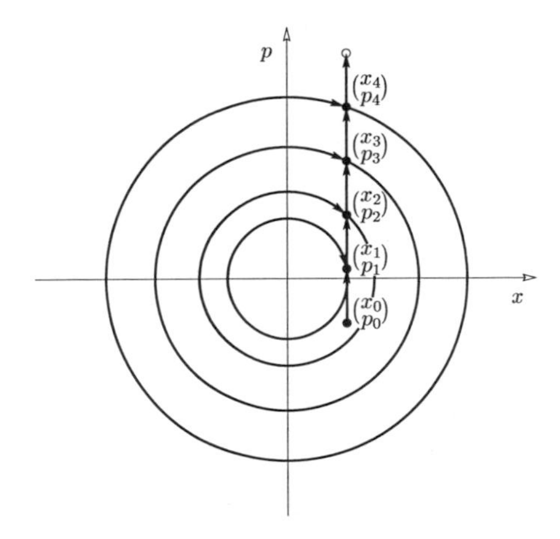

Figure 1.2: The same as figure 1.1, but for $T = 2\pi$, i.e. $q = 1$, and $(x_0, p_0)^t = (2, -1.5)^t$.

1.1.4 Generalized Web Maps

The choice of $\sin kx$ for the kick function as in equation (1.7) is not the only one that leads to the emergence of stochastic webs. Similarly, there are other systems than the harmonic oscillator that generate stochastic webs under the influence of appropriate periodic impulsive driving forces. It is the purpose of the present subsection to give some background information about how the web map (1.24) can be generalized in such a way that it still produces stochastic webs, and to provide some according references.

In the present study I concentrate the attention on the prototype of stochastic webs, which is generated by the harmonic oscillator submitted to sinusoidal kicks. The classical dynamics of this model has been studied extensively by ZASLAVSKY and co-workers; their combined effort has culminated in the review [ZSUC88] and the monograph [ZSUC91]; see also the references therein. Some other contributions to this field are [LW89, Low91, Low92, SJM92], where mainly the diffusive properties of the system are studied.

In a natural generalizing step one may pass from sinusoidal kick functions to more general periodic ones, where the latter include sinusoidal kicks as a special case. This path has been taken for example in [YP92, Hov92, Lam93, Low96, DK96]. Among the kick functions studied are the square wave and the saw tooth functions, as well as other piecewise linear functions. In all these cases variations of stochastic webs can be found. In [DA95] the sine function is again chosen for the kick, but here with a shift in the argument: $\sin k(x - x_0)$; depending on the choice of x_0, the stochastic web may or may not persist.

Even the condition of periodicity of the kick function can be dropped as in [Jun95], without necessarily destroying the web.

Another straightforward generalization of the web map concerns its extrapolation into phase spaces of higher dimensions (typically not more than 4). See for example [ZZN+89, Zas91, ASZ91, Jun95, Hau97]. As their lower-dimensional counterparts, these higher dimensional maps are most frequently studied with respect to their diffusive properties.

A more exotic discussion can be found in [LS87]. There, the relativistic analogue of the web map (1.24) is studied, with the result that the particle (in the underlying model system) can only be accelerated up to a critical energy. In other words, the corresponding web is finite. Still, the phase space structures close to the origin are the same as those generated by the map (1.24). The connection between the relativistic and nonrelativistic cases is further investigated in [HA99, AH00].

A limit beyond which no generalization can be pursued without spoiling the stochastic web is illustrated in [Vec95], where the effect of smoothing the δ-functions in the Hamiltonian (1.12) is investigated. The web is shown to be completely destroyed if the duration of the perturbation is arbitrarily short but finite, as opposed to being δ-shaped.

Until now I have only discussed the kicked *harmonic oscillator*.[3] Omitting the harmonic oscillator potential, it is also possible to consider a *free* particle under the influence of periodic kicks. This approach is chosen, for example, in [SK91] and [Sch93] for particles moving in one- and three-dimensional position space, respectively. Finally, rather than neglecting the oscillator potential in the Hamiltonian (1.12) one can also *add* a term, for example a centrifugal term that is obtained when discussing the radial motion of a particle in three dimensions [Hip94]. In all these cases stochastic webs can be found.

Summarizing, it is quite obvious that the emergence of stochastic webs is a much more general phenomenon than it appears at first sight; the kicked harmonic oscillator with sinusoidal kicks as in equation (1.17) is thus found to be a typical representative of a much larger class of systems. Having made the above remarks about the conditions of existence of these webs, I now turn to a detailed description of these objects of the present study.

1.2 The Stochastic Web

Two examples of stochastic webs that are generated by the web map (1.24) are shown in figure 1.3. The figures each show the first 10^5 iterates of the web map M_q for a single initial value. In figure 1.3a, for $q = 6$ ($T = \pi/3$), the iterates form a regular web-like pattern, which in this case is called a *kagome lattice* [ZSUC91]. The complete kagome lattice, which covers the entire phase plane, is obtained by iterating the map "infinitely often". As I discuss in more detail in the following subsections, the dynamics of the web map on this lattice is (weakly) chaotic, thus giving rise to the name *stochastic web*.

Certain (approximate) symmetry properties of the stochastic web for $q = 6$ are obvious: in this particular case the web is characterized by translational symmetry in two transversal directions (e.g. given by $(2\pi, \pm 2\pi/\sqrt{3})^t$) and by rotational symmetry with respect to several classes of 2-fold rotations, in addition to several reflection and glide reflection symmetries. In fact it turns out that the underlying *skeleton* of the web is even invariant under 2-fold, 3-fold and 6-fold

[3] It may be mentioned in passing that even the well-known Hénon maps — in their area-preserving, dissipative and logistic limit variants [Hén76, Hén83] — can be identified as Poincaré maps of a kicked harmonic oscillator with suitable kick functions [Hea92]. But in these particular cases no webs emerge.

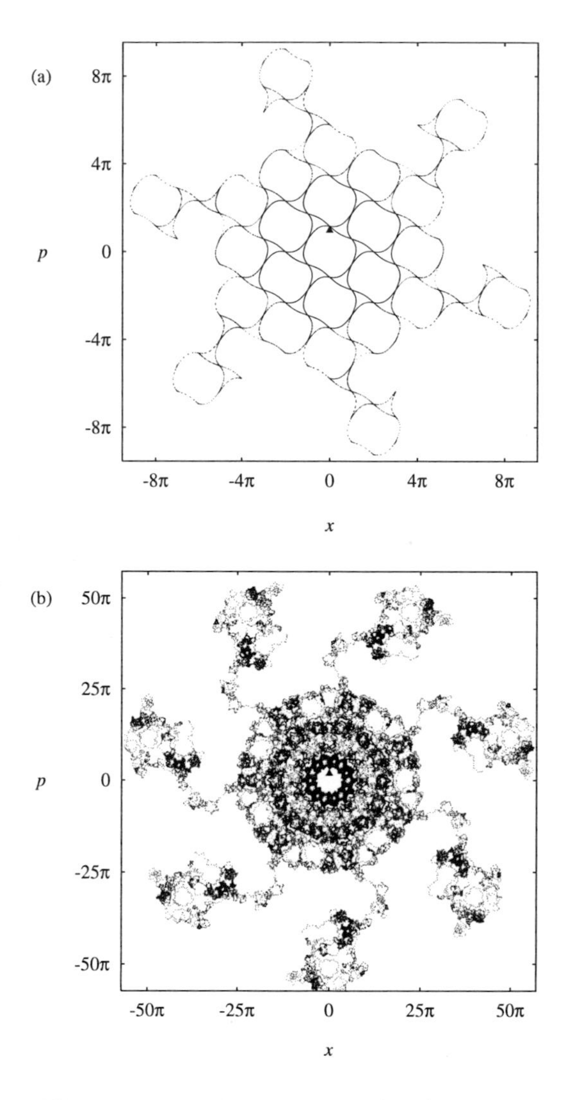

Figure 1.3: Phase portraits of the web map (1.24) in two cases of resonance. (a) A periodic stochastic web for $T = \pi/3$: the first 10^5 iterates of M_6 for the single initial value $(x_0, p_0)^t = (0, 3.62802)^t$ are plotted; (b) an aperiodic stochastic web for $T = 2\pi/7$: the same as in figure (a), but for M_7 and $(x_0, p_0)^t = (0, 9.5)^t$. $V_0 = 1.0$ for both figures; the initial values are marked by ▲.

rotations about suitably chosen centres of rotation. See subsection 1.2.2 for a rigorous definition of the skeletons of stochastic webs, and for more information on the symmetry groups of these skeletons.

In figure 1.3b, for $q = 7$ ($T = 2\pi/7$), the situation is different: although this phase portrait still reveals approximate rotational invariance with respect to the origin (rotations through the angle $2\pi/7$), there is no translational invariance as seen for example in figure 1.3a. In this sense, the phase portrait in figure 1.3b is less regular than that in figure 1.3a. Webs in phase space with just rotational symmetry — like the web for $q = 7$ — are called *aperiodic* stochastic webs, as opposed to the *periodic* stochastic webs exhibiting both rotational *and* translational symmetry, an example being the web for $q = 6$.

Having introduced periodic and aperiodic stochastic webs, the importance of the resonance condition (1.23) may now be illustrated by considering a value of $T = 2\pi/q$ with noninteger q. Figure 1.4 shows a phase portrait of the POINCARÉ map (1.21) for $T = 1.0$. Although this value of T is fairly close to $\pi/3 \approx 1.05$, the phase portraits for these two values of T are significantly different (cf. figure 1.3a). In the nonresonance case the phase space structures reveal much less regularity:

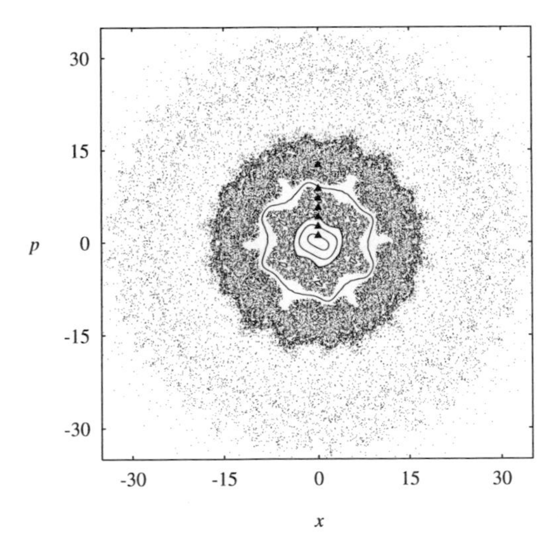

Figure 1.4: A phase portrait of the POINCARÉ map (1.21) for $T = 1.0$, i.e. in a case of nonresonance. The first $5 \cdot 10^3$ iterates of each of the initial values $(0, p_0)^t$ with $p_0 = 1.0/2.5/4.0/5.5/7.0/8.5$ are plotted, and $(0, 12.5)^t$ is iterated $3.5 \cdot 10^4$ times. $V_0 = 1.0$; the initial values are marked by ▲.

the whole phase plane — with the exception of a region near the origin where some invariant lines persist — forms a single dynamically connected chaotic region without an obvious inner structure, apart from variations in the density of points which are probably due to cantori.

In the following subsection 1.2.1 I present an argument for the fact that periodic webs, characterized by both rotational and translational symmetries, can develop not for all $q \in \mathbb{N}$ but only for a few specific values of q, namely for

$$q \in \mathcal{Q} := \{1, 2, 3, 4, 6\}; \tag{1.30}$$

in the present study I mainly concentrate the attention on these particular cases.

I then proceed in subsection 1.2.2 to the discussion of the skeleton, or "backbone", of the web and derive an equation that determines this overall structure of the web. Finally, in subsections 1.2.3 and 1.2.4 I briefly give an overview on what is known about the classical unbounded diffusive dynamics within the channels of the web that form around its skeleton; in particular I discuss the typical energy growth of a diffusing particle and the width of the channels of diffusive dynamics which is directly controlled by the kick strength V_0. The results presented in subsections 1.2.2 and 1.2.3 are mainly based on [ZSUC91].

1.2.1 Symmetries of the Web

In the theory of stochastic webs one is mainly interested in the symmetrical dynamical patterns that evolve in the phase plane as a result of the dynamics of the kicked harmonic oscillator. The types of symmetry that can be of interest here are determined by the symmetry groups admitted by the Hamiltonian (1.17). Its first part describes a rotation with unit angular velocity in the phase plane; as this rotation is interrupted by the kicks stroboscopically, with time constant $2\pi/q$, one can expect rotational symmetries through the angle $2\pi/q$ (and integer multiples thereof), i.e. q-fold rotation invariances. The second part of the Hamiltonian (1.17), describing the kicks, exhibits translational invariance with respect to shifts

$$x \longmapsto x + 2n\pi, \quad n \in \mathbb{Z}, \tag{1.31}$$

in x-direction. As a result, one has to look for phase space patterns that show both these rotational and translational symmetries at the same time, where (by the rotations) the translation invariance is not necessarily restricted to the x-direction any more.

Before taking into account the characteristic properties of the kicked harmonic oscillator (1.17), it is useful to determine what kinds of both rotationally and translationally invariant tilings of a two-dimensional plane are possible at all [Wey82]. See also [Lam93, LQ94, Jun95] for a more general exposition of this topic. The symmetry groups describing such tilings are the *planar space groups* or *wallpaper groups* [GS87].[4]

Consider a set of points $\mathcal{P}$ in the plane that is invariant both with respect to translations by the arbitrary period a in some direction and with respect to rotations through the angle α about the points of $\mathcal{P}$. Those α that fit into this setting are to be identified now. Depending on the value of α, there may be a single or two transversal directions of translation invariance.

Let A and B be elements of $\mathcal{P}$. Submitting these points to rotations through α about the points B and A, respectively, the points A' and B' are obtained (see figure 1.5), and because of the stipulated symmetries of $\mathcal{P}$, the *crystallographic condition* [Lax74, Che89]

$$\cos\alpha = \frac{1-n}{2} \quad \text{with suitable } n \in \mathbb{Z} \tag{1.32}$$

must be satisfied. This implies that $n \in \{-1, 0, 1, 2, 3\}$ and therefore $\alpha = 2\pi/q$ with $q \in \mathcal{Q}$. The case $q = 1$ is trivial in the sense that here no rotation is performed at all. Similarly, $q = 2$ is not really interesting as the corresponding rotations are through the angle π, such that the resulting pattern in the plane just consists of parallel lines of equidistant points. (Compare with the discussion

[4] *Space groups* is the general term used with respect to periodic (translation invariant) and rotation invariant patterns in arbitrary dimensions. Because of the most important application of these groups, in crystallography, they are also referred to as the *crystallographic groups* [Lax74].

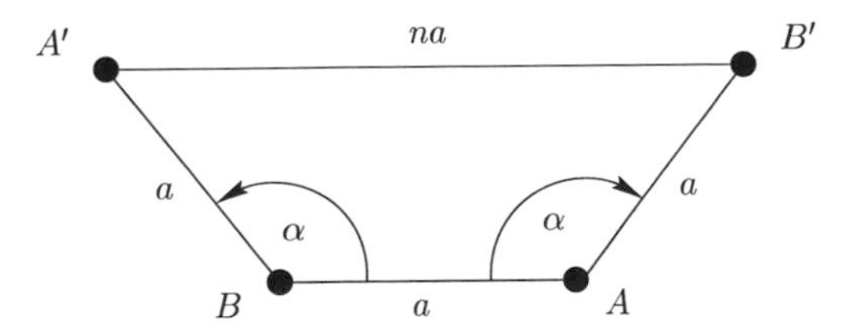

Figure 1.5: Schematic sketch of points A, B, A', B' belonging to the set $\mathcal{P}$ of points in the plane that is invariant with respect to translations by the period a and rotations through the angle α. n is a suitable integer. For this figure, $n = 2$ has been chosen as an example, corresponding to $\alpha = 2\pi/3$.

of the maps M_1 and M_2 in subsection 1.1.3. — In planar crystallography the corresponding symmetry groups are called *strip groups* or *frieze groups*.) In order to obtain webs with translational symmetry in two transversal directions and nontrivial rotational symmetries by means of the kicked harmonic oscillator I therefore restrict the larger part of the following investigation to kicks with the period

$$T = \frac{2\pi}{q}, \quad q \in \tilde{\mathcal{Q}} := \{3, 4, 6\} \subset \mathcal{Q}. \tag{1.33}$$

The phase space structures with $q \in \mathcal{Q}$ are the periodic stochastic webs introduced on page 19; frequently used synonymous terms are "uniform stochastic webs" or "stochastic webs with crystal symmetry". The aperiodic stochastic webs obtained for $q \notin \mathcal{Q}$ are also referred to as "stochastic webs with quasicrystal symmetry".

Although the webs for $q \in \mathcal{Q}$ and $q \notin \mathcal{Q}$ share their general web-like structure, they are fundamentally different beyond the issue of periodicity: The width of the channels of the latter typically decreases with the distance from the origin of the phase plane, whereas the periodicity of the first guarantees that the width of the channels is the same around every mesh of the web. Numerous examples of webs with quasicrystal symmetry can be found in the literature; see for example [CSUZ87, SUZ88, ZSUC88, CSUZ89, ZSUC91] and references therein. Aperiodic webs play an important role in the theory of *quasicrystals* in solid state physics [HG94, Zum97] and are related to many other fields, for example to the problem of tilings of the plane which finds its application both in arts (see figure 1.6 for an example) and science [Kep19, GS87]. The idea to explain symmetries of a planar pattern by identifying an appropriate model system which dynamically generates that pattern is originally due to KEPLER [Kep11].

1.2.2 The Skeleton of the Web

The iteration of the web map (1.24) for $q = 4$ ($T = \pi/2$) typically yields phase portraits like those shown in figure 1.7. In figures 1.7a through 1.7e one can observe the stochastic web for M_4 evolving under the iteration of a single initial value. Figure 1.7f shows how the irregular part of the phase portrait — the *channels* that develop around the skeleton of the web — and the regular part — the *meshes* or *cells* of the web — are intertwined with each other: each mesh of the web is densely filled[5] with invariant lines, i.e. cross-sections of invariant

[5]This statement is not to be taken too literally. In general, the interior of each mesh appears to be filled by just invariant lines only when studied on a coarse scale. A more thorough investigation typically exhibits narrow chains of islands and the corresponding stochastic layers, especially in the vicinity of the stochastic regions of the stochastic web itself [CSU+87, LL92]. The stochastic channels, on the other hand, are in general also interspersed with islands (some of them arbitrarily small) of regular dynamics.

tori of the underlying four-dimensional dynamics (in a generalized phase space spanned by x, p, t and H). Only a few of these invariant lines are shown here. The dynamics in the channels is chaotic, whereas the invariant lines within the meshes indicate regular, nonchaotic behaviour. A situation with such a mixed phase portrait with coexisting regions of regular and chaotic dynamics is often characterized as *soft* or *weak chaos* [Gut90, SUZ88]. The origin of phase space is a fixed point of M_4; this is already clear from equation (1.29). Figure 1.7f indicates that all other elliptic points $(l\pi, m\pi)^t$, with $l+m$ even, are periodic with period four.

Figure 1.8, where phase portraits of the web map for several values of the kick amplitude V_0 are shown, allows the conclusion that the overall structure of the web is the same, regardless of the value of V_0 — the main consequence of increasing the value of V_0 being a broadening of the channels of the web.

The stochastic webs for $q = 4$ shown in figure 1.8 are (approximately) characterized by translational symmetry in two linearly independent directions

Figure 1.6: *Zon en Maan* by M. C. ESCHER (1948). [*Sun and Moon*, © 2003 Cordon Art — The Netherlands. All rights reserved. Used by permission.] Taken from [Esc94], where a number of similar works can be found. The artist elaborates on the problem of tiling the plane using an aperiodic but still somewhat regular pattern. *Roughly* speaking, this pattern exhibits translational and rotational invariance with $q = 3$ and $q = 6$ ($q = 3$ if one distinguishes between birds of different colours, $q = 6$ otherwise).

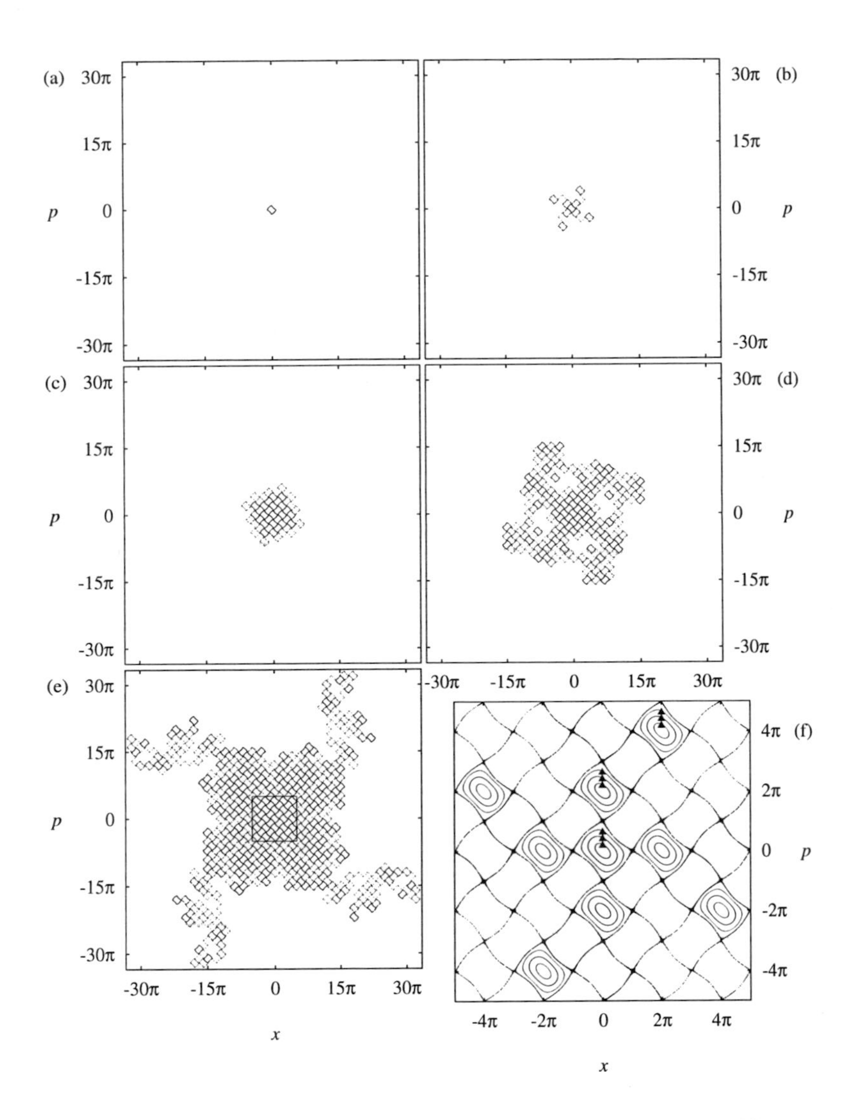

Figure 1.7: Time evolution of the stochastic web for $q = 4$ ($T = \pi/2$) and $V_0 = 1.0$. In figures (a) through (e) the initial value $(x_0, p_0)^t = (0, 3.14)^t$ is iterated $10^3/3\cdot10^3/10^4/3\cdot10^4/10^5$ times, respectively. Figure (f) is a magnification of the square in the centre of figure (e); in addition, the first 10^3 iterates of the points marked by ▲ are plotted.

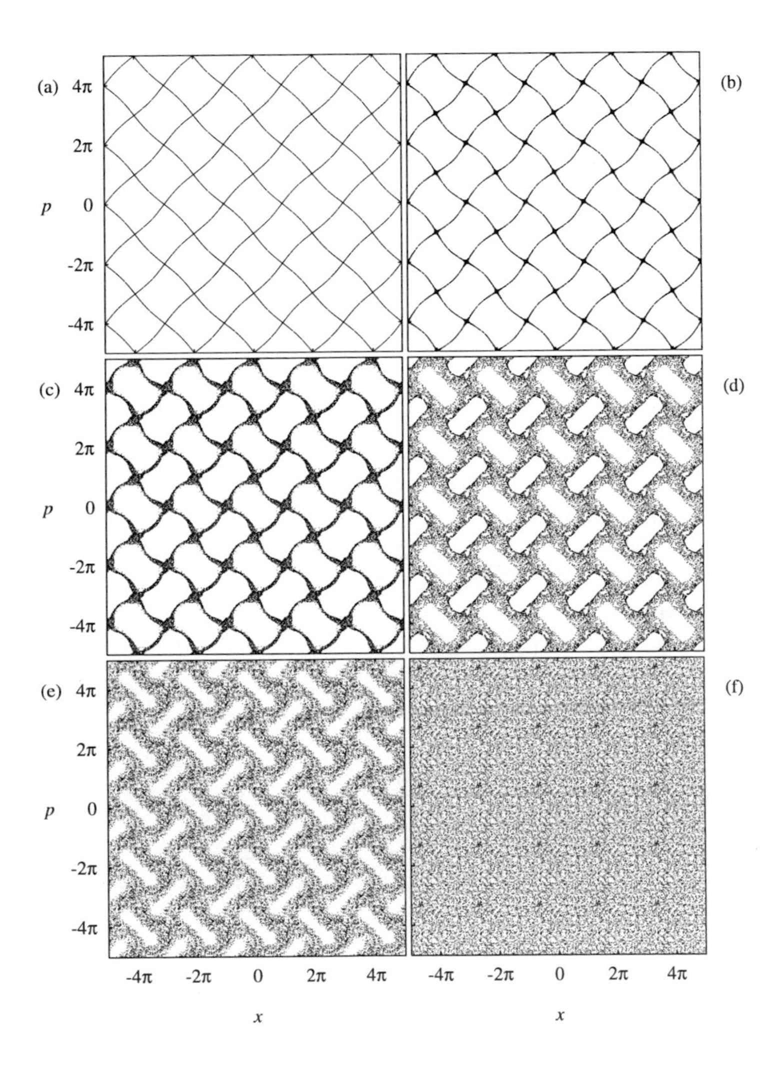

Figure 1.8: The stochastic webs for $q = 4$ ($T = \pi/2$) and several values of the kick amplitude V_0. In figures (a) through (f) V_0 takes on the values 0.5/1.0/1.5/2.0/2.5/5.0, respectively.

(e.g. given by $(2\pi, 0)^t$ and $(0, 2\pi)^t$) and by two rotational symmetries with respect to 2-fold and 4-fold rotation axes, in addition to several (glide) reflection symmetries. More can be said about the symmetries of the *skeleton* of this web — see below.

I now proceed to the explanation of these characteristics of the web dynamics that — for the resonant values $T = 2\pi/q$ with $q \in \mathcal{Q}$ — manifest themselves in figures 1.3a, 1.7 and 1.8.

The overall structure of the web, its *skeleton*, can be explained by splitting the Hamiltonian (1.17) into two parts, such that the first gives an analytical description of the skeleton and the second can be treated as a perturbation to the first.

In a preliminary step the Hamiltonian (1.17) is submitted to a pair of successive canonical transformations, the first replacing the original coordinates x, p with polar coordinates (which are identical to action-angle variables of the unperturbed harmonic oscillator except for a sign in the angle), while the second switches to a rotating frame of reference with coordinates ϑ, J. The combination of both transformations can be described by the time-dependent generating function

$$F_2(x, J, t) \;=\; c_1 \left(\frac{x}{2}\sqrt{2J - x^2} + J \arcsin \frac{x}{\sqrt{2J}} \right) + \left(c_2 - t\right) J \tag{1.34a}$$

with

$$(c_1; c_2) \;=\; \begin{cases} (+1; 0) & p > 0,\; x > 0 \\ (-1; \pi) \quad \text{for} & p < 0 \\ (+1; 2\pi) & p > 0,\; x < 0, \end{cases} \tag{1.34b}$$

which is of GOLDSTEIN's F_2-type [Gol80]. The old and new variables are related via

$$J \;=\; \frac{1}{2}\left(x^2 + p^2\right) \tag{1.35a}$$

$$\tan(\vartheta + t) \;=\; \frac{x}{p}, \tag{1.35b}$$

and the new Hamiltonian is obtained as

$$H(\vartheta, J, t) \;=\; V_0 \cos\left(\sqrt{2J}\sin(\vartheta + t)\right) \sum_{n=-\infty}^{\infty} \delta(t - nT). \tag{1.36}$$

Here, for simplicity, I use the same symbol H for the Hamiltonian before and after the transformation. Since the new frame of reference is made to rotate with unit angular velocity — which is just the angular velocity of the unperturbed oscillator after the scaling (1.15) — only the kick contribution of the original Hamiltonian remains in (1.36).

Setting $n =: k + lq$ with $k \in \{1, 2, \ldots, q\}$, $l \in \mathbb{Z}$, and using the resonance condition (1.23) I have

$$H(\vartheta, J, t) = V_0 \sum_{k=1}^{q} \cos\left\{\sqrt{2J}\sin\left(\vartheta + \frac{2k\pi}{q}\right)\right\} \sum_{l=-\infty}^{\infty} \delta\left(t - 2l\pi - \frac{2k\pi}{q}\right), \tag{1.37}$$

because in this frame of reference all the summands contribute only at the times of kicks. Then, using the spectral decomposition of the III-function (cf. equation (1.2)),

$$\mathrm{III}(t) = \sum_{l=-\infty}^{\infty} \delta(t - 2l\pi) = \frac{1}{2\pi}\left(1 + 2\sum_{l=1}^{\infty}\cos lt\right), \tag{1.38}$$

the Hamiltonian is split into two components in a natural way:

$$H(\vartheta, J, t) = \mathcal{H}_q(\vartheta, J) + \mathcal{V}(\vartheta, J, t) \tag{1.39a}$$

with $\mathcal{H}_q$ being the time averaged full Hamiltonian,

$$\begin{aligned} \mathcal{H}_q(\vartheta, J) &= \overline{H(\vartheta, J, t)} \\ &= \frac{V_0}{2\pi} \sum_{k=1}^{q} \cos\left\{\sqrt{2J}\sin\left(\vartheta + \frac{2k\pi}{q}\right)\right\}, \end{aligned} \tag{1.39b}$$

representing an autonomous (and therefore integrable) system which is submitted to the action of the time-dependent perturbation

$$\mathcal{V}(\vartheta, J, t) = \frac{V_0}{\pi} \sum_{k=1}^{q} \cos\left\{\sqrt{2J}\sin\left(\vartheta + \frac{2k\pi}{q}\right)\right\} \sum_{l=1}^{\infty} \cos\left\{l\left(t - \frac{2k\pi}{q}\right)\right\}. \tag{1.39c}$$

Both $\mathcal{H}_q$ and $\mathcal{V}$ depend on both parameters q and V_0. But in addition to separating the averaged Hamiltonian and the perturbation term, the splitting (1.39) also essentially separates the effects of the parameters: q takes a decisive role for $\mathcal{H}_q$

where it alone determines the shape of the skeleton of the web, whereas changing V_0 does not influence the shape of the skeleton at all.[6] For $\mathcal{V}$, on the other hand, V_0 is the more important parameter that controls the strength of the perturbation to the time averaged Hamiltonian. It turns out that it is mainly V_0 that determines the width of the diffusive channels that form around the skeleton of the web. I return to this issue in subsection 1.2.3, after having discussed the skeleton itself.

For the present purpose I need $\mathcal{H}_q$ expressed in terms of the original phase space variables x, p. Substitution of the transformation (1.35) into the Hamiltonian (1.39b) yields

$$\mathcal{H}_q(x,p,t) = \frac{V_0}{2\pi}\sum_{k=1}^{q}\cos\left\{x\cos\left(\frac{2k\pi}{q}-t\right)+p\sin\left(\frac{2k\pi}{q}-t\right)\right\}, \tag{1.40}$$

where again I use the same symbol $\mathcal{H}_q$ for the Hamiltonian as a function both of ϑ, J and x, p, t. Since this expression is unpleasant in that it contains the time t explicitly, I restrict the attention to just the times of kicks,

$$t_n := nT = \frac{2n\pi}{q}, \quad n \in \mathbb{Z}. \tag{1.41}$$

This is not a severe restriction because $\mathcal{H}_q$ is supposed to explain features of the web map, which is defined at kick times only. With the unit vectors

$$\vec{e}_{q,k} := \begin{pmatrix} \cos\dfrac{2k\pi}{q} \\ \sin\dfrac{2k\pi}{q} \end{pmatrix}, \quad k = 1,\ldots,q, \tag{1.42}$$

the averaged Hamiltonian now reads

$$\mathcal{H}_q(x,p,t_n) = \frac{V_0}{2\pi}\sum_{k=1}^{q}\cos\left\{\begin{pmatrix} x \\ p \end{pmatrix}\cdot\vec{e}_{q,k}\right\}, \tag{1.43}$$

which does not depend on n any more; this means that the averaged dynamics given by $\mathcal{H}_q$ is the same for all times of kicks. Using equation (1.43), the *skeletons*

[6] It is interesting, though, that there exists a V_0-dependence of the web-generating Hamiltonian $\mathcal{H}_q$ at all, regardless of the scaling that is used. This, a consequence of the transformation (1.35) reducing the Hamiltonian to its kick contribution, indicates that the web-like structure in phase space is the *combined* result of both the electric and magnetic components of the field (1.1). In other words, the structure stems from the *interaction* of the corresponding translational and rotational symmetries. Despite $\mathcal{H}_q$ being V_0-dependent, the *geometry* of the web generated by $\mathcal{H}_q$ does not depend on this parameter, since V_0 enters as a mere multiplicative factor into equation (1.39b).

of the stochastic webs can now be identified as level lines $\mathcal{H}_q(x,p,t_n) = \text{const.}(q)$, where the constant has to be chosen suitably for each q.

In subsection 1.2.3 I demonstrate that $\mathcal{V}$ can indeed be treated as a perturbation to $\mathcal{H}_q$. This means that for $V_0 \to 0$ the dynamics of the web map is confined to the neighbourhood of surfaces — i.e. lines in the (x,p)-plane — of constant $\mathcal{H}_q(x,p,t_n)$. For each sufficiently small V_0 the skeleton of the web is then given by some specific contour lines of the time averaged Hamiltonian (1.43). The term "specific" is important here: the $\mathcal{H}_q$-level has to be chosen in such a way that the corresponding contour lines form an infinitely extended structure, in fact the skeleton of the stochastic web. The other levels are important, too; they approximate the bounded quasiperiodic motion in the meshes of the web and accordingly have the topological structure of circles in the phase plane. Figures 1.9–1.11 show contour plots for some of the more important $\mathcal{H}_q$.

The case $q = 4$ $(T = \pi/2)$ may serve to exemplify the above statements. For this value of q the time averaged Hamiltonian

$$\mathcal{H}_4(x,p,t_n) = \frac{V_0}{\pi}(\cos x + \cos p) \tag{1.44}$$

is of the *muffin-tin* type and describes the classical analogue of the HARPER model [Har55, Ket92]; its level lines are plotted in figure 1.10b. The $\mathcal{H}_4$-contours

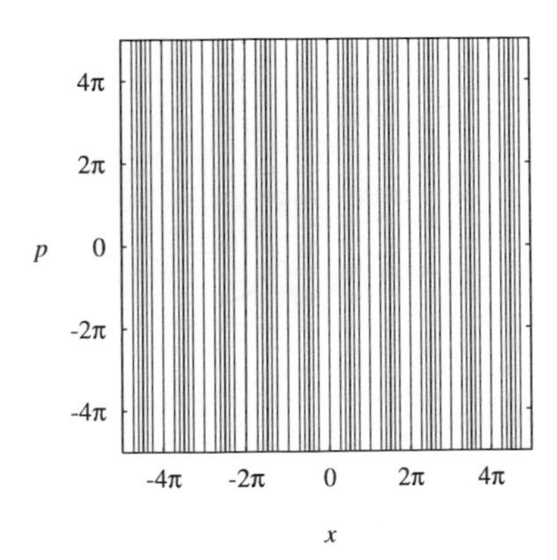

Figure 1.9: Contour plot of the time averaged Hamiltonian $\mathcal{H}_q(x,p,t_n)$ for $q \in \mathcal{Q}\backslash\tilde{\mathcal{Q}}$, i.e. for $q = 1$ and $q = 2$ ($T = 2\pi$ and $T = \pi$, respectively) — cf. equation (1.46) below. Level lines corresponding to 7 different levels of $\mathcal{H}_q$ are shown.

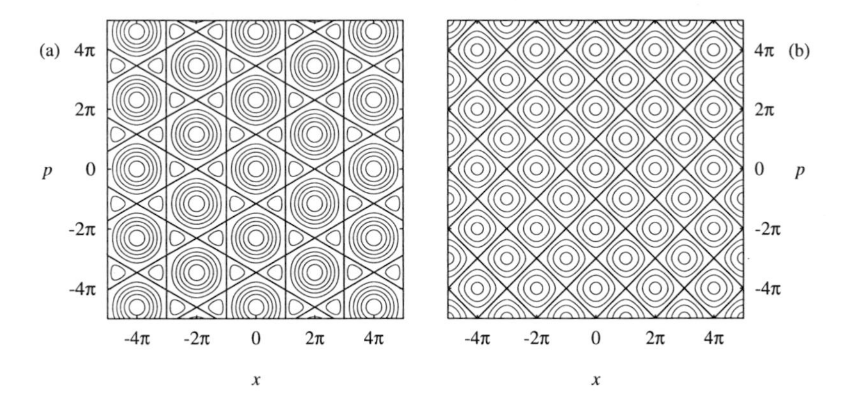

Figure 1.10: Contour plots of the time averaged Hamiltonian $\mathcal{H}_q(x,p,t_n)$ for $q \in \tilde{\mathcal{Q}}$. (a) $q = 3$ ($T = 2\pi/3$). $q = 6$ ($T = \pi/3$) generates the same contours; this is explained in equation (1.47) below. (b) $q = 4$ ($T = \pi/2$). In each plot level lines corresponding to 7 different values of $\mathcal{H}_q$ are shown. The skeletons of the webs — given by $\mathcal{H}_3 = -V_0/2\pi$ and $\mathcal{H}_4 = 0$, respectively — are drawn as thick lines; the other $\mathcal{H}_q$-level lines are drawn as thin lines.

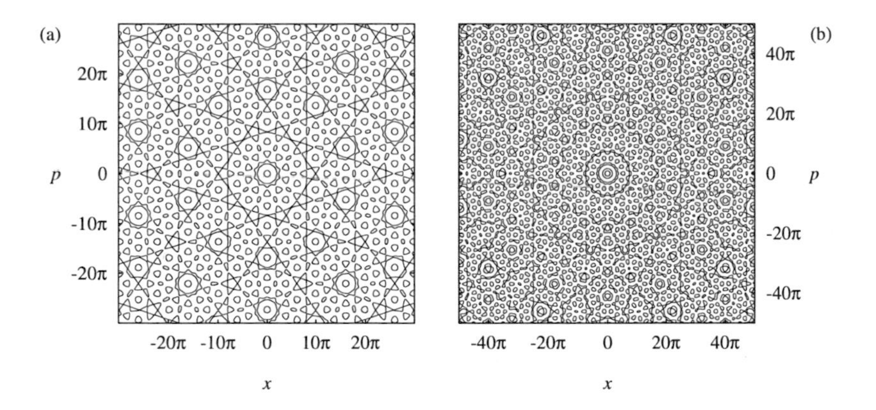

Figure 1.11: Contour plots of the time averaged Hamiltonian $\mathcal{H}_q(x,p,t_n)$ for $q \notin \mathcal{Q}$. (a) $q = 5$ ($T = 2\pi/5$); (b) $q = 7$ ($T = 2\pi/7$). In each plot a single $\mathcal{H}_q$ contour is shown: $\mathcal{H}_5 = V_0/2\pi$ and $\mathcal{H}_7 = -V_0/2\pi$, respectively.

are dominated by the square grid defined by

$$p = \pm x + (2k+1)\pi, \quad k \in \mathbb{Z}, \tag{1.45}$$

corresponding to $\mathcal{H}_4 = 0$. This square grid is the skeleton of the stochastic web for $q = 4$.

Figure 1.10b is to be compared with the phase portraits in figures 1.7 and 1.8 that have been generated by iteration of M_4. Going backwards from figure 1.8f to 1.8a, V_0 tends to zero, and accordingly the stochastic webs more and more approach the rectangular grid of figure 1.10b. The quasiperiodic regular dynamics in the meshes of the web, depicted in figure 1.7f, is also (more or less) well approximated by the contour lines of $\mathcal{H}_4$, as can be seen in figure 1.12, where magnifications of figures 1.7f and 1.10b have been superimposed. Summarizing, the time averaged Hamiltonian $\mathcal{H}_q$ works fine to explain the structure of the web's skeleton. Note that for increasing values of V_0 one observes not only a broadening of the channels of the web, but also a transition from the straight channels — described by the web-skeletons (1.45, 1.48) for $q = 4$ and $q = 3/6$, respectively (the skeleton (1.48) is given below when discussing the level lines

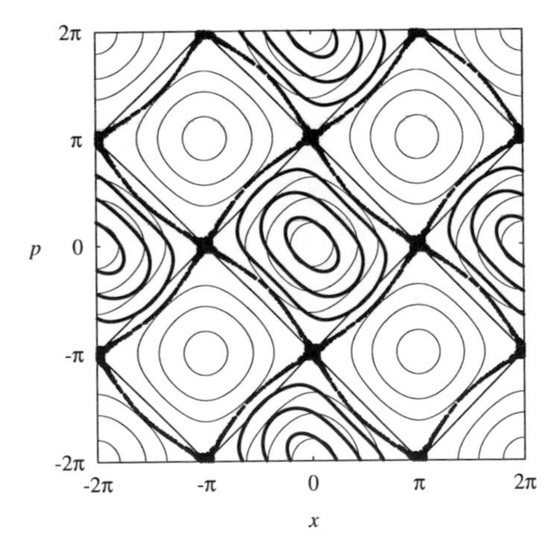

Figure 1.12: Superposition of the enlarged central portions of figures 1.7f and 1.10b. The stochastic web for $q = 4$ ($T = \pi/2$) and some of its accompanying invariant lines are shown using fat dots (which mark the iterates of the web map M_4), while the continuous lines indicate the skeleton of the web and some other $\mathcal{H}_4$-level lines.

of $\mathcal{H}_3$) — to channels with increasingly wavy boundaries, as the approximation $H \approx \mathcal{H}_q$ becomes worse.

In contrast to the inevitably qualitative description of the symmetry of the stochastic webs displayed in figures 1.3a and 1.7/1.8, the symmetry groups of the *skeletons* of the webs can be determined exactly, because with equation (1.43) a closed formula for the pattern they are forming is available. For the skeleton of stochastic webs with $q = 4$ (figure 1.10b), the symmetry group is $p4m$ (using the "international notation" for planar space groups [GS87]). It is characterized by two classes of 4-fold rotation symmetries (here: the centres of rotation being the elliptic points $(l\pi, m\pi)^t$, $l+m$ even, in the centres of the meshes and the points of channel crossings: $(l\pi, m\pi)^t$, $l+m$ odd) and a class of 2-fold rotation symmetries (rotations about $((l+1/2)\pi, (m+1/2)\pi)^t$), in addition to the obvious translation and (glide) reflection symmetries.

Some short remarks about the other values of $q \in \mathcal{Q}$ might be in order. Equation (1.43) again illustrates the triviality of the cases $q = 1$ and $q = 2$ ($T = 2\pi$ and $T = \pi$). For these q there is no p-dependence any more and

$$\mathcal{H}_1(x, p, t_n) = \frac{V_0}{2\pi} \cos x \tag{1.46a}$$

$$= \frac{1}{2} \mathcal{H}_2(x, p, t_n) \tag{1.46b}$$

such that the phase space structures effectively become one-dimensional rather than two-dimensional, as already discussed in the previous subsection and displayed in figure 1.9.

Similarly, $q = 3$ and $q = 6$ ($T = 2\pi/3$ and $T = \pi/3$) are also intertwined with each other,

$$\mathcal{H}_3(x, p, t_n) = \frac{V_0}{\pi} \left\{ \cos \frac{x}{2} \left(\cos \frac{x}{2} + \cos \frac{\sqrt{3}\, p}{2} \right) - \frac{1}{2} \right\} \tag{1.47a}$$

$$= \frac{1}{2} \mathcal{H}_6(x, p, t_n), \tag{1.47b}$$

and thus yield the same web skeleton, namely the level lines given by $\mathcal{H}_3(x, p, t_n) = -V_0/2\pi$:

$$x = (2k+1)\pi, \quad p \in \mathbb{R},\ k \in \mathbb{Z} \qquad \text{or} \tag{1.48a}$$

$$p = \frac{\pm x + (2k+1)2\pi}{\sqrt{3}}, \quad k \in \mathbb{Z}; \tag{1.48b}$$

see figure 1.10a. The symmetry group of this skeleton is $p6m$ and includes three classes of rotational symmetries, namely a 2-fold, a 3-fold and a 6-fold rotational symmetry (the rotations are about the vertices and the centres of the triangles, and about the centres of the hexagons in figure 1.10a, respectively), in addition to the obvious translation and (glide) reflection symmetries.

The skeletons for $q \notin \mathcal{Q}$ show a more complicated structure. In figures 1.11a and 1.11b contours for a single $\mathcal{H}_q$-level each are shown; these levels are chosen in such a way that the contour lines cover as many hyperbolic points in the stochastic layer as possible, thereby coming fairly close to the underlying separatrix (cf. [ZSUC88]). Here, as in figure 1.3b, the aperiodic organization of phase space is clearly visible.

Essentially, the method I have used here to identify the webs' skeletons is an averaging procedure (see the derivation of the splitting (1.39)). LOWENSTEIN [Low92] shows that this can be viewed as just the first step of an iterative scheme which is similar to the BIRKHOFF-GUSTAVSON normalization method [Gus66, Eng93, ESE95] and, step by step, yields higher order approximations for the Hamiltonian (1.17). Using this scheme it is possible to systematically derive improved expressions describing the separatrices of the webs and thereby explain their waviness that develops for larger values of V_0, as can be observed, for example, in figure 1.8.

1.2.3 The Influence of the Kick Amplitude

In this subsection I demonstrate the influence of the kick amplitude V_0 for the case $q = 4$; similar results can be obtained for $q = 3$ and $q = 6$. The following discussion holds for small values of V_0. The idea is to derive a *separatrix mapping* that describes the dynamics of the kicked harmonic oscillator in the vicinity of the separatrices and allows to estimate the width of the channels of diffusive dynamics.

The first step is to identify the $\mathcal{V}$ of equation (1.39c) as a high frequency perturbation to the time averaged Hamiltonian $\mathcal{H}_4$ (1.44). At kick times $t_n = n\pi/2$, $n \in \mathbb{Z}$, one has

$$\sqrt{2J}\sin\left(\vartheta + \frac{k\pi}{2}\right) = \sqrt{2J}\sin\Big((\vartheta + t_n) + \frac{\pi}{2}(k-n)\Big), \tag{1.49}$$

which gives, with the help of equations (1.35), for every fourth kick ($n = 4m$, $m \in \mathbb{Z}$)

$$x\cos\frac{k\pi}{2} + p\sin\frac{k\pi}{2},$$

such that the action-angle variables ϑ, J have been exchanged for the original x, p. $\mathcal{V}$ now takes the form

$$\mathcal{V}(x,p,t) = \frac{V_0}{\pi}\sum_{k=1}^{4}\cos\left(x\cos\frac{k\pi}{2}+p\sin\frac{k\pi}{2}\right)\sum_{l=1}^{\infty}\cos\left\{l\left(t-\frac{k\pi}{2}\right)\right\}, \quad (1.50)$$

which holds for every fourth kick time $t_n = t_{4m}$ only. This is a useful feature of equation (1.50), as it facilitates the discussion of the dynamics near *one* of the four separatrices enclosing each phase space cell, rather than near all four of them: while for the dynamics of the web map M_4, corresponding to the full Hamiltonian (1.17), it takes four iterates to return into the neighbourhood of an initial value near a separatrix, the same neighbourhood is reached by a single iteration of M_4^4, or equivalently by the discrete dynamics generated by the Hamiltonian $\mathcal{H}_4 + \mathcal{V}$, where $\mathcal{V}$ in the form (1.50) is used.

From the last formula it becomes clear that $\mathcal{V}$ can indeed be treated as a high frequency perturbation of the HARPER Hamiltonian $\mathcal{H}_4$: the time dependence of $\mathcal{V}$ is given by cosine terms the largest period of which is 2π, whereas the smallest period of $\mathcal{H}_4$ is $2\pi^2/V_0$ — as is shown below on page 38 — and can thus be made as large as desired in the limit $V_0 \to 0$.

In [LL73] it is argued that in a first approximation the higher frequency terms of a perturbation expanded as in equation (1.50) can be neglected, although they contribute with roughly the same weight ($\sim V_0$) as the small frequency terms. Therefore I can drop all cosine terms with $l \geq 3$ and get the approximant

$$\mathcal{V}(x,p,t) \approx \frac{2V_0}{\pi}(\cos x - \cos p)\cos 2t, \quad (1.51)$$

where it turns out to be useful to keep the explicit t-dependence, although strictly speaking this formula applies for $t = 2m\pi$ only.

Now consider a typical orbit of the web map in the stochastic region of the web. For some $n \in \mathbb{N}$ the corresponding orbit point $(x_n, p_n)^t$ will be close to and just below of the midpoint of the HARPER separatrix $p(x) = \pi - x$ with $0 < x < \pi$, which is displayed in the upper right quadrant of figure 1.13.[7] For notational convenience I define $P_i := (x_n, p_n)^t$ with some $i \in \mathbb{Z}$. At P_i, $\mathcal{H}_4$ takes on a certain value $\mathcal{E}_i \approx 0$ since the separatrices are characterized by $\mathcal{H}_4(x,p) = 0$. Let P_i be the initial value (at time τ_i) for the successive application of the fourth iterate of the iteration of the web map (1.29), M_4^4. The resulting orbit in the (x,p)-plane is

[7] Any other separatrix $p(x) = \pm x + (2k+1)\pi$ with $l\pi < x < (l+1)\pi$ and $k, l \in \mathbb{Z}$ could be considered as well. It depends on the choice of k and l whether the rotation is clockwise or anticlockwise. The cell centred around the origin of the (x,p)-plane exhibits anticlockwise rotation, and neighbouring cells have opposite directions of revolution. See figure 1.13.

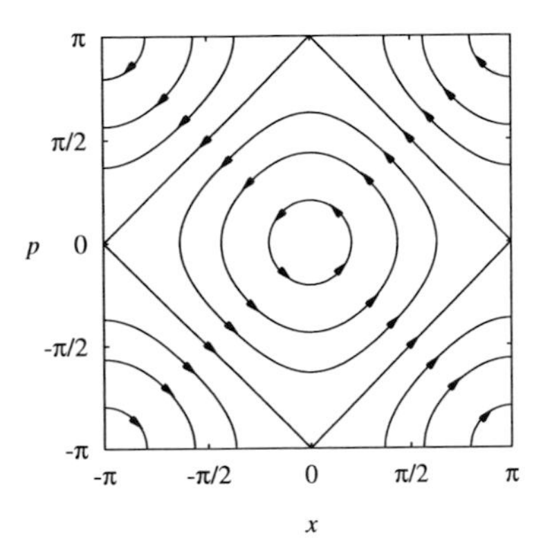

Figure 1.13: Dynamics of the HARPER system: asymptotic motion on the HARPER separatrices given by equation (1.45) towards the hyperbolic points $(l\pi, m\pi)^t$, $l + m$ being odd, and periodic (anticlockwise or clockwise) motion within the cells. The complete phase space picture is obtained by periodic continuation of the shown interval $[-\pi, \pi]^2$.

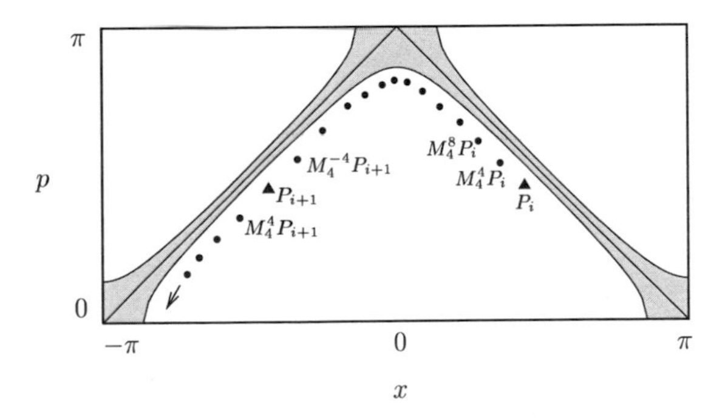

Figure 1.14: Schematic sketch of the dynamics of M_4^4 near the separatrix. The point P_i, shown as ▲, is mapped onto P_{i+1} by M_4^{4j} for some $j \in \mathbb{N}$: $P_{i+1} = M_4^{4j} P_i$; iterates of P_i under M_4^4, other than P_i and P_{i+1}, are drawn as •. The straight lines are the separatrices of the HARPER system (1.44), while the shaded area indicates the stochastic region for the full Hamiltonian (1.17). The separatrix mapping is defined by the variation of $\mathcal{E}$ and τ under the mapping $P_i \longmapsto P_{i+1}$; see equation (1.52).

quasiperiodic if P_i is not too near to the separatrix (cf. figure 1.7f). First the orbit follows the separatrix anticlockwise, until it comes close enough to the hyperbolic point $(0,\pi)^t$, where the orbit turns left and follows the perpendicular separatrix $p(x) = \pi + x$ with $-\pi < x < 0$; this is shown schematically in figure 1.14. At some time $\tau_{i+1} = \tau_i + 2j\pi$ (where j is the according number of iterations of M_4^4) the orbit reaches the point $P_{i+1} = M_4^{4j} P_i$, which is defined as that iterate of P_i under M_4^4 that comes closest to the centre of the second separatrix. This point again gives rise to a certain value of $\mathcal{H}_4$, $\mathcal{E}_{i+1}$. The desired separatrix mapping is now given by the change of the value of $\mathcal{H}_4$ during one such quarter revolution and by the corresponding (approximate) quarter period:

$$\mathcal{E}_{i+1} = \mathcal{E}_i + \Delta\mathcal{E}(\mathcal{E}_i, \tau_i; V_0) \tag{1.52a}$$

$$\tau_{i+1} = \tau_i + \Delta\tau(\mathcal{E}_i, \tau_i; V_0). \tag{1.52b}$$

In order to determine an approximate expression for $\Delta\mathcal{E}$, I first derive an explicit solution of the HARPER dynamics

$$\dot{x} = \frac{\partial \mathcal{H}_4}{\partial p} = -\frac{V_0}{\pi}\sin p \tag{1.53a}$$

$$\dot{p} = -\frac{\partial \mathcal{H}_4}{\partial x} = \frac{V_0}{\pi}\sin x \tag{1.53b}$$

on the separatrix $p = \pi - x$ for $0 < x < \pi$. Such a solution is given by

$$\begin{aligned} x(t) &= 2\arctan\exp\left(-\frac{V_0}{\pi}(t-\tau_i)\right) \\ p(t) &= \pi - x(t), \end{aligned} \tag{1.54}$$

where τ_i is chosen in such a way that $x(\tau_i) = p(\tau_i) = \pi/2$, corresponding to the midpoint of the separatrix. In figure 1.13 the corresponding trajectory connects the limiting points $(\pi, 0)^t$ and $(0,\pi)^t$ of the separatrix, for t going from $-\infty$ to ∞.

The rate of change of the value of $\mathcal{H}_4$ during the $\mathcal{V}$-perturbed dynamics can then approximately be calculated as

$$\frac{\mathrm{d}\mathcal{H}_4}{\mathrm{d}t} \approx \frac{\partial \mathcal{H}_4}{\partial x}\dot{x} + \frac{\partial \mathcal{H}_4}{\partial p}\dot{p}, \tag{1.55}$$

where the partial derivatives are to be taken from equations (1.53), and $\dot{x}, \dot{p}$ stem from the Hamiltonian equations according to

$$H(x,p,t) \approx \frac{V_0}{\pi}(\cos x + \cos p) + \frac{2V_0}{\pi}(\cos x - \cos p)\cos 2t, \tag{1.56}$$

which is used as an approximant to the full Hamiltonian (1.17) at times t_{4m}. This gives

$$\frac{\mathrm{d}\mathcal{H}_4}{\mathrm{d}t} \approx -\frac{4V_0^2}{\pi^2}\sin x(t)\sin p(t)\cos 2t, \tag{1.57}$$

which holds for all t_n, $n \in \mathbb{Z}$, again. Using the solution (1.54) for $x(t)$ and $p(t)$, the right hand side of equation (1.57) can be approximated by

$$-\frac{4V_0^2}{\pi^2}\sin^2\left(2\arctan\exp\left(-\frac{V_0}{\pi}(t-\tau_i)\right)\right)\cos 2t = -\frac{4V_0^2}{\pi^2}\frac{\cos 2t}{\cosh^2\left(\frac{V_0}{\pi}(t-\tau_i)\right)}, \tag{1.58}$$

such that by integration along the whole separatrix I finally obtain[8]

$$\Delta\mathcal{E}(\mathcal{E}_i,\tau_i;V_0) \approx \int_{-\infty}^{\infty}\frac{\mathrm{d}\mathcal{H}_4}{\mathrm{d}t}\,\mathrm{d}t \approx -\frac{8\pi\cos 2\tau_i}{\sinh\frac{\pi^2}{V_0}}. \tag{1.59}$$

Thus, while an exact formula for $\Delta\mathcal{E}$ certainly depends on $\mathcal{E}_i$, in the framework of this first approximation $\Delta\mathcal{E}$ is independent of $\mathcal{E}_i$.

It remains to determine the time interval $\Delta\tau$ in equation (1.52b). Within the cells, i.e. away from the separatrices, the equations of motion of the unperturbed HARPER system can be integrated using Jacobian elliptic functions. With

$$c := \cos x(t) + \cos p(t), \tag{1.60}$$

which remains constant on each integral curve of the HARPER system, and

$$\tilde{x}(t) := \cos x(t) \tag{1.61}$$

equation (1.53a) can be transformed into

$$\dot{\tilde{x}}^2 = \left(\frac{V_0}{\pi}\right)^2\left(1-\tilde{x}^2\right)\left(1-(c-\tilde{x})^2\right). \tag{1.62}$$

[8] More explicitly, $\Delta\mathcal{E}$ has to be calculated in two steps: first one integrates along the separatrix $p=\pi-x$ from $x=\pi/2$ to $x=0$; then the perpendicular separatrix $p=\pi+x$ is followed from $x=0$ to $x=-\pi/2$. A closer investigation shows that this procedure can be replaced by integrating along $p=\pi-x$ from $x=\pi$ up to $x=0$; this corresponds to the time integral from $-\infty$ to ∞ using the on-separatrix solution (1.54), as in equation (1.59).

This is essentially the characteristic differential equation satisfied by the elliptic function $\mathrm{cd}(t;k)$ [DV73], where the parameter k is determined by the value of c:

$$k(c) = \frac{2-|c|}{2+|c|}. \tag{1.63}$$

With $|c| < 2$ the solution of the characteristic equation gives

$$\begin{aligned} \cos x(t) &= \frac{c}{2} + \left(1 - \frac{|c|}{2}\right) \mathrm{cd}\left(\left(1 + \frac{|c|}{2}\right)\frac{V_0}{\pi}t;\ k(c)\right) \\ \cos p(t) &= c - \cos x(t). \end{aligned} \tag{1.64}$$

For real arguments, as in the present case, $\mathrm{cd}(t;k)$ is periodic with period $4K$, K being a complete elliptic integral of the first kind:

$$K(k) = \int_0^{\frac{\pi}{2}} \frac{\mathrm{d}\varphi}{\sqrt{1-k^2\sin^2\varphi}}. \tag{1.65}$$

This integral is bounded below by $\pi/2$. (The value $\pi/2$ corresponds to the case $|c| = 2$, characterizing the centres of the phase space cells.) Therefore the period of the HARPER solution $\big(x(t), p(t)\big)^t$,[9]

$$T_{\text{HARPER}} = \frac{8K\big(k(c)\big)}{\left(1+\frac{|c|}{2}\right)\frac{V_0}{\pi}}, \tag{1.66}$$

is not smaller than $2\pi^2/V_0$, which tends to infinity for $V_0 \to 0$. This justifies the above treatment of $\mathcal{V}$ as a high frequency perturbation of $\mathcal{H}_4$ (see pages 34f).

Approximate expressions for the integral (1.65) can be found in [AS72] and yield here, in the vicinity of a separatrix:

$$K\big(k(c)\big) \approx \frac{1}{2}\log\frac{8}{|c|} \quad \text{for} \quad c \to 0. \tag{1.67}$$

Taking into account that for the separatrix mapping only a quarter of a full revolution has to be considered, equations (1.66) and (1.67) finally give the desired expression for $\Delta\tau$,

$$\Delta\tau(\mathcal{E}_i, \tau_i; V_0) \approx \frac{\pi}{V_0}\log\frac{8V_0}{\pi|\mathcal{E}_{i+1}|}; \tag{1.68}$$

[9] Note that the period of $(x(t), p(t))^t$ is twice the period of $\cos x(t)$, $\cos p(t)$.

when solving the upcoming equation (1.71) it turns out that it is technically more convenient to use $\mathcal{E}_{i+1}$ here as a replacement for c rather than $\mathcal{E}_i$. The $\mathcal{E}_i$- and τ_i-dependence in this formula comes from $\mathcal{E}_{i+1}(\mathcal{E}_i, \tau_i; V_0)$ — see equations (1.52a) and (1.59).

With equations (1.59) and (1.68), the separatrix mapping (1.52), approximating the dynamics of M_4 near the separatrices, is completely specified. Note that the separatrix mapping can also be obtained in a different way: up to terms of order V_0^2, the fourth iterate of the web map (1.29) is given by

$$\begin{pmatrix} x \\ p \end{pmatrix} \longmapsto \begin{pmatrix} x - 2V_0 \sin p \\ p + 2V_0 \sin x \end{pmatrix} = M_4^4 \begin{pmatrix} x \\ p \end{pmatrix} + \mathcal{O}(V_0^2). \tag{1.69}$$

This map can be generated using the *kicked* HARPER *Hamiltonian*

$$\tilde{H}(x, p, t) = \frac{V_0}{\pi} \left\{ \cos x + \cos p \sum_{n=-\infty}^{\infty} \delta\left(\frac{t}{2\pi} - n \right) \right\}, \tag{1.70}$$

which finds its application in solid state physics, for example in the theory of electrons in certain magnetic fields [KSD92, Dan95, FGKP95], alongside its unkicked counterpart, the HARPER Hamiltonian $\mathcal{H}_4$. Submitting $\tilde{H}$ to manipulations similar to those of the present subsection, formulae (1.59) and (1.68) can be derived once again.

Using the separatrix mapping, I now can proceed to the estimation of the width of the channels of diffusive dynamics. As a criterion for the border between regular dynamics within the meshes and stochastic dynamics in the channels,

$$\max_{\tau_i} \left| \frac{\partial \tau_{i+1}}{\partial \tau_i} - 1 \right| \approx 1 \tag{1.71}$$

may be used, since this expression characterizes the region of phase space where the value of τ_i begins to change significantly under iteration of the separatrix mapping. Solving relation (1.71) for $|\mathcal{E}_{i+1}|$ and renaming it $|\mathcal{E}_{\text{border}}|$, I get

$$|\mathcal{E}_{\text{border}}| \approx \frac{16\pi^2}{V_0 \sinh \frac{\pi^2}{V_0}}, \tag{1.72}$$

such that with equation (1.44) I finally obtain for the width w of the channels:

$$w(V_0) \approx \frac{32\sqrt{2}\pi^3}{V_0^2} e^{-\frac{\pi^2}{V_0}} \quad \text{for} \quad V_0 \to 0. \tag{1.73}$$

This expression scales with $V_0^{-2}e^{-\frac{\pi^2}{V_0}}$ and, not surprisingly, tends to zero in the limit $V_0 \to 0$.

Figure 1.15 illustrates this behaviour: for several values of V_0 stochastic webs are obtained numerically by iterating the web map M_4 for a large number of times; then the channel widths of these webs are measured — by determining the largest distance of any two points of the web which approximately lie on the line $p = x$ and near the point $(\pi/2, \pi/2)^t$ — and compared with the numbers given by the approximating formula (1.73). The agreement between the analytical formula and the numerical data is reasonably good for $V_0 \lesssim 1$ and improves for $V_0 \to 0$. Figure 1.15 also indicates that for $V_0 \lesssim 0.3$ the precision of the computer algorithm does not suffice any more to produce accurate numerical results, because in this parameter range the web gets *very* thin — the width of the channels shrinks to less than 10^{-9} here. Similar results hold for the other values of $q \in \tilde{Q}$.

In this subsection I have discussed how the kick strength V_0 of the kicked harmonic oscillator determines the shape of its phase portrait. In particular it has been shown that the periodic kicking acts in a way which is typical for perturbed

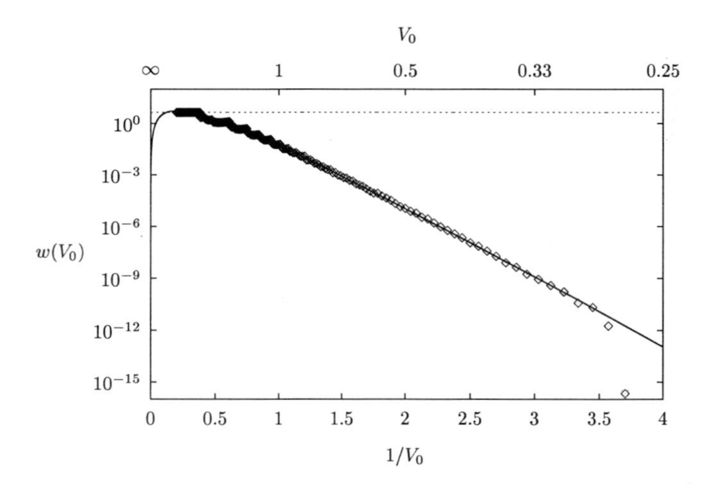

Figure 1.15: The width $w(V_0)$ of the channels of irregular dynamics for $q = 4$ $(T = \pi/2)$ as a function of the kick amplitude V_0. The data points marked by $\diamond$ have been obtained by numerically computing (with up to $2{\cdot}10^9$ iterations of M_4) the stochastic web for the corresponding value of V_0 and measuring the width of the channels where they are the thinnest, i.e. near the centres of the separatrices. The continuous line is the graph of the approximating formula (1.73); the dotted horizontal marks the distance $\sqrt{2}\pi$ of the centres of two neighbouring meshes of the web.

systems: starting from an integrable system at $V_0 = 0$ (here: the harmonic oscillator) the kicking (with $V_0 \neq 0$) renders the system nonintegrable, and the area of the phase space region of irregular, chaotic dynamics grows with increasing perturbation parameter V_0.

Having investigated the *shape* and the *size* of the channels of irregular motion in this subsection and the previous one, in the next subsection I turn to the discussion of the most characteristic *dynamical* aspect of the irregular motion.

1.2.4 Diffusive Energy Growth in the Channels

The time evolution of the stochastic web shown in figure 1.7 allows the conclusion that with a growing number of iterations of the web map an increasingly large region of phase space is covered by orbit points. This observation can be cast into a more exact form by considering the energy

$$E_n := \frac{1}{2}\left(p_n^2 + x_n^2\right), \quad n \in \mathbb{N}_0, \tag{1.74}$$

for a particle moving within the stochastic web. Obviously, E_n is the well-defined energy only *between* the kicks, namely between the $(n-1)$-st and the n-th kick. E_n grows quadratically with the distance of the particle from the origin of the phase plane.

Figure 1.16 shows how, for $q = 4$, the energy develops as a function of time for some typical orbits and for two different values of V_0. Two entirely different types of motion can be distinguished: first, if the initial condition $(x_0, p_0)^t$ of the orbit is chosen from one of the regular regions where invariant lines persist, then E_n remains bounded as the orbit revolves on its corresponding invariant line. The orbits with $p_0 = 2.9$ in figure 1.16a and $p_0 = 1.0$ in figure 1.16b are of this type, giving rise to the nearly horizontal lines at $E_n \approx 0$. Second, for initial conditions within the channels of the web, unbounded motion is possible and dominates the dynamics. The other orbits in figure 1.16 correspond to initial conditions of this class. This type of motion manifests itself by sharp increases and declines in the energy. Between the times of rapidly changing energy there are time intervals for which the energy remains roughly constant. The length of these time intervals varies and typically decreases with increasing values of V_0.

For orbits of the second type, on the average the energy grows with time; this is the most prominent feature of the dynamics in the channels of the web. For computational convenience this time average can be replaced by an ensemble average. Rather than considering a very long orbit for a single initial condition I generate a family of orbits, corresponding to a Gaussian distribution of initial values, and compute the accordingly averaged energy at time $nT - 0$, denoted by $\langle E\rangle_n^{\mathrm{cl}}$.[10] Using an ensemble average also allows for a more straightforward comparison with the quantum dynamics of the system, since in quantum mechanics HEISENBERG's uncertainty relation rules out the possibility of considering δ-shaped initial distributions (corresponding to single classical initial values) in phase space.

In figure 1.17 I employ Gaussian distributions of initial conditions which are centred around the initial conditions of the preceding figure. The Gaussians used

[10] The superscript ^{cl} is used to distinguish the classical ensemble average $\langle\cdot\rangle^{\mathrm{cl}}$ from the quantum expectation value $\langle\cdot\rangle$.

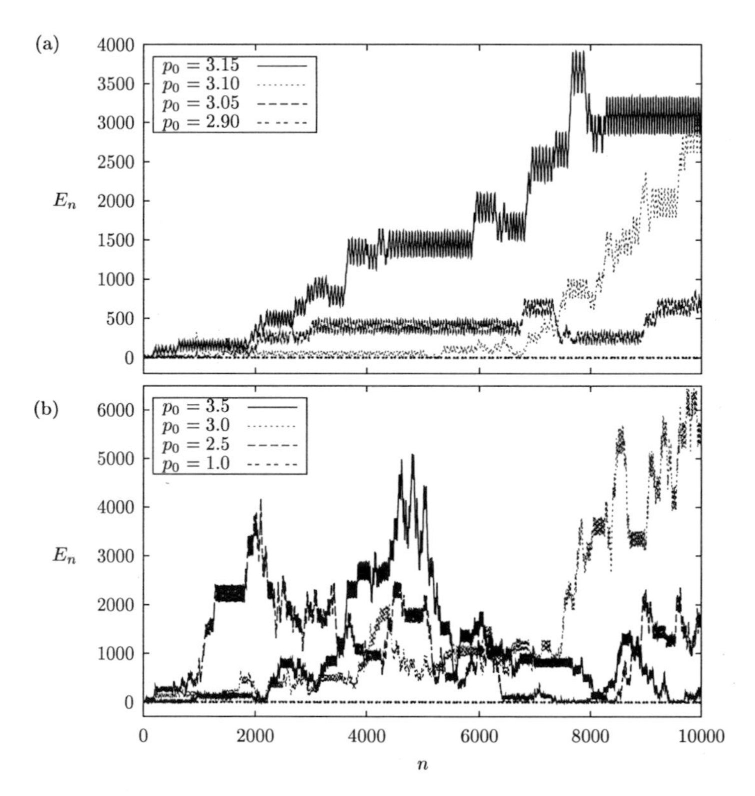

Figure 1.16: The energy E_n of the kicked harmonic oscillator for $q = 4$ ($T = \pi/2$) versus the number n of kicks. (a) $V_0 = 1.0$; (b) $V_0 = 2.0$. The initial values are $(0, p_0)^t$ with the values of p_0 given in the insets.

here are of half width 0.1, both in x- and p-direction. From these curves it is quite clear that the gross oscillations seen in figure 1.16 cancel as a result of the averaging over many orbits. Initial distributions centred in regular regions of phase space yield a slower increase of the energy at first, but due to the tails of the Gaussian distributions even in these cases the averaged energy is not bounded any more.

After averaging it becomes obvious that — apart from a brief ($n \lesssim 1000$) transient stage of subdiffusive motion — the growth of the energy is characterized by *normal* or EINSTEIN *diffusion* [Ein06], i.e. the averaged energy grows as a

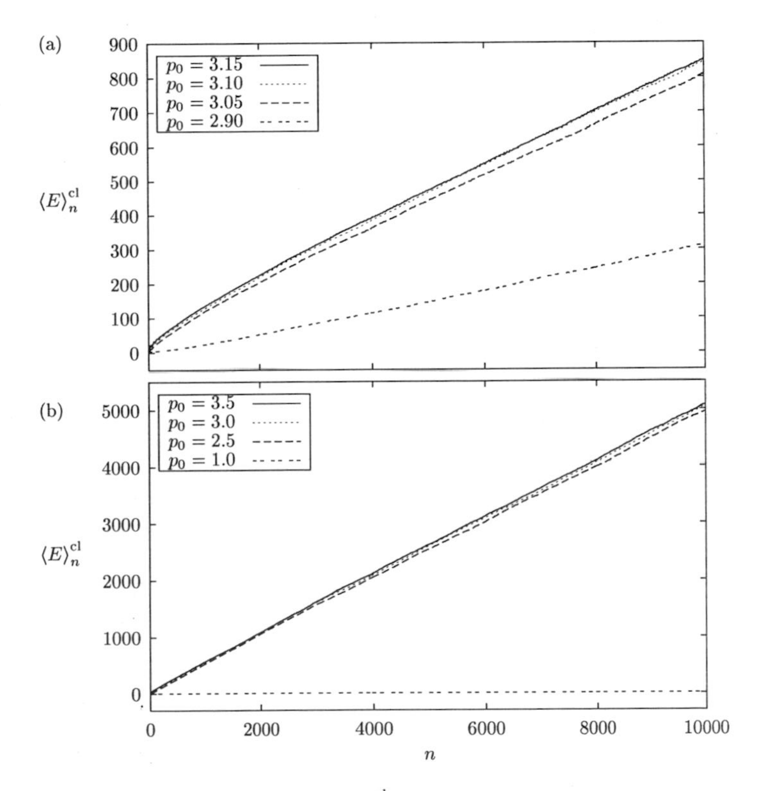

Figure 1.17: The averaged energy $\langle E\rangle_n^{\rm cl}$ of the kicked harmonic oscillator for $q = 4$ ($T = \pi/2$) versus the number n of kicks. (a) $V_0 = 1.0$; (b) $V_0 = 2.0$. For each graph 10^4 initial values are used which are distributed according to a Gaussian with half width 0.1, centered around the same initial values as in figure 1.16.

linear function of time,

$$\langle E\rangle_n^{\rm cl} \;\approx\; D(V_0)\, n + \text{const.} \quad \text{for} \quad n \to \infty, \tag{1.75}$$

with a diffusion coefficient $D(V_0)$. As indicated by the above figures, $D(V_0)$ should be expected to be a function of the amplitude V_0 of the kicks, but independent of n.

For larger values of V_0 a formula for $D(V_0)$ can be derived analytically, which holds (with deviations which are discussed below) for all web maps with $q \in \tilde{\mathcal{Q}}$. From equation (1.21) it follows that for a single orbit the energy change inflicted by the n-th kick is determined by

$$E_{n+1} \;=\; E_n + V_0 p_n \sin x_n + \frac{V_0^2}{2}\sin^2 x_n. \tag{1.76}$$

Averaging over x_n and p_n then gives

$$\langle E\rangle_n^{\rm cl} \;\approx\; \langle E\rangle_0^{\rm cl} + \frac{V_0^2}{4} n, \tag{1.77}$$

such that I have for the diffusion coefficient:

$$D(V_0) \;\approx\; \frac{V_0^2}{4}. \tag{1.78}$$

This derivation, which is largely analogous to the *random phase approximation* for the kicked rotor [Chi79, CCIF79] (see also subsection 5.1.1), relies on the straightforward averaging over x_n and p_n. Roughly, it can be taken to be justified if unbounded and unhindered motion into all directions of the phase plane is possible, i.e. if $q \in \tilde{\mathcal{Q}}$, and if, in addition, the channels are wide enough, i.e. if V_0 is large enough. For *aperiodic* webs, on the other hand, the dynamics is subdiffusive, because the nonperiodicity implies that in general the channels of the webs are not wide enough everywhere for unrestricted diffusion through all the channels (cf. [Hip94]).

The quality of the approximation leading to equation (1.78) is checked in figure 1.18 for the case of $q = 4$. For several values of V_0 I have calculated $\langle E\rangle_n^{\rm cl}$ and obtained $D(V_0)$ using a least squares fit;[11] these values of the diffusion coefficient are compared with the graph of formula (1.78). By and large, agreement of the two can be observed. But it is also a striking feature of figure 1.18 that an additional oscillation of the diffusion coefficient is found to be superimposed to the expected

[11] More sophisticated methods for determining the speed of diffusion can be found in the literature; see for example [Hip94] and references therein.

parabola. Following [KM90] such oscillations of the diffusion coefficient can be attributed to autocorrelations of the orbits, due to small islands of stability in the channels of the web, while strictly speaking the ("quasilinear") result (1.78) holds for Markovian dynamics only [CM81, Rei98]. For a more detailed analysis of the parameter dependence of the diffusion coefficient for the kicked harmonic oscillator see [DH95, DH97].

1.3 The Complementary Case: Nonresonance

While in the present chapter for the most part the *resonance* cases of the kicked harmonic oscillator have been discussed, in this section I briefly turn to the complementary case of *nonresonance*, where condition (1.22) is not satisfied. This case is of particular importance for the discussion in chapter 5.

It has been shown in section 1.2 that in order to dynamically construct stochastic webs by means of the POINCARÉ map (1.21), the resonance condition (1.23) needs to be satisfied, thereby restricting the kick period T to a discrete set of values. What is more, the *periodic* stochastic webs leading to unrestricted diffusion in phase space are obtained for the few values of T specified by equation (1.30) only.

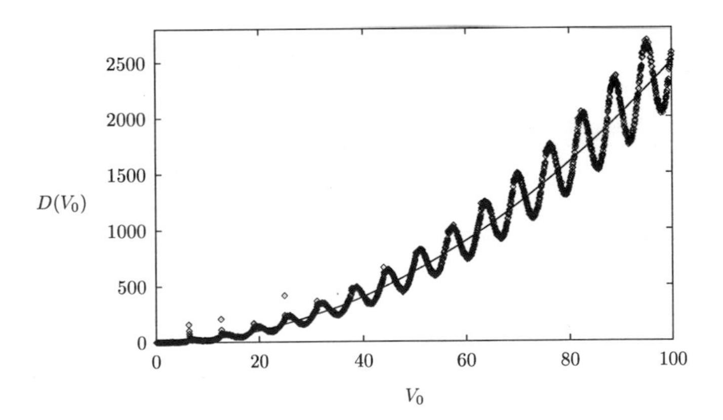

Figure 1.18: Diffusion coefficients for $q = 4$ ($T = \pi/2$). The $\diamond$ indicate least squares approximations of $D(V_0)$ which are determined numerically, based on ensembles consisting of $5{\cdot}10^4$ points which are iterated 10^4 times; the continuous line is the graph of the function (1.78).

The following typical example demonstrates that for *non*resonant T, typically the energy grows diffusively as well.

Similar to figure 1.17, figure 1.19 shows the time development of the ensemble averaged energy, but in this case for the nonresonant value of $T = 1.0$. Gaussian ensembles of initial conditions centered around some of the initial conditions $(0, p_0)^t$ used for the phase portrait of figure 1.4 are used to compute the averaged

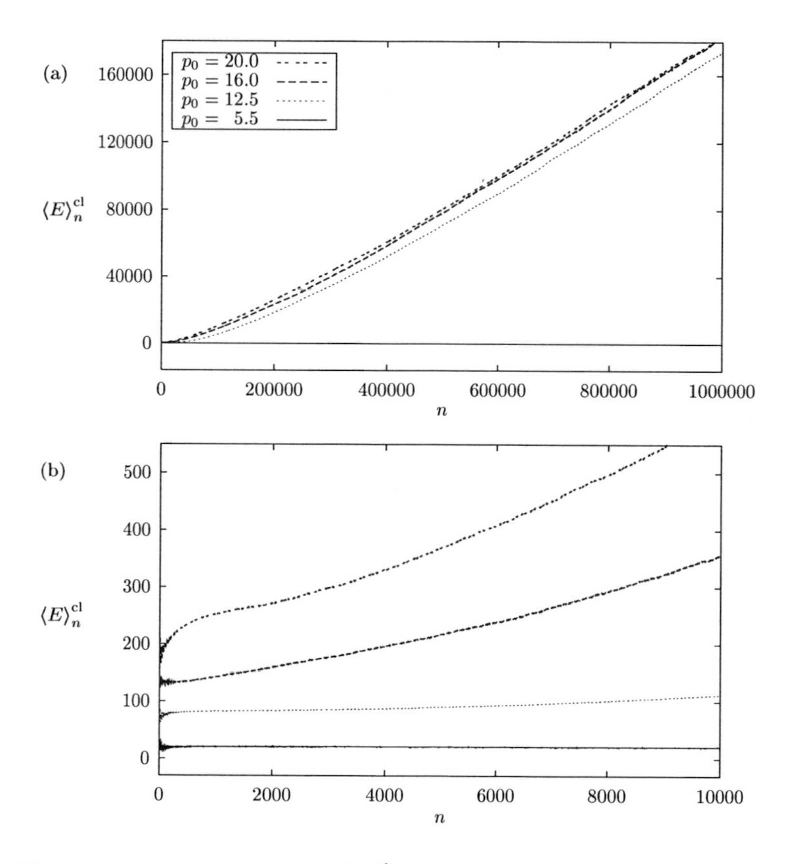

Figure 1.19: The averaged energy $\langle E \rangle_n^{\mathrm{cl}}$ of the kicked harmonic oscillator versus the number n of kicks, in a case of nonresonance: $T = 1.0$ and $V_0 = 1.0$, corresponding to the phase portrait in figure 1.4. For each of the points $(0, p_0)^t$, with p_0 given in the inset, 10^4 Gaussian distributed initial conditions centered around $(0, p_0)^t$ are used to compute the averaged energy. (b) is a magnification of (a).

energy in this case. For the initial condition with $p_0 = 5.5$ (close enough to the origin to be encircled by the outer invariant line shown in figure 1.4) there is no energy growth beyond a certain saturation value, as should be expected. But for initial conditions outside of the boundary given by the outermost invariant line, for example for $p_0 = 12.5$, the energy grows unboundedly as in the case of a periodic stochastic web. The same effect, but even more clearly, can be observed for initial conditions placed further outward, for example with $p_0 = 16.0$ or $p_0 = 20.0$.

Figure 1.19a also indicates that, depending on the values of the parameters, it may take a large number of kicks (here: $n \gtrsim 60000$) before it becomes clear that asymptotically the energy growth is linear with time. For larger values of V_0, typically this regime of diffusive energy growth is reached after a smaller number of kicks.

The existence of diffusive energy growth is confirmed by considering the rate of growth of $\langle E \rangle_n^{\mathrm{cl}}$ analytically. As in subsection 1.2.4, the "random x_n, p_n approximation" can be applied when unbounded motion is granted, such that the result of equation (1.78) is obtained again. Therefore, in the case of nonresonance for large enough values of V_0, or for $(x_0, p_0)^t$ far enough from the origin, the rate of growth of the energy should be expected to increase quadratically with V_0, as in the case of resonance.

Chapter 2

Quantum Mechanics of the Kicked Harmonic Oscillator

Züricher Lokalaberglauben.

WOLFGANG PAULI commenting on ERWIN SCHRÖDINGER's interpretation of quantum mechanics [HvMW79].

In chapter 1 the kicked harmonic oscillator has been introduced, and typical properties of this model system have been discussed within the framework of *classical mechanics*. On the other hand, it is the main objective of this study to investigate the *quantum mechanics* of the kicked harmonic oscillator and to compare that system's differing dynamical properties with respect to these two dynamical theories. To this end a quantum description of the model, consistent with the classical formulation of the previous chapter, is supplied in the present chapter.

In section 2.1 the Hamiltonian, scaled in a similar way as its classical counterpart, is given and a quantum map for the quantum states that takes the role of the classical POINCARÉ map is derived. For this derivation, based on the periodicity of the Hamiltonian in time, I employ FLOQUET theory, some results of which are also an important prerequisite for the discussion of the theory of ANDERSON-like localization in chapter 5. The derivation of the quantum map explicitly relies on the δ-shaped time-dependence of the kicks, as chosen in equation (1.12) for the excitation of the harmonic oscillator. A different approach has been chosen, for example, in [VT99] by replacing the δ-functions with rectangularly peaked functions of finite width. While this simplifies the resulting time evolution operator in the sense that the dynamics becomes piecewise autonomous, at the same time this approach rules out the existence of stochastic webs which are in the focus of attention here — cf. the discussion in subsection 1.1.4, in particular with respect to [Vec95]. Therefore, in order to study the quantum counterparts of stochastic webs, keeping the δ-shaped kicks is essential. What is more, although they make the dynamics nonautonomous in a nontrivial way, the δ-kicks have the advantage of allowing to compute the quantum kick dynamics in a comparatively simple way.

Another way of quantizing the problem is to consider the equation of motion not for the quantum states, but for the relevant operators, i.e. for $\hat{x}$ and $\hat{p}$, or for the annihilation and creation operators $\hat{a}$, $\hat{a}^\dagger$. This leads, in section 2.2, to two more versions of the quantum map, this time in the HEISENBERG picture rather than the SCHRÖDINGER picture.

In section 2.3 I then argue that it is difficult to discuss the classical limit of the quantum dynamics and its semiclassical approximation based on the analytic formulae for the quantum maps alone. This provides the motivation for the numerical analysis of the system in the following chapters.

2.1 The Quantum Map

The quantum dynamics of the unscaled kicked harmonic oscillator with the Hamiltonian function (1.12) is governed by the SCHRÖDINGER equation

$$\hat{H}\,|\psi(t)\rangle \;=\; i\hbar\frac{\partial}{\partial t}\,|\psi(t)\rangle\,, \tag{2.1}$$

where $\hat{H}$ is the explicitly time-dependent Hamiltonian operator

$$\hat{H} \;=\; \hat{H}(\hat{x},\hat{p},t) \;=\; \frac{1}{2m_0}\hat{p}^2 + \frac{1}{2}m_0\omega_0^2\hat{x}^2 + V_0T\cos k\hat{x}\sum_{n=-\infty}^{\infty}\delta(t-nT) \tag{2.2}$$

and $|\psi(t)\rangle$ the quantum state at time t. As in the classical case I switch to dimensionless variables by means of the scaling transformation (1.15). In the quantum mechanical context the classical momentum scaling can be interpreted as the implicit definition of a scaled PLANCK constant $\hbar$, since equation (1.15b) and the usual definition of the momentum operator in the position representation,

$$\hat{p} \;=\; \frac{\hbar}{i}\frac{\partial}{\partial x}, \tag{2.3}$$

combined with the position scaling (1.15a) — which plays the same role classically and quantum mechanically — yield the scaling

$$\frac{k^2}{m_0 \omega_0} \hbar \longmapsto \hbar, \tag{2.4}$$

resulting in a dimensionless $\hbar$. The definition (2.3) of the momentum operator holds both before and after scaling; in particular, $\hbar$ is retained after scaling, albeit in scaled form — in contrast to the scaling used elsewhere [BRZ91]. Using the scaled PLANCK constant and the parameters V_0 and T, scaled according to equations (1.16), I obtain the Hamiltonian operator in its dimensionless form:

$$\boxed{\hat{H} \;=\; -\frac{\hbar^2}{2}\frac{\partial^2}{\partial x^2} + \frac{1}{2}x^2 + V_0 \cos x \sum_{n=-\infty}^{\infty} \delta(t - nT).} \tag{2.5}$$

This Hamiltonian is to be used in conjunction with the SCHRÖDINGER equation (2.1) where the scaled version of $\hbar$ is employed.

By virtue of the scaling one is left with only the three dimensionless parameters V_0, T and $\hbar$. The first two of these describe the nature of the kick and have to be considered both in the classical and the quantum realms, whereas the third — and only the third — is a genuinely quantum mechanical parameter.

As discussed in the Introduction (pages 1ff), the main objective of the theory of quantum chaos is the investigation of the way in which the dynamics of the system changes when advancing from the quantum to the classical case, i.e. when passing from $\hbar \neq 0$ via the semiclassical $\hbar \approx 0$ to the limiting case $\hbar = 0$.

For this purpose, comparison of the two dynamical theories of classical and quantum mechanics, the scaling used here is more appropriate than the one used in [BRZ91], for example: there, the oscillator length $\sqrt{\hbar/m_0\omega_0}$ is used to scale lengths, which is a natural choice in the quantum context, but makes comparison with the classical case more difficult, as using this scale in classical mechanics does not make sense. This problem is avoided here by measuring lengths in units of k, given by the kick function, which is present in both dynamical theories in exactly the same way. As a result, in this scaling the only parameter involving quantum effects is $\hbar$, and the other two remaining parameters V_0 and T both play the same role classically and quantum mechanically.

2.1.1 Floquet Theory

The natural quantum analogue of the classical map (1.21) is obtained by considering the evolution of quantum states during one period T of the excitation.

Therefore I define, in close analogy with (1.19), the state immediately before the n-th kick as

$$|\psi_n\rangle \;:=\; \lim_{t\nearrow nT} |\psi(t)\rangle\,, \quad n \in \mathbb{Z}. \tag{2.6}$$

The *quantum map* is then given by the propagator $\hat{U}_n$ for a period of time of length T, acting on $|\psi_n\rangle$:

$$|\psi_{n+1}\rangle \;=\; \hat{U}_n\,|\psi_n\rangle\,. \tag{2.7}$$

$\hat{U}_n$ is a special case of the general time evolution operator $\hat{U}(t', t)$:

$$\hat{U}_n \;:=\; \lim_{t\nearrow nT} \hat{U}\,(t+T, t)\,, \tag{2.8}$$

where $\hat{U}(t', t)$ is defined to take the quantum state from time t to time t',

$$\hat{U}(t', t) :\; |\psi(t)\rangle \longmapsto |\psi(t')\rangle\,, \tag{2.9}$$

and satisfies

$$\hat{U}(t', t) \;=\; \hat{U}(t', t'')\,\hat{U}(t'', t) \quad \forall\, t'' \in \mathbb{R}, \tag{2.10}$$

as usual. By the Hermiticity of the Hamiltonian and the initial condition $\hat{U}(t, t) = \mathbb{1}$, the time evolution operator is unitary,

$$\left(\hat{U}(t', t)\right)^{-1} \;=\; \hat{U}(t, t') \;=\; \left(\hat{U}(t', t)\right)^{\dagger}. \tag{2.11}$$

For a system with a time-periodic Hamiltonian

$$\hat{H}(t+T) \;=\; \hat{H}(t) \quad \forall\, t \in \mathbb{R}, \tag{2.12}$$

as in the present case of equation (2.5), $\hat{U}(t', t)$ is also time-periodic in the sense of

$$\hat{U}(t'+T, t+T) \;=\; \hat{U}(t', t). \tag{2.13}$$

Therefore the propagator (2.8) is the same for all n,

$$\hat{U} \;:=\; \hat{U}_n \quad \forall\, n, \tag{2.14}$$

and the quantum map (2.7) simplifies to

$$|\psi_{n+1}\rangle \;=\; \hat{U}\,|\psi_n\rangle \tag{2.15}$$

for all iterations.

The time-T-propagator $\hat{U}$ is also known as the FLOQUET operator of the quantum system. This naming convention is due to the fact that, using a time-independent orthonormal basis $\{|\phi_n\rangle\}$ of HILBERT space for expanding $|\psi(t)\rangle$ into $\sum_n a_n(t)\,|\phi_n\rangle$, the time-dependent SCHRÖDINGER equation (2.1) can be transformed into a system of ordinary linear differential equations, the coefficients of which are T-periodic because $\hat{H}$ exhibits the same periodicity. For a finite basis this is the setting of the FLOQUET theorem which asserts existence and uniqueness of the solutions and explicitly states their functional dependence on t [Flo83, YS75]. In the present case the FLOQUET theorem does not apply as the HILBERT space is infinite-dimensional, but nevertheless the typical form of the FLOQUET solution and several other properties do carry over [Sal74]. Therefore, by analogy, the quantum theory of systems with time-periodic Hamiltonians is often also called FLOQUET theory.

Using the time ordering operator $\hat{\mathcal{Z}}$,

$$\hat{\mathcal{Z}}\left(\hat{A}(t)\,\hat{B}(t')\right) \;=\; \begin{cases} \hat{A}(t)\,\hat{B}(t') & \text{if} \quad t > t' \\ \hat{B}(t')\,\hat{A}(t) & \phantom{\text{if}} \quad t < t', \end{cases} \tag{2.16}$$

the FLOQUET operator can formally be written as [Sch02]

$$\hat{U} \;=\; \lim_{t \nearrow nT} \hat{\mathcal{Z}} \exp\left(-\frac{i}{\hbar}\int_t^{t+T} \mathrm{d}t'\,\hat{H}(t')\right), \tag{2.17}$$

but it is difficult to evaluate this expression for general time-dependent Hamiltonians. For autonomous $\hat{H}$, on the other hand, $\hat{U}$ is found in its usual form as an exponential of the Hamiltonian, $e^{-\frac{i}{\hbar}\hat{H}T}$; this can be used for the free (i.e. unkicked) propagation part of the Hamiltonian (2.5). The propagator for the explicitly time-dependent (kick) part of the Hamiltonian (2.5) is found in the following subsection by direct integration of the SCHRÖDINGER equation. Before turning to the calculation of $\hat{U}$ for the full Hamiltonian (2.5), I now discuss some general properties of the FLOQUET operator that are of importance later on in chapter 5.

Consider the (normalized) eigenstates $|\phi_E\rangle$ of the FLOQUET operator $\hat{U}$ with respect to the eigenvalues λ_E:

$$\hat{U}\,|\phi_E\rangle \;=\; \lambda_E\,|\phi_E\rangle\,; \tag{2.18}$$

for the index $E \in \mathbb{R}$ see below. Using

$$\hat{U}(t,-0) := \lim_{t_0 \nearrow 0} \hat{U}(t,t_0), \tag{2.19}$$

time-dependent solutions

$$|\phi_E(t)\rangle := \hat{U}(t,-0)\,|\phi_E\rangle \tag{2.20}$$

of the SCHRÖDINGER equation (2.1) with respect to the initial condition $|\psi_0\rangle = |\phi_E\rangle$ can be constructed for all times t. I now discuss some properties of these solutions for general t, although later on mainly solutions for the stroboscopic times $nT - 0$ are needed.

Since $|\phi_E\rangle$ is an eigenstate of $\hat{U}$ with respect to the eigenvalue λ_E, $|\phi_E(t)\rangle$ is an eigenstate of $\hat{U}(t+T,t)$ with respect to the same eigenvalue λ_E:

$$\begin{aligned} \hat{U}(t+T,t)\,|\phi_E(t)\rangle &= \hat{U}(t+T,t)\,\hat{U}(t,-0)\,|\phi_E\rangle \\ &= \hat{U}(t+T,T-0)\,\hat{U}\,|\phi_E\rangle \\ &= \lambda_E\,|\phi_E(t)\rangle\,, \end{aligned} \tag{2.21}$$

where in the last step the periodicity (2.13) of $\hat{U}(t',t)$ has been used. Therefore the $|\phi_E(t)\rangle$ can all be labelled by the same index E, indicating the same eigenvalue for all t.

Because of the unitarity of $\hat{U}$ its eigenvalues are of unit modulus and can be written as

$$\lambda_E =: e^{-\frac{i}{\hbar}ET}. \tag{2.22}$$

The motivation for this formulation of the parameter dependence of the eigenvalues and for labelling the eigenstates by E becomes clear when investigating the time dependence of the $|\phi_E(t)\rangle$. By the definition

$$|\phi_E(t)\rangle =: e^{-\frac{i}{\hbar}Et}\,|u_E(t)\rangle \tag{2.23}$$

for the *reduced states* $|u_E(t)\rangle$, the trivial part of the time dependence of the $|\phi_E(t)\rangle$, corresponding to the time dependence of an energy eigenstate with respect to an autonomous system with energy E, is effectively separated off, and it remains to discuss the $|u_E(t)\rangle$. Obviously, these states inherit the periodicity (2.12) of the Hamiltonian:

$$|u_E(t+T)\rangle = |u_E(t)\rangle\,. \tag{2.24}$$

This property provides the basis for some of the considerations in chapter 5.

The definition (2.23) is tailored to make the description of the time dependence of the $|\phi_E(t)\rangle$ as similar to the dynamics of an autonomous system as possible. What is more, for stroboscopic times these two types of dynamics coincide,

$$|\phi_E(t+T)\rangle = e^{-\frac{i}{\hbar}ET}\,|\phi_E(t)\rangle\,, \tag{2.25}$$

which can be seen by combining equations (2.23) and (2.24); as required, equation (2.25) reproduces the result (2.21).

Since the parameter E plays a similar role as the energy eigenvalue of a time-independent system, E is called a *quasienergy* of the time-periodic Hamiltonian, and the FLOQUET states $|\phi_E(t)\rangle$ are referred to as its *quasienergy states* [Zel67]. For brevity, often the states $|u_E(t)\rangle$ are called (reduced) quasienergy states, too. The quasienergy is defined modulo $2\pi\hbar/T$ only, since it originates from the exponential in equation (2.22). Due to this nonuniqueness of the quasienergy it cannot be identified with any physical observable in a straightforward way, but note that, for an unscaled system, the quasienergy has the dimension of an energy. A discussion of the problems potentially arising from identifying the quasienergy with the conventional energy — and thereby linking the quasienergy spectrum directly with the resonance (emission/absorption) spectrum of the respective system — may be found in [DM98]. Normally, one restricts E to the interval $[0, 2\pi\hbar/T)$. By equation (2.21), the case of $E = 0$, i.e. $\lambda_E = 1$, corresponds to the quantum map's stationary states, for which not only the reduced $|u_0(t)\rangle$, but also the full FLOQUET states $|\phi_0(t)\rangle$ are periodic with period T.

The quasienergy states are characterized by several useful properties. It is easy to show that for $E_1 \neq E_2$ the quasienergy states $|\phi_{E_1}(t)\rangle$, $|\phi_{E_2}(t)\rangle$ are orthogonal: their scalar product is

$$\langle\phi_{E_1}(t)\,|\phi_{E_2}(t)\rangle = e^{\frac{i}{\hbar}(E_1-E_2)t}\,\langle u_{E_1}(t)\,|u_{E_2}(t)\rangle\,, \tag{2.26}$$

where the exponential is periodic with a period larger than T, whereas the period of $\langle u_{E_1}(t)\,|u_{E_2}(t)\rangle$ is smaller than or equal to T; this means that $\langle\phi_{E_1}(t)\,|\phi_{E_2}(t)\rangle$ must be zero, as every pair of solutions of the SCHRÖDINGER equation yields a constant scalar product. Furthermore, $\{\,|\phi_E(t)\rangle\,|\,0 \le E < 2\pi\hbar/T\,\}$ is a complete set [Zel67, Per93]. The last two properties combined imply that the set $\{|\phi_E(t)\rangle\}$ can be used in the conventional way as a basis for expanding arbitrary states of the system,

$$|\psi(t)\rangle = \sum_E A_E\,|\phi_E(t)\rangle\,, \tag{2.27}$$

with *constant* (i.e. time-independent) expansion coefficients $A_E \in \mathbb{C}$ [KW96].

Summarizing, with respect to a time-periodic Hamiltonian the quasienergies and the quasienergy states play much the same role as the energy eigenvalues and the stationary energy eigenstates do with respect to a time-independent Hamiltonian [Sam73]. This analogy includes the observation that in the same way as any solution of the time-independent SCHRÖDINGER equation can be expanded in terms of energy eigenstates with constant coefficients, the same can be accomplished using quasienergy states in the FLOQUET case.

2.1.2 The Floquet Operator of the Kicked Harmonic Oscillator

Having discussed the properties of the FLOQUET operator from a general point of view, I now proceed to the investigation of the specific $\hat{U}$ defined by equations (2.8) and (2.14) with the Hamiltonian (2.5) of the kicked harmonic oscillator.

As its classical counterpart M in equations (1.20), $\hat{U}$ can be decomposed into a contribution $\hat{U}_{\text{kick}}$, describing the kick, and the propagator of the free harmonic oscillator dynamics for time T, $\hat{U}_{\text{free}}$:

$$\hat{U} = \hat{U}_{\text{free}} \hat{U}_{\text{kick}}, \tag{2.28}$$

corresponding to the Hamiltonians

$$\hat{H}_{\text{free}} = \frac{1}{2}\left(\hat{p}^2 + \hat{x}^2\right) \tag{2.29a}$$

$$\hat{H}_{\text{kick}} = V_0 \cos \hat{x} \sum_{n=-\infty}^{\infty} \delta(t - nT). \tag{2.29b}$$

The free propagator — as many more expressions to follow — is most conveniently expressed in terms of the ladder operators

$$\hat{a} := \frac{1}{\sqrt{2\hbar}}\left(\hat{x} + i\hat{p}\right) \tag{2.30a}$$

$$\hat{a}^\dagger := \frac{1}{\sqrt{2\hbar}}\left(\hat{x} - i\hat{p}\right), \tag{2.30b}$$

which can be used to write

$$\hat{H}_{\text{free}} = \hbar\left(\hat{a}^\dagger \hat{a} + \frac{1}{2}\right) \tag{2.31}$$

and thus

$$\hat{U}_{\text{free}} = e^{-\frac{i}{\hbar}\hat{H}_{\text{free}}T} \tag{2.32a}$$

$$= e^{-iT\left(\hat{a}^\dagger\hat{a}+\frac{1}{2}\right)}. \tag{2.32b}$$

Note that this means that the free propagator in the harmonic oscillator eigenrepresentation solely depends on T and not on $\hbar$ (naturally it does not depend on V_0 either): $\hat{U}_{\text{free}} = \hat{U}_{\text{free}}(T)$; cf. equation (2.44) below.

The computation of the kick propagator requires slightly more effort because of the explicit time-dependence of the kick part of the Hamiltonian. Dividing the SCHRÖDINGER equation in coordinate representation by the corresponding wave function $\psi(x,t) := \langle x\,|\psi(t)\rangle$, and integrating over time at the n-th kick I obtain

$$i\hbar\int\limits_{nT-\varepsilon}^{nT+\varepsilon}\frac{1}{\psi}\frac{\partial\psi}{\partial t}\,\mathrm{d}t = \int\limits_{nT-\varepsilon}^{nT+\varepsilon}\frac{\hbar}{\psi}\left(\hat{a}^\dagger\hat{a}+\frac{1}{2}\right)\psi\,\mathrm{d}t + V_0\cos x\int\limits_{nT-\varepsilon}^{nT+\varepsilon}\delta(t-nT)\,\mathrm{d}t. \tag{2.33}$$

Taking the limit $\varepsilon \to 0$, the first term on the right hand side vanishes, such that

$$|\psi(nT+0)\rangle = e^{-\frac{i}{\hbar}V_0\cos\hat{x}}\,|\psi(nT-0)\rangle \tag{2.34}$$

and therefore

$$\hat{U}_{\text{kick}} = e^{-\frac{i}{\hbar}V_0\cos\hat{x}} \tag{2.35a}$$

$$= e^{-\frac{i}{\hbar}V_0\cos\left(\sqrt{\frac{\hbar}{2}}\left(\hat{a}^\dagger+\hat{a}\right)\right)}. \tag{2.35b}$$

In the position representation the kick propagator thus depends on the quotient of V_0 and $\hbar$ only: $\hat{U}_{\text{kick}} = \hat{U}_{\text{kick}}(V_0/\hbar)$.

Summarizing, this gives for the full FLOQUET operator

$$\hat{U} = e^{-iT\left(\hat{a}^\dagger\hat{a}+\frac{1}{2}\right)}\,e^{-\frac{i}{\hbar}V_0\cos\left(\sqrt{\frac{\hbar}{2}}\left(\hat{a}^\dagger+\hat{a}\right)\right)}, \tag{2.36}$$

and the full quantum map for the kicked harmonic oscillator is obtained as

$$\boxed{|\psi_{n+1}\rangle = e^{-iT\left(\hat{a}^\dagger\hat{a}+\frac{1}{2}\right)}\,e^{-\frac{i}{\hbar}V_0\cos\hat{x}}\,|\psi_n\rangle\,.} \tag{2.37}$$

In the last equation a "mixed" notation, using both $\hat{a}, \hat{a}^\dagger$ and $\hat{x}$, is used, not only because this is the most concise form of the quantum map, but also because it is in this form that the quantum map (2.37) is evaluated in the following subsection.

Note that equation (2.37) cannot be simplified significantly by restricting the discussion to values of T that satisfy a resonance condition (1.23) as in the classical case (cf. equations (1.25–1.29)). This statement also holds with respect to the explicit expressions for the matrix elements of $\hat{U}$ that I derive in subsection 2.1.3; there, T may take any value as well. For the comparison of classical and quantum results on stochastic webs, one has to concentrate on resonant values of T. The way in which the choice of T controls the existence of quantum mechanical periodic stochastic webs is discussed in chapter 4; for the discussion of ANDERSON localization in chapter 5, nonresonant values of T are in the focus of attention.

2.1.3 Matrix Elements of the Floquet operator

It is the objective of the next chapters to study the time evolution of initial wave packets $|\psi_0\rangle$ under the iteration of the quantum map. One way to do this is to expand the initial state into a series of eigenstates of the harmonic oscillator and then use the corresponding matrix representation of $\hat{U}$ for the iteration. In this subsection I supply the formulae that are needed for this approach. Similar results may be found in [BRZ91].

The number states $|m\rangle$, i.e. the eigenstates of the unkicked quantum harmonic oscillator given by the Hamiltonian (2.29a), are the solutions of the eigenvalue equation

$$\hat{H}_{\text{free}}\,|m\rangle \;=\; \hbar\left(m+\frac{1}{2}\right)|m\rangle\,, \quad m \in \mathbb{N}_0\,. \tag{2.38}$$

It is well known that in the position representation these eigenstates can be written as

$$\langle x\,|m\rangle \;=\; \frac{1}{\sqrt[4]{\pi\hbar}\,\sqrt{2^m m!}}\,e^{-\frac{x^2}{2\hbar}}\,\mathrm{H}_m\!\left(\frac{x}{\sqrt{\hbar}}\right), \tag{2.39}$$

where the H_m are HERMITE polynomials — see for example [Gra89]. Using $\{|m\rangle\}$ as a basis, any $|\psi_n\rangle$ can be expanded into

$$|\psi_n\rangle \;=\; \sum_{m=0}^{\infty} a_m^{(n)}\,|m\rangle \tag{2.40}$$

(with the expansion coefficients $a_m^{(n)} \in \mathbb{C}$), which evolves, after one iteration of the quantum map, into

$$|\psi_{n+1}\rangle = \sum_{m,m'=0}^{\infty} U_{mm'}\, a_{m'}^{(n)}\, |m\rangle\,, \tag{2.41}$$

where the

$$U_{mm'} := \langle m|\hat{U}|m'\rangle \tag{2.42}$$

are the matrix elements of the FLOQUET operator. In terms of the coefficients $a_m^{(n)}$ this may also be formulated as

$$a_m^{(n+1)} = \sum_{m'=0}^{\infty} U_{mm'}\, a_{m'}^{(n)}. \tag{2.43}$$

In other words, the matrix $U = (U_{mm'})$ is the propagator of the quantum map in the eigenrepresentation of the free harmonic oscillator. It remains to derive an explicit expression for the matrix elements in order to complete the scheme given by equation (2.43) for iterating the quantum map.

In the harmonic oscillator eigenrepresentation the free propagator (2.32b) is diagonal:

$$\left\langle m\middle|\hat{U}_{\text{free}}\middle|m'\right\rangle = e^{-iT\left(m+\frac{1}{2}\right)}\,\delta_{mm'}. \tag{2.44}$$

Using this expression and the splitting (2.28), $U_{mm'}$ can be written as

$$U_{mm'} = e^{-iT\left(m+\frac{1}{2}\right)}\,K_{mm'}, \tag{2.45}$$

where the *kick matrix elements* are defined as

$$K_{mm'} := \left\langle m\middle|\,\hat{U}_{\text{kick}}\,\middle|m'\right\rangle \tag{2.46a}$$

$$= \left\langle m\middle|\,e^{-\frac{i}{\hbar}V_0\cos\hat{x}}\,\middle|m'\right\rangle. \tag{2.46b}$$

Due to the oscillator eigenstates in the position representation (2.39) being real-valued functions, the kick matrix elements are symmetric:

$$K_{mm'} = K_{m'm}. \tag{2.47}$$

A FOURIER expansion of the kick propagator $e^{-\frac{i}{\hbar}V_0 \cos\hat{x}}$ (cf. [AS72]) gives

$$K_{mm'} = \sum_{l=-\infty}^{\infty} (-i)^l \, \mathrm{J}_l\left(\frac{V_0}{\hbar}\right) \left\langle m \left| e^{il\hat{x}} \right| m' \right\rangle, \tag{2.48}$$

with the BESSEL functions J_l. The matrix element $\left\langle m \left| e^{il\hat{x}} \right| m' \right\rangle$ on the right hand side can be evaluated by means of the formula (see [GR00]):

$$\int_{-\infty}^{\infty} e^{-x^2} \, \mathrm{H}_m(x+y) \, \mathrm{H}_{m'}(x+z) \, \mathrm{d}x = m'! \, 2^m \sqrt{\pi} \, z^{m-m'} \, \mathrm{L}_{m'}^{(m-m')}(-2yz), \tag{2.49}$$

holding for $m \geq m'$; the $\mathrm{L}_{m'}^{(m-m')}$ are generalized LAGUERRE polynomials. With equation (2.49) I get

$$\left\langle m \left| e^{il\hat{x}} \right| m' \right\rangle = \sqrt{\frac{m'!}{m!}} \left(i\sqrt{\frac{\hbar}{2}} l \right)^{m-m'} e^{-\frac{\hbar l^2}{4}} \, \mathrm{L}_{m'}^{(m-m')}\left(\frac{\hbar l^2}{2}\right), \quad m \geq m', \tag{2.50}$$

and using the elementary property $\mathrm{J}_{-l}(x) = (-1)^l \, \mathrm{J}_l(x)$ of the BESSEL functions, I finally obtain

$$K_{mm'} = \begin{cases} \mathrm{J}_0\left(\frac{V_0}{\hbar}\right) \delta_{mm'} + 2 \sum_{l=1}^{\infty} (-i)^{l+m'-m} \times & \\ \times \, \mathrm{J}_l\left(\frac{V_0}{\hbar}\right) \sqrt{\frac{m'!}{m!}} \left(\sqrt{\frac{\hbar}{2}} l \right)^{m-m'} \times & \\ \times \, e^{-\frac{\hbar l^2}{4}} \, \mathrm{L}_{m'}^{(m-m')}\left(\frac{\hbar l^2}{2}\right) & m-m' \text{ even}, \, m \geq m' \\ K_{m'm} & \text{for} \quad m-m' \text{ even}, \, m \leq m' \\ 0 & m-m' \text{ odd}. \end{cases} \tag{2.51}$$

Note that $U_{mm'} = K_{mm'} = 0$ for $m - m'$ odd may be viewed as a *selection rule* that — with equation (2.46b) — follows directly from the fact that the kick potential (1.18) has been chosen as $V(x) = V_0 \cos x$; every other kick potential with the same symmetry property — i.e. being an even function of x — would lead to the same selection rule. As a result, the dynamics in the harmonic oscillator eigenrepresentation decomposes into two parts, taking place in the dynamically disconnected subspaces of HILBERT space that are given by the states $|n\rangle$ with odd and even indices, respectively.[1]

With equations (2.45) and (2.51) the matrix elements of the FLOQUET operator are completely specified. For the above derivation no assumption concerning the (resonant or nonresonant) value of T had to be made; also note that T enters $U_{mm'}$ via the elementary exponential $e^{-iT(m+1/2)}$ in equation (2.45) only and does not interfere with the more complicated calculation of $K_{mm'}$.

The rest can be left to a computer with large memory. The way in which formula (2.51) can be evaluated efficiently within a computer program is discussed in chapter 3.

2.2 The Classical Limit

It is instructive to use the quantum mechanical system introduced in the previous section to rederive the classical POINCARÉ map (1.21). This is most conveniently done in the HEISENBERG picture, rather than in the SCHRÖDINGER picture that has been used in section 2.1 to formulate the quantum map. In the present section, and only here, I use the indices $_{\mathrm{H}}$ and $_{\mathrm{S}}$ in order to distinguish between HEISENBERG and SCHRÖDINGER operators. (All operators used elsewhere without explicit reference to a particular picture are SCHRÖDINGER operators.)

Using the time evolution operator $\hat{U}(t', t)$, the time-dependent annihilation operator $\hat{a}_{\mathrm{H}}(t)$ can be defined as

$$\hat{a}_{\mathrm{H}}(t) := \left(\hat{U}(t, nT - 0)\right)^{\dagger} \hat{a}_{\mathrm{S}} \, \hat{U}(t, nT - 0); \tag{2.52}$$

the time-independent annihilation operator $\hat{a}_{\mathrm{S}}$ has been defined in equation (2.30a), and the initial condition

$$\hat{a}_{\mathrm{H}}(nT - 0) := \hat{a}_{\mathrm{S}} \tag{2.53}$$

immediately before the n-th kick has been chosen. This choice facilitates the comparison with the classical POINCARÉ map further below.

The dynamics of $\hat{a}_{\mathrm{H}}(t)$ is governed by the HEISENBERG equation of motion [Mes91],

$$i\hbar \frac{\mathrm{d}\hat{a}_{\mathrm{H}}}{\mathrm{d}t} = \left[\hat{a}_{\mathrm{H}}, \hat{H}_{\mathrm{H}}\right] \tag{2.54}$$

[1] A similar manifestation of this symmetry of the kick potential arises in subsection 5.3.1 below, where the selection rule (5.88) leads to two discrete SCHRÖDINGER equations (5.94) that are interwoven with but independent of each other.

$[\cdot,\cdot]$ being the usual commutator. For the *kick dynamics* at the n-th kick I have

$$i\hbar\,\frac{\mathrm{d}\hat{a}_{\mathrm{H}}}{\mathrm{d}t} \;=\; V_0\left[\hat{a}_{\mathrm{H}}, e^{\frac{i}{\hbar}V_0\cos\hat{x}_{\mathrm{S}}}\cos\hat{x}_{\mathrm{S}}\,e^{-\frac{i}{\hbar}V_0\cos\hat{x}_{\mathrm{S}}}\right]\delta(t-nT), \tag{2.55}$$

where I have discarded the time-independent part of the Hamiltonian $\hat{H}_{\mathrm{S}}$ (2.5) and used the expression (2.35a) for $\hat{U}_{\mathrm{kick}} = \hat{U}(nT+0, nT-0)$. Equation (2.55) and the following expressions up to (2.59) hold for $t \in [nT-\varepsilon, nT+\varepsilon]$ in the limit of $\varepsilon \to 0$ only. Using equation (2.52) and well-known commutator algebra, the commutator on the right hand side can be written as

$$e^{\frac{i}{\hbar}V_0\cos\hat{x}_{\mathrm{S}}}\,[\hat{a}_{\mathrm{S}},\cos\hat{x}_{\mathrm{S}}]\,e^{-\frac{i}{\hbar}V_0\cos\hat{x}_{\mathrm{S}}},$$

and by direct computation it is easily shown that

$$[\hat{a}_{\mathrm{S}},\cos\hat{x}_{\mathrm{S}}] \;=\; -\sqrt{\frac{\hbar}{2}}\sin\hat{x}_{\mathrm{S}}. \tag{2.56}$$

Thus the HEISENBERG equation for the kick becomes

$$i\hbar\,\frac{\mathrm{d}\hat{a}_{\mathrm{H}}}{\mathrm{d}t} \;=\; -V_0\sqrt{\frac{\hbar}{2}}\,\sin\hat{x}_{\mathrm{S}}\,\delta(t-nT), \tag{2.57}$$

which, with

$$\hat{x}_{\mathrm{S}} \;=\; \hat{x}_{\mathrm{H}}(nT-0), \tag{2.58}$$

gives by integration over the kick

$$\hat{a}_{\mathrm{H}}(nT+0) \;=\; \hat{a}_{\mathrm{H}}(nT-0) + i\frac{V_0}{\sqrt{2\hbar}}\,\sin\hat{x}_{\mathrm{H}}(nT-0). \tag{2.59}$$

Solving the HEISENBERG equation *between* two consecutive kicks, i.e. for the unkicked harmonic oscillator, is somewhat easier. One has

$$i\hbar\,\frac{\mathrm{d}\hat{a}_{\mathrm{H}}}{\mathrm{d}t} \;=\; \left[\hat{a}_{\mathrm{H}}, \hbar\left(\hat{a}_{\mathrm{H}}^{\dagger}\hat{a}_{\mathrm{H}} + \frac{1}{2}\right)\right] \;=\; \hbar\hat{a}_{\mathrm{H}}, \tag{2.60}$$

and therefore

$$\hat{a}_{\mathrm{H}}((n+1)T-0) \;=\; \hat{a}_{\mathrm{H}}(nT+0)\,e^{-iT}, \tag{2.61}$$

such that the complete quantum map in terms of the ladder operators is obtained as

$$\hat{a}_{n+1} = \left(\hat{a}_n + i \frac{V_0}{\sqrt{2\hbar}} \sin \left(\sqrt{\frac{\hbar}{2}} \left(\hat{a}_n^\dagger + \hat{a}_n \right) \right) \right) e^{-iT}, \tag{2.62}$$

where the notation

$$\hat{a}_n := \hat{a}_{\rm H}(nT - 0) \tag{2.63}$$

has been adopted for notational convenience; $\hat{x}_n$ and $\hat{p}_n$ (to be used below) are defined accordingly. By inversion of equations (2.30), substituting $\hat{a}_{n+1}$ and its Hermitian adjoint into the resulting expressions, and using equations (2.30) again, I finally get the quantum map for the position and momentum operators in the HEISENBERG picture:

$$\hat{x}_{n+1} = \sqrt{\frac{\hbar}{2}} \left(\hat{a}_{n+1}^\dagger + \hat{a}_{n+1} \right) = \hat{x}_n \cos T + (\hat{p}_n + V_0 \sin \hat{x}_n) \sin T \tag{2.64a}$$

$$\hat{p}_{n+1} = i\sqrt{\frac{\hbar}{2}} \left(\hat{a}_{n+1}^\dagger - \hat{a}_{n+1} \right) = -\hat{x}_n \sin T + (\hat{p}_n + V_0 \sin \hat{x}_n) \cos T. \tag{2.64b}$$

An equivalent way of deriving this form of the quantum map is to consider

$$\hat{H}_{\rm H} = \frac{1}{2} \left(\hat{p}_{\rm H}^2 + \hat{x}_{\rm H}^2 \right) + V_0 \cos \hat{x}_{\rm H} \sum_{n=-\infty}^{\infty} \delta(t - nT), \tag{2.65}$$

determine the canonical HEISENBERG equations of motion for $\hat{x}_{\rm H}$ and $\hat{p}_{\rm H}$,

$$\dot{\hat{x}}_{\rm H} = \frac{\partial \hat{H}_{\rm H}}{\partial \hat{p}_{\rm H}} = \hat{p}_{\rm H} \tag{2.66a}$$

$$\dot{\hat{p}}_{\rm H} = -\frac{\partial \hat{H}_{\rm H}}{\partial \hat{x}_{\rm H}} = -\hat{x}_{\rm H} + V_0 \sin \hat{x}_{\rm H} \sum_{n=-\infty}^{\infty} \delta(t - nT), \tag{2.66b}$$

and solve them in the same way as the classical equations of motion in subsection 1.1.3. In this way the discussion of the ladder operators as above is avoided; on the other hand the most condensed form (2.62) of the quantum map in the HEISENBERG picture is not obtained as a byproduct, but has to be derived afterwards, if desired.

Equations (2.62) and (2.64) contain exactly the same information as their SCHRÖDINGER picture counterpart (2.37), but especially equations (2.64) are better suited than the SCHRÖDINGER picture quantum map to make contact with the classical POINCARÉ map, as I show immediately.

Substituting *formally* in the quantum map (2.64) the expectation values $\langle\hat{x}\rangle_n$, $\langle\hat{p}\rangle_n$ at time nT for the operators $\hat{x}_n$, $\hat{p}_n$ results in just the equations (1.21) of the classical POINCARÉ map (where the expectation values take the places of the corresponding classical observables):

$$\langle\hat{x}\rangle_{n+1} \approx \langle\hat{x}\rangle_n \cos T + \Big(\langle\hat{p}\rangle_n + V_0 \sin\langle\hat{x}\rangle_n\Big) \sin T \tag{2.67a}$$

$$\langle\hat{p}\rangle_{n+1} \approx -\langle\hat{x}\rangle_n \sin T + \Big(\langle\hat{p}\rangle_n + V_0 \sin\langle\hat{x}\rangle_n\Big) \cos T. \tag{2.67b}$$

This clear formal analogy between the classical and quantum maps is one of the advantages of the HEISENBERG picture. Note that the transition from the quantum mechanically exact equations (2.64) to (2.67) is an approximation that cannot be justified in general terms. In fact it is generally a quite bad approximation the quality of which decreases with growing $\Delta\hat{x}$ and $\Delta\hat{p}$.

The same approximation may be accomplished by replacing the operators $\hat{a}$, $\hat{a}^\dagger$ in equation (2.62) by their associated c-numbers α, α^* and taking into account the relationship of these with the classical observables x, p (or equivalently with the action-angle variables J, ϑ of equation (1.35)):

$$\hat{a} \;\widehat{=}\; \alpha = \frac{1}{\sqrt{2\hbar}}(x + ip) = \sqrt{\frac{J}{\hbar}}\, e^{i\vartheta}. \tag{2.68}$$

Decomposing the resulting expression into its real and imaginary parts, equations (2.67) are again obtained [Zas85].

The EHRENFEST equations of motion of the kicked harmonic oscillator, on the other hand, are obtained by taking the expectation values on both sides of equations (2.64) which gives the following, slightly different result:

$$\langle\hat{x}\rangle_{n+1} = \langle\hat{x}\rangle_n \cos T + \Big(\langle\hat{p}\rangle_n + V_0 \langle\sin\hat{x}\rangle_n\Big) \sin T \tag{2.69a}$$

$$\langle\hat{p}\rangle_{n+1} = -\langle\hat{x}\rangle_n \sin T + \Big(\langle\hat{p}\rangle_n + V_0 \langle\sin\hat{x}\rangle_n\Big) \cos T, \tag{2.69b}$$

the difference being marked by the fact that in general $\langle\sin\hat{x}\rangle_n \neq \sin\langle\hat{x}\rangle_n$. Equations (2.69) are quantum mechanically exact, in contrast to the approximative expressions (2.67). In the classical limit both the position and momentum distribution become δ-peaked, as they describe a single classical point particle. In that limit $\langle \sin \hat{x} \rangle_n$ and $\sin \langle \hat{x} \rangle_n$ become equal and the EHRENFEST equations coincide with (2.67), thus becoming formally equivalent to their classical counterparts (1.21).

However, the way in which the limit of classical behaviour is reached is not obvious at all, among other reasons due to the fact that the expectation values are taken with respect to quantum states (solutions of the quantum map) about which not much is known a priori. In the following subsection I address this question from another point of view, by considering the explicit parameter dependence of the FLOQUET operator.

2.3 Discussion of the Parameter Dependence

In this section the parameter dependence of the quantum map is discussed qualitatively, with some emphasis on the semiclassical limit. Because of the observation stated in the last paragraph of the previous section, the equations of that section are not well suited for this purpose. Therefore, in the present section mainly the FLOQUET operator in the form of equations (2.37) and (2.45, 2.51) is investigated.

Generally speaking, $U_{mm'}$ depends on the three parameters T, V_0 and $\hbar$. As mentioned above, only $\hat{U}_{\text{free}}$ (as opposed to $\hat{U}_{\text{kick}}$) is T-dependent; similarly, only $\hat{U}_{\text{kick}}$ (and therefore $K_{mm'}$) depends on the kick strength V_0. From inspection of (2.51) it is clear that V_0 enters the formulae via the quotient $V_0/\hbar$ only, an obvious consequence of the form (2.35a) of the kick propagator. In this sense, the quantum map in essence depends on the three parameters T, $V_0/\hbar$ and $\hbar$, rather than T, V_0 and $\hbar$.

The intricacy with respect to the parameter $\hbar$ makes the analysis of the semiclassical limit $\hbar \to 0$ particularly difficult: all nontrivial terms in equation (2.51) depend on $\hbar$ in a nontrivial way. This is not an artificially created consequence of the scaling (1.15, 2.4) used here, but a general phenomenon that occurs within all the different scalings that up to now have been used for the analysis of the kicked harmonic oscillator (see for example [BRZ91, SS92]).

The first thing to note about the quantum map (2.37) is that for $V_0/\hbar \to 0$ the kick propagator becomes the identity operator and the quantum map is essentially determined by just the rotation given by the free harmonic oscillator propagator. This becomes more clear by considering equations (2.45) and (2.51); they show that in that limit $K_{mm'}$, and thus $U_{mm'}$, become diagonal, such that the contributing harmonic oscillator eigenstates each perform their respective rotation dynamics in quantum phase space (cf. section A.6 of the appendix) without mutual perturbation.

There are several ways to obtain the case of $V_0/\hbar \to 0$; here I discuss two of them. First, one can keep $\hbar$ fixed and let V_0 tend to zero. This scenario agrees with the discussion of the dynamics of the classical POINCARÉ map (1.21) near the origin of phase space when the kick strength is decreased and approaching zero.

Second, when keeping V_0 fixed and increasing $\hbar$, i.e. moving away from the semiclassical towards the full quantum regime, another scenario with $V_0/\hbar \to 0$ is obtained which is quite different from the first: clearly the evaluation of the series in equation (2.51) becomes increasingly difficult with growing $\hbar$, as large degree LAGUERRE polynomials and powers of large arguments have to be evaluated and summed up. The numerical intricacy of the evaluation of the matrix elements for large values of $\hbar$ makes it quite impossible to follow this approach by direct calculation on a computer.

The reverse direction, the semiclassical limit with V_0 fixed and $\hbar$ tending to zero, which leads to $V_0/\hbar \to \infty$, is even more difficult to analyze, because for semiclassically small values of $\hbar$ many more terms contribute to the series in equation (2.51), due to the exponential $\exp(-\hbar l^2/4)$. This mathematical observation can also be understood from a physical point of view: for expanding a quantum state in the same part of phase space a much larger basis of (e.g. harmonic oscillator) eigenstates is needed for smaller values of $\hbar$, because the structures to be resolved become smaller, too. The only clear observation is that in this case there is no argument by which the kick propagator attains a simple form or even becomes trivial (in a similar way as for $V_0/\hbar \to 0$). So in the semiclassical limit there seems to be no way to make the desired contact with the classical POINCARÉ map *analytically*.

Summarizing, in order to study the correspondence between classical and quantum dynamics — especially in the semiclassical approximation — one has to revert to *numerical* means. After some preparations in chapter 3, this approach is followed in chapters 4 and 5 in the cases of resonance and nonresonance, respectively.

Chapter 3

Numerical Methods

Erst die natürlichen Betrachtungen gemacht,
ehe die subtilen kommen.

Κέρας 'Αμαλθείας (276)
GEORG CHRISTOPH LICHTENBERG

In the previous chapter it has been shown that in order to study the quantum dynamics of the kicked harmonic oscillator the time-dependent SCHRÖDINGER equation (2.1) needs to be solved for the Hamiltonian (2.5). Equivalently, the quantum map (2.37) is to be iterated. I now discuss some numerical methods that may be used for this purpose.

In section 3.1 two finite differences methods for solving the partial differential equation (2.1) in the position representation are presented; in section 3.2 I discuss how the the FLOQUET operator can be evaluated numerically in the position representation, and in section 3.3 some technical aspects of iterating the quantum map in the eigenrepresentation of the harmonic oscillator as outlined in subsection 2.1.3 are addressed.

The main result of this chapter is that from a numerical point of view the method best suited for studying the dynamics (including long-time effects) of the quantum kicked harmonic oscillator is the method described in subsection 2.1.3 and section 3.3. This is discussed in section 3.4.

3.1 Finite Differences

In the literature a multitude of finite differences methods for solving partial differential equations can be found, see for example [Smi85, HS96] and references therein. I do not attempt to even only try to give a complete overview of this field here; rather I want to present two typical finite difference approaches to the problem of solving the SCHRÖDINGER equation, and I discuss some of the limitations of approaches of this kind. The first method presented here relies on a discretization of time only, whereas for the second both the time and position variables are being discretized.

3.1.1 An Algorithm Based on the Trotter Formula

Consider the quantum map (2.37). When discussing states in the position representation, application of the kick propagator as in equation (2.34) is easily accomplished, because $\hat{U}_{\text{kick}}$ depends on $\hat{x}$ only, and not on $\hat{p}$. On the other hand, the application of the free propagator,

$$|\psi((n+1)T-0)\rangle = \hat{U}_{\text{free}}\,|\psi(nT+0)\rangle\,, \tag{3.1}$$

poses a greater problem. In the form of equation (2.44) it is straightforward to let $\hat{U}_{\text{free}}$ act on states expanded in terms of the harmonic oscillator eigenstates only — I come back to this point later in section 3.3. For now, I want to discuss the application of the free propagator to states in the position representation.

Clearly, the exponential (2.32a) can be evaluated by expanding it into its TAYLOR series,

$$\hat{U}_{\text{free}} = \sum_{m=0}^{\infty}\frac{1}{m!}\left(-\frac{i}{\hbar}\hat{H}_{\text{free}}T\right)^{m}, \tag{3.2}$$

allowing to evaluate the mapping (3.1) term by term, but this approach is problematic from a practical point of view. Namely, in practice the infinite sum must be truncated at some finite value of m, such that the resulting approximant to the unitary $\hat{U}_{\text{free}}$ may become nonunitary. Unitarity of the propagator used for the calculation is a natural prerequisite for obtaining an *unconditionally stable* [DR96] numerical algorithm, because in this way conservation of the norm of the states is automatically built into the method. Unconditional stability being granted implies that the numerical error — mainly due to discretization with respect to position and time — that builds up in the course of many iterations of the quantum map remains controllable.

Therefore, from the point of view of unconditional stability, using a plain TAYLOR expansion should be excluded from the considerations, especially when it is intended to study long-time evolution of quantum states. Nevertheless, in [Dro95] a procedure is described that allows to improve on expansions like equation (3.2), but truly long-time dynamics still seems to remain beyond reach even with this refined approach.

The basic idea for constructing an unconditionally stable algorithm is to decompose $\hat{U}_{\text{free}}$ into a product where all factors are already unitary — and where each factor can be diagonalized by a suitably simple transformation. In this way the propagation over a full period T gets replaced with successive propagations over smaller time steps, such that a discretization with respect to time is introduced into the method.

Above all, the problems with constructing a unitary approximant to $\hat{U}_{\text{free}}$ are caused by the fact that $\hat{H}_{\text{free}}$ is a sum of two noncommuting contributions, namely the kinetic and potential terms $\hat{p}^2/2$ and $\hat{x}^2/2$. Since $[\hat{p}^2, \hat{x}^2] \neq 0$, a simple BAKER-CAMPBELL-HAUSDORFF (BCH) decomposition into two unitary factors is not possible:

$$e^{-\frac{i}{\hbar}\hat{H}_{\text{free}}T} \neq e^{-\frac{iT}{2\hbar}\hat{p}^2} e^{-\frac{iT}{2\hbar}\hat{x}^2} \tag{3.3}$$

(cf. equation (A.4) in appendix A for more on BCH formulae).

An alternative and more suitable approach is based on the TROTTER *product formula*. For two operators $\hat{A}$, $\hat{B}$ and $z \in \mathbb{C}$, the exponential of the sum of the operators can be expressed as an infinite ordered product of exponentials of the individual operators:

$$e^{z(\hat{A}+\hat{B})} = \lim_{s\to\infty} \left(e^{\frac{z}{s}\hat{A}} e^{\frac{z}{s}\hat{B}} \right)^s ; \tag{3.4}$$

for a proof see [Sch81]. Using the TROTTER formula, for sufficiently large $s \in \mathbb{N}$ one may approximate:

$$e^{\frac{z}{s}(\hat{A}+\hat{B})} \approx e^{\frac{z}{s}\hat{A}} e^{\frac{z}{s}\hat{B}} . \tag{3.5}$$

For purely imaginary z, the error induced by this approximation, due to s being finite, can be estimated using [DR87]

$$\left\| e^{\frac{z}{s}(\hat{A}+\hat{B})} - e^{\frac{z}{s}\hat{A}} e^{\frac{z}{s}\hat{B}} \right\| \leq \frac{|z|^2}{2s^2} \left\| [\hat{A}, \hat{B}] \right\| , \tag{3.6}$$

where $\|\cdot\|$ is a suitable operator norm. Note that, using the estimate (3.6), for commuting $\hat{A}$, $\hat{B}$ the simplest BCH result $e^{\hat{A}+\hat{B}} = e^{\hat{A}}e^{\hat{B}}$ is confirmed again.

In the present context of the kicked harmonic oscillator, I obtain the following approximation for the free propagator,

$$\hat{U}_{\text{free}} = \left(e^{-\frac{iT}{2\hbar s}\hat{p}^2} e^{-\frac{iT}{2\hbar s}\hat{x}^2} \right)^s + \mathcal{O}\left(\frac{T^2}{s}\right), \tag{3.7}$$

which can be made as accurate as desired by increasing s. Note that the upper bound of the error does not depend explicitly on $\hbar$ any more, such that the quality of the algorithm, as given by the error term $\mathcal{O}(T^2/s)$, does not vary with $\hbar$. Unitarity of the approximating propagator is granted by $\hat{p}^2$, $\hat{x}^2$ being Hermitian.

As opposed to $\hat{U}_{\text{free}}$ in the form of equation (3.2), application of its approximant (3.7) is straightforward. For each of the s time steps per period T, expressions of the following type have to be evaluated:

$$\begin{aligned} &\left\langle x \left| e^{-\frac{iT}{2\hbar s}\hat{p}^2} e^{-\frac{iT}{2\hbar s}\hat{x}^2} \right| \psi \right\rangle = \\ &\int_{-\infty}^{\infty} \mathrm{d}p' \int_{-\infty}^{\infty} \mathrm{d}x' \left\langle x \left| e^{-\frac{iT}{2\hbar s}\hat{p}^2} \right| p' \right\rangle \left\langle p' \left| e^{-\frac{iT}{2\hbar s}\hat{x}^2} \right| x' \right\rangle \langle x' | \psi \rangle , \end{aligned} \tag{3.8}$$

i.e. the $\hat{x}$-dependent exponential is multiplied to the initial state in the position representation, the result is transformed into the momentum representation, where the $\hat{p}$-dependent exponential is easily applied; finally, the resulting expression is transformed back into the position representation.

In this way, the free harmonic oscillator dynamics of arbitrary states is obtained essentially by successively switching between the position and momentum representations; this can be done quite effectively using the technique of *fast* FOURIER *transformation* (FFT) [CT65, EMR93]. Nevertheless, the need for applying the FFT *very often* in order to obtain the desired accuracy is the limiting bottleneck for this algorithm.[1]

[1] In [See95] a stroboscopically kicked (and otherwise free) particle is studied quantum mechanically using a similar algorithm of successive transformations between the position and momentum representations (with modifications as described in [HM87] which regrettably are not applicable to the present case of the cosine kicked harmonic oscillator where the kick potential is not of finite range). Similarly, in subsection 5.1.1 and in [Lan94] the quantum dynamics of the cosine-kicked rotor is discussed. In these systems, only a *single* pair of FOURIER transformations per kick period is needed, because between the kicks the (free) dynamics in the momentum representation is trivial. The present subsection 3.1.1 illustrates the way in which the situation becomes considerably more intricate when replacing the truly free dynamics between the kicks with the "free" harmonic oscillator dynamics.

The full kicked harmonic oscillator dynamics is then obtained by alternately applying the simple $\hat{U}_{\text{kick}}$ of equation (2.34) and the TROTTER-approximated $\hat{U}_{\text{free}}$ as described above.

Effectively, the TROTTER algorithm implies a discretization of the dynamics with respect to time; the period of time T of the unperturbed harmonic oscillator dynamics is split into s sufficiently small time steps of length

$$\delta t := \frac{T}{s}. \tag{3.9}$$

This may be used to implement an adaptive step size control: each time the approximate propagator (3.7) is applied, s can be tuned to be small enough to make the algorithm not too time-consuming on the one hand, and large enough to achieve the desired accuracy on the other hand; the latter can be checked by applying (3.7) twice each time, for example using s and $2s$, and comparing the two resulting iterated states. (The same applies to the number of nodes used for storing $\langle x \,|\psi\rangle$ and $\langle p \,|\psi\rangle$, but in contrast to s this number cannot be changed easily in the course of the calculation, but rather has to be specified before the algorithm is started.)

3.1.2 The Goldberg Algorithm

In the TROTTER-based algorithm, the position and momentum variables remain nondiscretized; only through the FFT — which essentially is just a sophisticated formulation of the *discrete* FOURIER *transformation* using a discrete set of nodes in position and momentum space — effectively x and p become discretized, too.

Now I describe a more typical finite differences approach to solving the SCHRÖDINGER equation for the free harmonic oscillator in the position representation, where both space and time are discretized in equidistant steps from the beginning. The solution $\langle x \mid \psi(t)\rangle$ on the interval $[x_{\text{min}}, x_{\text{max}}]$ is constructed at the nodes

$$\begin{aligned} x_j &= x_{\text{min}} + j\,\delta x, \qquad j = 0, \ldots, j_{\text{max}} \\ x_{j_{\text{max}}} &= x_{\text{max}}\,, \end{aligned} \tag{3.10}$$

and the time steps

$$t_n^{(m)} = nT + m\,\delta t, \qquad m = 0, \ldots, s \tag{3.11}$$

are used. For simplicity, only the dynamics in the time interval $\big[nT,(n+1)T\big)$ is considered here, such that the subscript n can be dropped. The wave function at site x_j and time $t^{(m)}$, and the potential contribution to $\hat{H}_{\text{free}}$ at x_j are denoted as

$$\begin{aligned} \psi_j^{(m)} &:= \langle x_j \,|\, \psi\left(t^{(m)}\right)\rangle && (3.12)\\ v_j &:= \frac{1}{2}x_j^2, && (3.13) \end{aligned}$$

respectively.

In a first step, $\hat{U}_{\text{kick}}$ is applied to the discretized wave function:

$$\psi_j^{(0)} \longmapsto e^{-\frac{i}{\hbar}V_0 \cos x_j}\,\psi_j^{(0)}. \tag{3.14}$$

Then, the new $\psi_j^{(0)}$ is propagated for a period of time of length T using the GOLDBERG algorithm [Koo86, KW96] that I describe now in some more detail.

With the discrete first order approximation for the second derivative [SB00]

$$\left\langle x_j \middle| \frac{\partial^2}{\partial x^2} \middle| \psi\left(t^{(m)}\right)\right\rangle = \frac{1}{(\delta x)^2}\left(\psi_{j-1}^{(m)} - 2\psi_j^{(m)} + \psi_{j+1}^{(m)}\right) + \mathcal{O}\left((\delta x)^2\right), \tag{3.15}$$

the application of the free harmonic oscillator Hamiltonian to $\psi_j^{(m)}$ can be written as

$$\begin{aligned} \hat{H}_{\text{free}}\,\psi_j^{(m)} &:= -\frac{\hbar^2}{2(\delta x)^2}\left(\psi_{j-1}^{(m)} - 2\psi_j^{(m)} + \psi_{j+1}^{(m)}\right) + v_j\,\psi_j^{(m)} \\ &= \left\langle x_j \middle| \hat{H}_{\text{free}} \middle| \psi\left(t^{(m)}\right)\right\rangle + \mathcal{O}\left((\hbar\,\delta x)^2\right). \end{aligned} \tag{3.16}$$

Using this notation, the propagation over the time interval δt is approximately

$$\psi_j^{(m+1)} = e^{-\frac{i}{\hbar}\hat{H}_{\text{free}}\,\delta t}\,\psi_j^{(m)}. \tag{3.17}$$

As in the previous subsection it remains to replace the exponential with a suitable approximant, and again using the TAYLOR formula (3.2) is not a good choice, as it does not yield the desired unitary expression. Rather, the GOLDBERG algorithm calls for the CAYLEY form [PTVF94] of the approximant:

$$e^{-\frac{i}{\hbar}\hat{H}_{\text{free}}\,\delta t} = \frac{\mathbb{1} - \frac{i}{2\hbar}\hat{H}_{\text{free}}\,\delta t}{\mathbb{1} + \frac{i}{2\hbar}\hat{H}_{\text{free}}\,\delta t} + \mathcal{O}\left(\left(\frac{\delta t}{\hbar}\right)^3\right). \tag{3.18}$$

Clearly, the quotient on the right hand side is unitary, thereby in principle allowing for an unconditionally stable algorithm. But note that the approximation (3.16) for the application of $\hat{H}_{\text{free}}$ can still spoil the stability of the algorithm. The CAYLEY approximant is correct up to second order in $\delta t/\hbar$, one power more than would have been supposed at first sight, as both the numerator and the denominator in equation (3.18) are first order approximations only. On the other hand, this very desirable feature of the approximant is accompanied by the less desirable feature that it results in a set of *implicit* difference equations to be solved — see equation (3.20) below.

With the abbreviations

$$\tilde{v}_j := \frac{2(\delta x)^2}{\hbar^2} v_j + 2 \tag{3.19a}$$

$$\lambda := \frac{4(\delta x)^2}{\hbar\,\delta t}, \tag{3.19b}$$

combination of equations (3.16–3.18) gives the iteration scheme

$$\psi_{j-1}^{(m+1)} + (i\lambda - \tilde{v}_j)\,\psi_j^{(m+1)} + \psi_{j+1}^{(m+1)} = -\psi_{j-1}^{(m)} + (i\lambda + \tilde{v}_j)\,\psi_j^{(m)} - \psi_{j+1}^{(m)}. \tag{3.20}$$

It is to be solved for the $\psi_j^{(m+1)}$ and essentially requires the inversion of a tridiagonal system of linear equations. This is seen more easily in vectorial notation. Equation (3.20) is equivalent to

$$A_+\vec{\psi}^{(m+1)} = A_-\vec{\psi}^{(m)}, \tag{3.21}$$

for which the vectors

$$\vec{\psi}^{(m)} := \begin{pmatrix} \psi_1^{(m)} \\ \vdots \\ \psi_{j\text{max}}^{(m)} \end{pmatrix} \tag{3.22}$$

and the symmetric $(j_{\text{max}}, j_{\text{max}})$ matrices

$$A_\pm := \begin{pmatrix} i\lambda \mp \tilde{v}_1 & \pm 1 & 0 & \cdots & 0 & \pm c \\ \pm 1 & i\lambda \mp \tilde{v}_2 & & & & 0 \\ 0 & & & \ddots & \ddots & \vdots \\ \vdots & \ddots & \ddots & \ddots & & 0 \\ 0 & & & & & \pm 1 \\ \pm c & 0 & \cdots & 0 & \pm 1 & i\lambda \mp \tilde{v}_{j\text{max}} \end{pmatrix} \tag{3.23}$$

have been defined. The value of the parameter c depends on the (DIRICHLET) boundary conditions to be used.

$c = 0$: This case corresponds to the boundary conditions

$$\psi_0^{(m)} = \psi_{j\max+1}^{(m)} = 0. \tag{3.24a}$$

For all times t considered, the wave function is assumed to decay rapidly enough with $|x|$, i.e. it is assumed that $\langle x \,|\psi(t)\rangle = 0$ for all $x \notin [x_{\min}, x_{\max}]$. For $c = 0$, the matrices $A_\pm$ are indeed tridiagonal.

Effectively, depending on the initial quantum state and the dynamics that evolves from it, this type of boundary condition poses a restriction on the maximum period of time for which the algorithm may be applied meaningfully: if the (physical) wave packet flows apart, as one should expect for the quantum dynamics in a case of resonance, where classically a stochastic web develops, then after sufficiently long time — which can be short — the wave packet may be delocalized enough to violate the (numerical) boundary condition (3.24a). From this point of view, the interval $[x_{\min}, x_{\max}]$ should be chosen as large as practically possible, in order to minimize cut-off errors.

$c = 1$: In this case of *periodic boundary conditions*, only

$$\psi_0^{(m)} = \psi_{j\max}^{(m)} \tag{3.24b}$$

is required, possibly taking on nonzero values. The $A_\pm$ are not tridiagonal any more in this case, but the only additional nonzero matrix elements $(A_+)_{1,j\max}$ and $(A_+)_{j\max,1}$ do not cause a significant complication of the numerical algorithm for inverting A_+.

Using this type of boundary condition should be considered if it has been shown *beforehand* that the evolving wave packet exhibits spatial periodicity (but cf. the corresponding remarks in section 3.4 below). In such a case, the interval $[x_{\min}, x_{\max}]$ can be chosen to cover the smallest possible periodicity interval in position space.

The right hand side of equation (3.21) is easily computed from the already known $\psi_j^{(m)}$. Then $\vec{\psi}^{(m+1)} = A_+^{-1} A_- \vec{\psi}^{(m)}$ is determined by inversion of the (essentially) tridiagonal linear system given by the matrix A_+; for that purpose, several efficient algorithms are available [PTVF94, Sto99].

As in the case of the TROTTER-based method, the time increment δt can be changed adaptively at each time step in order to control the numerical accuracy of the algorithm. Also, as in the TROTTER algorithm, changing the spatial nodes (and thus δx) in the course of the calculation is much more difficult and prone to cause additional numerical error.

Regarding the initial choice of the parameters δt and δx, it is natural to choose these parameters small enough such that the smallest oscillation of interest (both with respect to time and space) of the system can be resolved. Furthermore, these parameters can be chosen in such a way that the errors induced by time and space discretization are balanced — this interdependence is made plausible by equation (3.19b) which shows that reducing δx leads to similar numerical problems as increasing δt; more information on this issue can be found in [KW96] and references therein.

3.2 Numerical Evaluation of the Propagator in the Position Representation

In [BR95] a numerical "brute force approach" to determining the kicked harmonic oscillator dynamics is described. Using the present terminology and scaling, this approach can be formulated as a discretization of the integral form of the quantum map (2.37) in the position representation,

$$\langle x|\psi_{n+1}\rangle = \int_{-\infty}^{\infty} \mathrm{d}x' \, G_{\text{free}}(x,T;x',0) \left\langle x' \middle| e^{-\frac{i}{\hbar}V_0 \cos x'} \middle| \psi_n \right\rangle, \tag{3.25}$$

expressing the time evolution over one period using the position representation propagator

$$G_{\text{free}}(x,T;x',0) = \frac{1}{\sqrt{2\pi i\hbar \sin T}} \exp\left\{ i\,\frac{(x^2+x'^2)\cos T - 2xx'}{2\hbar \sin T} \right\} \tag{3.26}$$

of the unkicked harmonic oscillator (2.29a) [FH65]. In contrast to the methods discussed in section 3.1 these expressions by construction take into account the FLOQUET nature of the system.

Using the x-discretization as given by equations (3.10), the integral expression (3.25) can be approximated by

$$\langle x_j|\psi_{n+1}\rangle \approx \frac{x_{\max}-x_{\min}}{j_{\max}} \sum_{j'=1}^{j_{\max}} G_{\text{free}}(x_j,T;x_{j'},0)\, e^{-\frac{i}{\hbar}V_0 \cos x_{j'}} \langle x_{j'}|\psi_n\rangle, \tag{3.27}$$

which is once again denoted more concisely in vectorial form,

$$\vec{\psi}_{n+1} = U^{(\mathrm{pr})}\, \vec{\psi}_n, \tag{3.28}$$

with the vectors

$$\vec{\psi}_n := \begin{pmatrix} \langle x_1 | \psi_n \rangle \\ \vdots \\ \langle x_{j_{\max}} | \psi_n \rangle \end{pmatrix} \tag{3.29}$$

and the $(j_{\max}, j_{\max})$ propagator matrix $U^{(\mathrm{pr})}$ in the position representation, with the matrix elements

$$U^{(\mathrm{pr})}_{jj'} := \frac{x_{\max} - x_{\min}}{j_{\max}}\, G_{\mathrm{free}}\big(x_j, T; x_{j'}, 0\big)\, e^{-\frac{i}{\hbar} V_0 \cos x_{j'}}, \quad 1 \leq j, j' \leq j_{\max}. \tag{3.30}$$

$U^{(\mathrm{pr})}$ takes the same role with respect to the position representation as the propagator U of equation (2.42) does with respect to the eigenrepresentation of the harmonic oscillator. Note that $U^{(\mathrm{pr})}$, which describes the propagation over a full period of the excitation, is the same for all n, such that it suffices to determine its matrix elements once and store them for repeated use.

In order to obtain a numerically stable algorithm, $U^{(\mathrm{pr})}$ must be required to be unitary,

$$\sum_{j'=1}^{j_{\max}} \left| U^{(\mathrm{pr})}_{jj'} \right|^2 \stackrel{!}{=} 1 \qquad \forall j, \tag{3.31}$$

whence with equations (3.26) and (3.30)

$$\frac{(x_{\max} - x_{\min})^2}{j_{\max}} \stackrel{!}{=} 2\pi\hbar\, |\sin T| \tag{3.32}$$

can be concluded; the parameters $x_{\min}$, $x_{\max}$, $j_{\max}$ of the numerical algorithm, on the left hand side of equation (3.32), have to be chosen as specified by the physical parameters $\hbar$, T on the right hand side. The kick amplitude V_0 does not interfere with the unitarity of $U^{(\mathrm{pr})}$.

In addition to the condition (3.32), a boundary condition of the type (3.24a) also needs to be satisfied, i.e. the interval $[x_{\min}, x_{\max}]$ must be chosen large enough to allow the discrete mapping (3.27) to approximate the original (3.25) well enough. This is granted if the wave function $\langle x | \psi_n \rangle$ decays rapidly enough with x reaching out to the boundaries of the interval considered.

These two restrictions combined are the reason why the algorithm given by the mapping (3.27) is of limited practical use when the long-time dynamics of spreading states is to be studied: for delocalized states spreading widely along the x-axis, very large values of $j_{\max}$ are required by the condition (3.32), in particular when $\hbar$ takes on small values. This means that the memory requirements on the computer grow rapidly, and the evaluation of each iteration of (3.27) quickly becomes more time-consuming. For example, the values $-x_{\min} = x_{\max} = 30$, $\hbar = 0.01$ and $T = \pi/2$ lead to $j_{\max} \approx 57000$, making it practically impossible to store the matrix elements $U^{(\mathrm{pr})}_{jj'}$ on a typical workstation and slowing down the speed of the computation considerably. It is important to keep in mind this limitation of the algorithm when working with it in practice.

Conversely, if the dynamics is followed for times nT not too large, such that the initial wave packets have not spread too much, then equation (3.27) provides a simple and efficient way to evaluate the quantum map (2.37), especially for larger values of $\hbar$.

3.3 The Propagator in the Eigenrepresentation of the Free Harmonic Oscillator

In this section I discuss some of the aspects of the practical implementation of the iteration scheme outlined in subsection 2.1.3.

From a formal point of view the methods formulating $\hat{U}$ in the position representation (section 3.2) and the eigenrepresentation of the harmonic oscillator (subsection 2.1.3 and the present section) obviously are very similar — for example, both equation (2.43) and equation (3.28) are explicit linear mappings of vectors of complex numbers.

What is more, the dimensions of the matrices U and $U^{(\mathrm{pr})}$ used for the iterations are of the same order of magnitude, too. This can be seen as follows. The dimension $j_{\max}$ to be used for $U^{(\mathrm{pr})}$ has been derived in the previous subsection; on the other hand, the dimension needed for U is determined by the maximum index $m_{\max}$ of the set $\{\, |m\rangle \,|\, 0 \le m \le m_{\max} \}$ necessary for resolving quantum phase space structures at a distance of $x_{\max}$ from the origin. Roughly speaking, $|\langle x\,|m\rangle|^2$ is peaked around the corresponding classical turning points at $\pm\sqrt{\hbar(2m+1)}$ and decays exponentially for larger values of $|x|$.[2] This means

[2] Note that the same applies with respect to the p-direction. The quantum phase space portion covered by $\{\, |m\rangle \,|\, 0 \le m \le m_{\max} \}$ is a circular disc with the approximate area $\pi\hbar(2m_{\max}+1)$, i.e. it grows linearly with $m_{\max}$, the growth rate being given by the square of the oscillator length ($\sqrt{\hbar}$ in the scaling used here).
See figures A.1–A.3 in appendix A for a graphical demonstration of this property of the harmonic oscillator eigenstates.

that, as a rule of the thumb, the dimension of U needed for describing the dynamics on the interval $[-x_{\max}, x_{\max}]$ is not very much larger than $x_{\max}^2/2\hbar$. This estimate is close to the value of $j_{\max}$ determined in equation (3.32).

There are some decisive differences between the matrices used for the two methods, as well. First of all, with respect to identical dimensions of the matrices, storing the matrix elements of U only requires a quarter of the memory needed for storing the matrix elements of $U^{(\mathrm{pr})}$: in practice, the kick matrix elements $K_{mm'}$ of equation (2.51) — rather than the full FLOQUET matrix elements $U_{mm'}$ — are stored in the computer memory, such that the selection rule $K_{mm'} = 0$ for $m - m'$ odd and the symmetry property $K_{mm'} = K_{m'm}$ can be used — see equation (2.51). Then, each time they are needed during the iteration, the $U_{mm'}$ are trivially calculated from the $K_{mm'}$ using the splitting (2.45). Storing the $K_{mm'}$ requires computer memory for roughly $m_{\max}^2/4$ complex numbers, i.e. for the typical value of $m_{\max} = 6000$ some 138 MBytes are needed. The approximate CPU time per iteration of the quantum map then typically varies from 45 sec (on a Pentium II with 350 MHz) to 10 sec (Pentium 4 with 2 GHz).

Second, in contrast to the evaluation of (3.30), computation of the $K_{mm'}$ and thus of the $U_{mm'}$ is a nontrivial task. Equation (2.51) shows that the formula for the $K_{mm'}$ is composed of several terms (BESSEL functions, generalized LAGUERRE polynomials, factorials, exponentials) the absolute values of which can — and in practice do — differ greatly, such that all kinds of numerical cancellations, overflows and underflows are to be expected — and indeed do occur — when (2.51) is evaluated directly. Therefore, this calculation has to be implemented in a slightly more sophisticated way, basically by exploiting the recurrence relation

$$\mathrm{L}_{m'}^{(\delta m)}(z) = \frac{\delta m + 2m' - 1 - z}{m'} \mathrm{L}_{m'-1}^{(\delta m)}(z) - \frac{\delta m + m' - 1}{m'} \mathrm{L}_{m'-2}^{(\delta m)}(z), \quad m' \geq 2 \tag{3.33a}$$

$$\mathrm{L}_{0}^{(\delta m)}(z) = 1 \tag{3.33b}$$

$$\mathrm{L}_{1}^{(\delta m)}(z) = \delta m + 1 - z \tag{3.33c}$$

for the generalized LAGUERRE polynomials [AS72]. In this way, not only the abovementioned numerical problems are avoided, but also the computation of the whole matrix is sped up considerably.

Starting from the kick matrix elements in the form of equation (2.48) and using the definition

$$N_{mm'}(l) := \langle m | e^{il\hat{x}} | m' \rangle + \langle m | e^{-il\hat{x}} | m' \rangle \tag{3.34}$$

I obtain

$$K_{mm'} = \mathrm{J}_0\left(\frac{V_0}{\hbar}\right)\delta_{mm'} + \sum_{l=1}^{\infty}(-i)^l\,\mathrm{J}_l\left(\frac{V_0}{\hbar}\right)N_{mm'}(l). \tag{3.35}$$

The matrix elements $N_{mm'}(l)$ in this expression need to be studied with respect to their behaviour along the diagonals given by

$$\delta m := m - m' = \text{const.} \tag{3.36}$$

With equation (2.50), $N_{mm'}(l)$ can be rewritten in terms of a suitably normalized variant,

$$\tilde{\mathrm{L}}_{m'}^{(\delta m)}(z) := \frac{1}{\sqrt{(\delta m + m')(\delta m + m' - 1)\cdots(m'+1)}}\,\mathrm{L}_{m'}^{(\delta m)}(z), \tag{3.37}$$

of the generalized LAGUERRE polynomials:

$$N_{\delta m+m',m'}(l) = \begin{cases} 2\left(-\frac{\hbar l^2}{2}\right)^{\frac{\delta m}{2}} e^{-\frac{\hbar l^2}{4}}\,\tilde{\mathrm{L}}_{m'}^{(\delta m)}\left(\frac{\hbar l^2}{2}\right) & \delta m \text{ even} \\ & \text{for} \\ 0 & \delta m \text{ odd.} \end{cases} \tag{3.38}$$

In this expression, working with $\tilde{\mathrm{L}}_{m'}^{(\delta m)}$ (rather than $\mathrm{L}_{m'}^{(\delta m)}$) is advantageous, as it collects all m'-dependent terms. And using the indices δm, m' rather than m, m' is appropriate here, because in the next step the $N_{\delta m+m',m'}(l)$ are calculated by repeated application of a recurrence relation along the diagonals of the matrix, where m' specifies the individual matrix elements on the diagonal determined by δm. Note that equations (3.37) and (3.38) hold for $\delta m \geq 0$ only, because the same is true for equation (2.50). For $\delta m < 0$, by equation (2.51) the symmetry $K_{mm'} = K_{m'm}$ allows to determine the kick matrix elements with equations (3.35–3.38), too.

The $N_{\delta m+m',m'}(l)$ can now be calculated efficiently using the recurrence relation

$$\begin{aligned} N_{\delta m+m',m'}(l) &= \frac{\delta m + 2m' - 1 - \frac{\hbar l^2}{2}}{\sqrt{m'(\delta m + m')}}\,N_{\delta m+m'-1,m'-1}(l) \\ &\quad - \sqrt{\frac{(\delta m + m' - 1)(m'-1)}{m'(\delta m + m')}}\,N_{\delta m+m'-2,m'-2}(l), \quad m' \geq 2 \end{aligned} \tag{3.39a}$$

$$N_{\delta m,0}(l) \quad = \quad 2\left(-\frac{\hbar l^2}{2}\right)^{\frac{\delta m}{2}} e^{-\frac{\hbar l^2}{4}} \frac{1}{\sqrt{(\delta m)!}} \tag{3.39b}$$

$$N_{\delta m+1,1}(l) \quad = \quad \frac{\delta m + 1 - \frac{\hbar l^2}{2}}{\sqrt{\delta m + 1}} N_{\delta m,0}(l), \tag{3.39c}$$

which is obtained by evaluating the $\tilde{\mathrm{L}}_{m'}^{(\delta m)}\left(\hbar l^2/2\right)$ in equation (3.38) by means of the definition (3.37) and the recursion relation (3.33) of the generalized LAGUERRE polynomials.

The numerical problems mentioned above with respect to the direct evaluation of equation (2.51) are reduced to a minimum with this algorithm, because multiplying and adding up very large and very small numbers at the same time is avoided here as far as possible: the definition (3.37) avoids the factorials of equation (2.51), and the respective absolute values of the numerators and denominators in each of the fractions in the recursion (3.39a) are constructed to be as comparable as possible.

Similarly, when in the next chapters $\langle x\,|m\rangle$ needs to be evaluated numerically for larger values of m, this cannot be done using equation (2.39) directly. Rather, a scaled variant of the HERMITE polynomials is defined,

$$\tilde{\mathrm{H}}_m(z) := \frac{1}{\sqrt{2^m m!}} \mathrm{H}_m(z), \tag{3.40}$$

which can be evaluated efficiently and in a numerically sound way even for larger values of m via the recurrence relation

$$\tilde{\mathrm{H}}_m(z) \quad = \quad \sqrt{\frac{2}{m}}\, z\, \tilde{\mathrm{H}}_{m-1}(z) - \sqrt{\frac{m-1}{m}}\, \tilde{\mathrm{H}}_{m-2}(z), \quad m \geq 2 \tag{3.41a}$$

$$\tilde{\mathrm{H}}_0(z) \quad = \quad 1 \tag{3.41b}$$

$$\tilde{\mathrm{H}}_1(z) \quad = \quad \sqrt{2}\, z \tag{3.41c}$$

that follows immediately from the conventional recurrence relation

$$\mathrm{H}_m(z) \quad = \quad 2z\, \mathrm{H}_{m-1}(z) - 2(m-1)\, \mathrm{H}_{m-2}(z), \quad m \geq 2 \tag{3.42a}$$

$$\mathrm{H}_0(z) \quad = \quad 1 \tag{3.42b}$$

$$\mathrm{H}_1(z) \quad = \quad 2z \tag{3.42c}$$

for the HERMITE polynomials [AS72]. The harmonic oscillator eigenstates in the position representation are then obtained as

$$\langle x\,|m\rangle \;=\; \frac{1}{\sqrt[4]{\pi\hbar}}\,e^{-\frac{x^2}{2\hbar}}\,\tilde{\mathrm{H}}_m\!\left(\frac{x}{\sqrt{\hbar}}\right). \tag{3.43}$$

With the $N_{\delta m+m',m'}(l)$ computed as described above, the series in equation (3.35) can be evaluated term by term. In the parameter range considered for the numerical calculations in the following chapters, the series can typically be truncated at values of l not exceeding 200, while keeping the error induced by the truncation reasonably small. (For the maximum relative error due to this truncation, 10^{-14} was a somewhat typical value in the practical calculations.)

As in the previous sections, boundary conditions have to be imposed here, too. After each time step it is checked if the condition (3.24a) is still satisfied. In practice, for diffusing states violation of this boundary condition — due to the implicit cut-off error as a result of the finite number $m_{\max}$ of basis elements $|m\rangle$ used for expanding the states — is one of the two main sources of numerical error. The other is the unavoidable rounding error generated in the course of the numerical computation of the $K_{mm'}$.

3.4 Discussion of the Numerical Methods

Both the TROTTER and the GOLDBERG algorithms do not take into account the FLOQUET nature of the stroboscopically kicked system discussed here. Rather, these algorithms provide quite general methods for numerically treating the SCHRÖDINGER equation and are applicable to a much larger variety of systems. Although being efficient in this general sense — in particular the GOLDBERG algorithm should be expected to give good results due to the second order CAYLEY approximant (3.18), as opposed to the first order approximation of equation (3.7) of the TROTTER algorithm — more efficiency can be gained by using methods that take into account the particular properties of the kicked harmonic oscillator. The methods described in sections 3.2 and 3.3 do exactly that, for example by making use of the FLOQUET nature of the system when calculating the propagator for a full period of the excitation.

The finite differences methods are also limited by the disadvantageous feature that there the main computational effort has to be spent for each small time step of length δt and must thus be repeated very often, thereby effectively slowing down long-time computations. The methods described in sections 3.2 and 3.3 improve on the finite differences methods by computing the FLOQUET operators once and for all during a more or less lengthy calculation; having accomplished this, application of the FLOQUET operators is then reduced to simple and comparatively fast matrix multiplications.

With respect to the discussion of boundary conditions in subsection 3.1.2, one might at first come to the conclusion that, when studying the quantum analogues of the stochastic webs discussed in chapter 1, the GOLDBERG algorithm with periodic boundary conditions might be the best numerical method to use. But there are several counterarguments to this approach. First of all, the use of periodic boundary conditions automatically incorporates into the numerical algorithm spatial periodicity of the states that are to be computed. If one aims — as in this study — at confirming that the investigated system by itself dynamically develops periodic structures in position space (and phase space), then this periodicity should not already be an ingredient of the algorithm. Furthermore, *periodic* boundary conditions are of no use when studying the quantum analogues of *quasiperiodic* structures such as the aperiodic webs discussed in section 1.2, or when the case of *nonresonance* is considered for which no web-like structures should be expected.

On the other hand, once the periodicity of the states with respect to the x-coordinate has been established in some way, this observation can be used as a starting point for considerably speeding up the GOLDBERG algorithm, namely by using the smallest possible periodicity interval for $[x_{\text{min}}, x_{\text{max}}]$. In this way the number of nodes needed for representing the wave packet is minimized, thereby minimizing the numerical effort as well. Using periodic boundary conditions also has the advantage that in this way the algorithm avoids cut-off errors altogether that otherwise could arise when the boundary conditions (3.24a) are not satisfied any more, thus spoiling the norm-conservation of the computation.

The problem of cut-off errors does not only occur with respect to the GOLDBERG algorithm, as discussed on page 74, but with respect to all algorithms discussed here. Localized states — such as the typical initial states specified in the following chapters — can be well represented within all the algorithms. But more delocalized nonperiodic states that violate the condition (3.24a) cannot be described and propagated any better within the TROTTER-based algorithm, because the FOURIER transformations used there also require the wave functions (in both the position and momentum representations) to be well localized in the intervals considered. And in the framework of the eigenrepresentation of the harmonic oscillator, a spreading state after a sufficiently long period of time reaches out far enough such that any finite basis $\{|m\rangle \mid 0 \leq m \leq m_{\text{max}}\}$ does not suffice any more to meaningfully expand the state. This observation describes a general and natural restriction for the long-time numerical analysis of unbounded dynamics on a computer. Naturally, this problem does not arise if periodic boundary conditions can be used as discussed above.

If not much computer memory is available, then using one of the finite differences methods might be preferred again: they are characterized by very moderate memory requirements, since both the fast FOURIER transformation and the solving of a tridiagonal linear system need very little memory, as no huge $(j_{\max}, j_{\max})$ matrices need to be stored.

Finally, it is interesting to note the complementary role of the kick propagation within the different numerical methods discussed here. On the one hand there are the methods treated in sections 3.1 and 3.2, where the implementation of the kick is trivial, as these methods are based on the position representation, and the numerically more difficult part is the free harmonic oscillator propagation. On the other hand, the situation is reversed in the framework of the method using the eigenrepresentation of the harmonic oscillator, where the free propagation is easily accomplished, but the kick poses severe numerical difficulties, as discussed in section 3.3.

In the following chapter, the methods described in sections 3.1 and 3.3 are applied to the quantum kicked harmonic oscillator. It is finally established there that using the FLOQUET operator in the harmonic oscillator eigenrepresentation is the most efficient (i.e. fastest) and most general method, allowing to study the quantum analogues of all three interesting classical types of dynamics: periodic webs, aperiodic webs and dynamics in the case of nonresonance (leading to localized quantum motion). Most of the calculations in the following chapters are performed using that method. The position representation FLOQUET operator $U^{(\mathrm{pr})}$ of section 3.2 is not considered any more from here on: it has been shown above to be similar — in a certain technical sense — to the U of section 3.3, while needing more computer memory than the latter. But the decisive argument against $U^{(\mathrm{pr})}$ might be that at several points in the following chapters it is just technically more convenient to have the states expanded in terms of harmonic oscillator eigenfunctions.

Chapter 4

Quantum Stochastic Webs

"Well, in our *country," said Alice, still panting a little, "you'd generally get to somewhere else if you ran very fast for a long time as we've been doing."*
"A slow sort of country!" said the Queen. "Now, here, *you see, it takes all the running* you *can do, to keep in the same place."*

Through the Looking Glass
LEWIS CARROL

Having discussed some methods for numerically iterating the quantum map of the kicked harmonic oscillator in the previous chapter, the present chapter is dedicated to describing the results of extensive application of these methods in the case of resonance that classically leads to the emergence of stochastic webs — see chapter 1. The case of nonresonance is dealt with in chapter 5.

In section 4.1 I show that, as a result of the quantum dynamics, *web-like structures in quantum phase space* develop which are very similar to the classical stochastic webs. As in the classical case, the emergence of such webs turns out to be associated with *diffusive energy growth.* These numerical findings are given a theoretical explanation in section 4.2.

4.1 Numerical Indications of Quantum Stochastic Webs

First, in subsection 4.1.1, from the numerical methods discussed in chapter 3 that method that is best suited to the present study is finally identified: the algorithm using the propagator in the eigenrepresentation of the harmonic oscillator, as discussed in section 3.3. This method is then used to generate, for several initial states $|\psi_0\rangle$ and for a number of different values of the parameters $T \in \mathcal{Q}$, $\hbar$ and V_0, long time series of quantum states $|\psi_n\rangle$. These states are analyzed in terms of patterns in quantum phase space (subsection 4.1.1) and energy growth (subsection 4.1.2).

4.1.1 The Quantum Skeleton

The classical stochastic webs discussed in chapter 1 are objects within classical phase space, which is spanned by both the position *and* momentum variables. This is an inherently classical concept, as in quantum mechanics either the position *or* the momentum representation can be used, but these representations are mutually exclusive. Therefore it is not a priori clear how the classical and quantum results are to be compared.

A solution for this problem of classical-quantum comparison is to consider *quantum phase space distribution functions.* In appendix A I describe in some detail how such distributions can be defined and used to compare a quantum state with a classical phase portrait. It turns out that many different phase space distributions can be defined, but the most important is the HUSIMI distribution $F^{\mathrm{H}}(x,p,t;\zeta)$. In many respects, this quasiprobability distribution function is as close as possible to the (LIOUVILLE) probability distributions in phase space obtained for classical systems. In the same way as $\langle x\,|\psi(t)\rangle$ in the position representation contains exactly the same information as $\langle p\,|\psi(t)\rangle$ in the momentum representation, $F^{\mathrm{H}}(x,p,t;\zeta)$ gives an equivalent description of the quantum state $|\psi(t)\rangle$. $\zeta \in \mathbb{R}$ is a numerical parameter that can be chosen as desired; here I use $\zeta = 1$ alone, in which case the HUSIMI distribution is also known as the *coherent state representation.* See appendix A for more information on the theory of quantum phase distribution functions.

I now begin the discussion of the numerical methods with the example of $\hbar = 1.0$, $V_0 = 1.0$ and the resonance given by $T = \pi/2$ ($q = 4$). This value of T classically leads to rectangular classical stochastic webs as shown in figures 1.7, 1.8, 1.10b and 1.12. The skeletons of the classical stochastic webs for $T = \pi/2$ are given by the square grid (1.45). In the figures of the present section for

$T = \pi/2$, this grid is displayed via thin lines, in addition to the contour lines of $F^{\mathrm{H}}(x, p, nT - 0; 1)$.

In figure 4.1, the result of a numerical simulation is shown for which the GOLDBERG finite differences algorithm of subsection 3.1.2 has been used. In the interval $[-10\pi, 10\pi]$, $\langle x | \psi_n \rangle$ is sampled at $j_{\max} = 16384$ equidistant points, and boundary conditions according to equation (3.24a) have been assumed. Figure 4.2 shows the same as the preceding figure, the difference being that here the dynamics has been obtained using the propagator of the quantum map in the eigenrepresentation of the harmonic oscillator, with $m_{\max} = 6000$ being the size of the basis used — see section 3.3. For both figures, the ground state $|0\rangle$ of the harmonic oscillator has been chosen as the initial state. In the figures, contour lines of the HUSIMI distributions are drawn at 2.5%, 10%, 20%, ..., 90%, 99% of the respective maximum values of $F^{\mathrm{H}}(x, p, nT - 0; 1)$ for each state.

Using the parameters stated above, the harmonic oscillator representation method runs much faster than the finite differences method: on a Pentium 4 with 2 GHz, the algorithms require approximately 10 sec / 150 sec per iteration of the quantum map, respectively. After a closer inspection of the numerical data it also turns out that the GOLDBERG algorithm produces a faster growing numerical error, manifesting itself in an increasing deviation of the norm of the numerically modelled quantum state from its nominal value of 1. In principle, this error can be controlled by adaptively using smaller time steps T/s, but only at considerable numerical cost. The second method also produces a numerical error, but at a distinctly smaller rate: after $n = 2000$ kicks, for example, the respective errors of the two runs shown in the figures are of the order of 10^{-3} and 10^{-5}, respectively.

This is reflected in the figures: the initial state $|0\rangle$ — the HUSIMI distribution corresponding to the ground state $|0\rangle$ of the harmonic oscillator is a two-dimensional Gaussian in phase space, centered around $(0, 0)^t$; see figures 4.1 ($n = 0$), 4.2 ($n = 0$) and A.1c — evolves into a web-like structure, as the more exact figure 4.2 nicely shows. The spreading of the wave packet is visible in figure 4.1 as well, but there it takes place much more slowly and in a much less pronounced fashion; the periodic pattern clearly displayed in figure 4.2 evolves much more slowly in figure 4.1, again indicating a numerical error of the GOLDBERG result. This error could be reduced by increasing $j_{\max}$ and s, and by spending more numerical effort on solving equation (3.21) in order to control cancellations and other numerical errors; but in any case the method would be slowed down further.

This behaviour of the GOLDBERG algorithm, suggesting that a considerably greater numerical effort has to be spent for obtaining sufficiently accurate results, has also been observed in several other calculations performed for other parameter combinations; it seems to be typical for this algorithm when applied to the quantum map of the kicked harmonic oscillator. Consequentially, the GOLD-

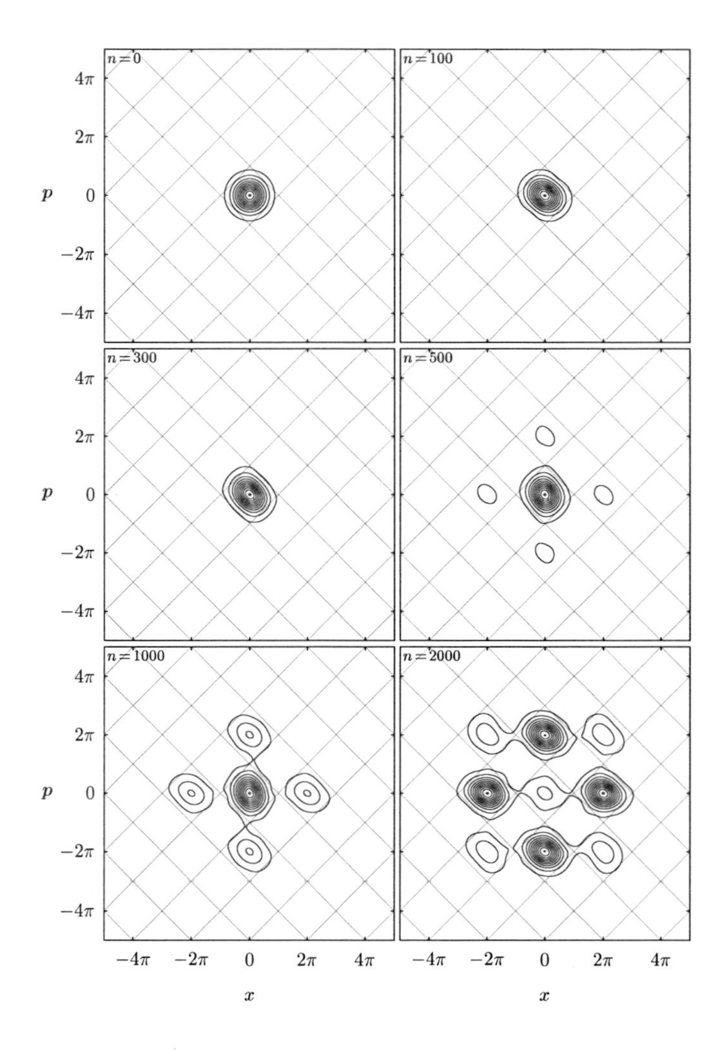

Figure 4.1: A sample result of the application of the GOLDBERG algorithm with periodic boundary conditions; contour plots of HUSIMI distributions of $|\psi_n\rangle$, i.e. after n iterations of the quantum map (2.37), in a case of resonance with respect to T. For each state, the contour lines are drawn at 2.5%, 10%, 20%, ..., 90%, 99% of the maximum value of $F^{\mathrm{H}}(x, p, nT - 0; 1)$, respectively.
Parameters: $T = \pi/2$, $\hbar = 1.0$, $V_0 = 1.0$. Initial state: $|\psi_0\rangle = |0\rangle$.

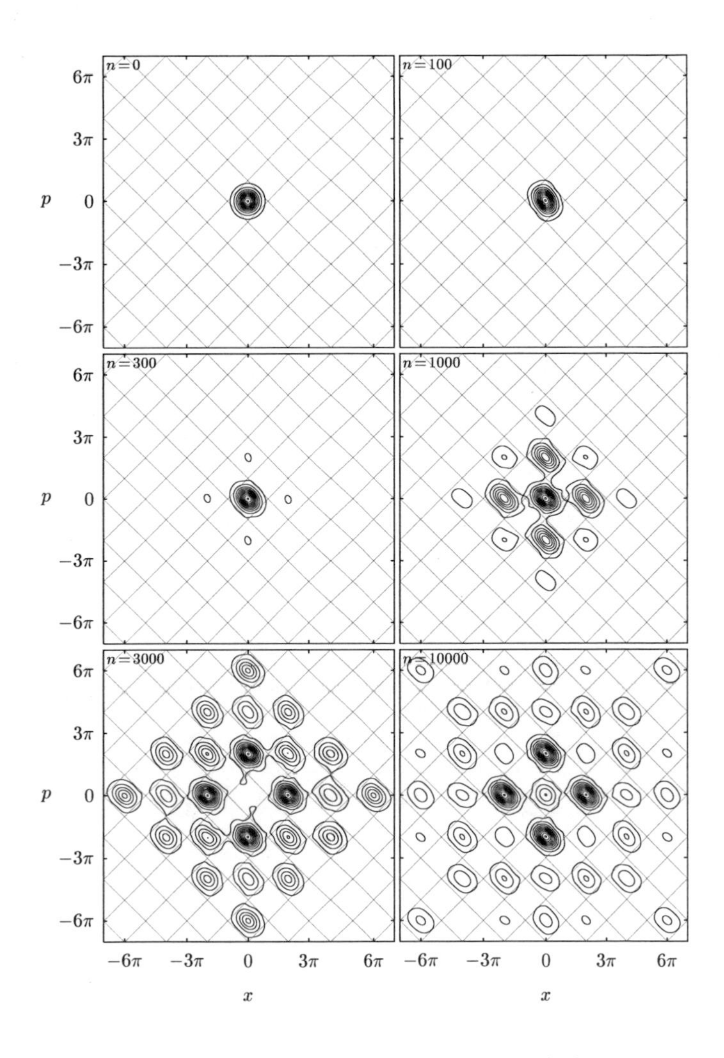

Figure 4.2: Contour plots of HUSIMI distributions of $|\psi_n\rangle$, obtained — as all numerical results from here on — using the propagator of the quantum map in the eigenrepresentation of the free harmonic oscillator. The contour lines are drawn at 2.5%, 10%, 20%, ..., 90%, 99% of the maximum value of $F^{\mathrm{H}}(x, p, nT - 0; 1)$. Parameters: $T = \pi/2$, $\hbar = 1.0$, $V_0 = 1.0$. Initial state: $|\psi_0\rangle = |0\rangle$.

BERG algorithm is not used further on in this study. All the rest of the numerical iterations of the quantum map shown in chapters 4 and 5 and in appendix C have been carried out using the superior algorithm, namely the method using the propagator in the eigenrepresentation of the harmonic oscillator.

If the computation of the quantum analogues of stochastic webs had been the only focus of this work, then using the GOLDBERG algorithm with *periodic boundary conditions* (3.24b) might have been an alternative approach, potentially faster in obtaining the complete web than the other methods, because periodicity in x-direction is already built into the algorithm. But this advantage is compensated by the disadvantage that solving equation (3.21) becomes numerically more difficult for periodic boundary conditions and easily leads to cancellations, overflows etc. In principle, these difficulties can be mastered — for example by solving equation (3.21) using iterative methods like SOR (successive overrelaxation) [SB00] — but again only at considerable numerical cost for each iteration of the quantum map.

In addition to the better numerical performance, the harmonic oscillator eigenrepresentation method also has the advantage that the same numerical algorithm can be used for all three relevant cases, i.e. for periodic and aperiodic stochastic webs and for the case of nonresonance. Of these, only the periodic webs allow the application of an algorithm relying on periodic boundary conditions, whereas the method used here can cope with all three cases.

Although this algorithm runs comparably fast, while producing quite accurate results, it should be noted that the computer time needed for these simulations can be quite long. As long as the achieved numerical accuracy allows it, in the following often the long-time dynamics for up to 10^5 kicks is studied. Typically, on a fast workstation, it takes up to and beyond ten days to complete such a simulation run. Series of simulations of this scope, where the parameters T, $\hbar$, V_0 each take on several different values, can be performed only if a larger number of fast workstations is available.

In figure 4.2, the dynamics can be iterated up to roughly 10^4 kicks, before the cut-off error reduces the numerical norm of $|\psi_n\rangle$ too much. For $n = 10^4$, the figure indicates that the HUSIMI distribution already exhibits a nearly periodic pattern within the square grid shown. Furthermore, the figure also suggests a 4-fold rotational symmetry, just as for the classical stochastic webs for $T = \pi/2$.[1] $F^{\mathrm{H}}(x, p, nT - 0; 1)$ tends to be concentrated in the meshes of the classical web, rather than in the channels, where the — classical and quantum mechanical — phase space density gets transported away along the classical skeleton (1.45) rapidly.

[1] Similar to the patterns in the classical figures, the rotational symmetry exhibited in figure 4.2 is only 2-fold if the *shape* and the orientation of the distribution contours in the phase space meshes is also taken into account. The skeleton (1.45) describes the 4-fold symmetry pattern, disregarding the shape of the contour lines.

Figure 4.2 leads to the conjecture that, as in the classical case, for $n \to \infty$ the web-like structure uniformly extends over the complete phase space and thus establishes (the meshes of) the *quantum stochastic web*, only the central portion of which is shown in the figure. The classical and the quantum stochastic webs for $T = \pi/2$ seem to be characterized by the same symmetries.

In figures 4.1 and 4.2, an initial state $|\psi_0\rangle$ corresponding to a HUSIMI distribution centered in a *mesh* of the stochastic web has been used, leading to a web-like structure which obviously is concentrated in the meshes of the web.[2] This makes it natural to ask for an initial state that is located somewhere in the *channels* of the web.

To this end, WEYL's unitary *displacement operator*, or translation operator,

$$\hat{D}(\alpha) \ := \ e^{\alpha \hat{a}^\dagger - \alpha^* \hat{a}}, \tag{4.1a}$$

with the ladder operators $\hat{a}$, $\hat{a}^\dagger$ of equations (2.30), or equivalently

$$\hat{D}(x', p') \ := \ e^{\frac{i}{\hbar}(p'\hat{x} - x'\hat{p})} \tag{4.1b}$$

needs to be considered; the parameters $x', p' \in \mathbb{R}$ are essentially the real and imaginary parts of $\alpha \in \mathbb{C}$:[3]

$$\alpha \ = \ \alpha(x', p') \ =: \ \frac{1}{\sqrt{2\hbar}} \,(x' + ip')\,. \tag{4.2}$$

$\hat{D}(x', p')$ acts by translating a state in quantum phase space by x' along the x-axis and by p' along the p-axis. This is discussed in some more detail in subsection A.3.1 of the appendix.

Using the translation operator, initial states like

$$|\psi_0\rangle \ = \ \hat{D}(0, p_0)\,|m\rangle \tag{4.3}$$

[2] Note that in contour plots of HUSIMI distributions showing quantum stochastic webs, typically not *all* meshes of the web show up with significantly nonzero $F^{\mathrm{H}}(x, p, nT - 0; 1)$. In figures 4.1 and 4.2, for example, the meshes at $((2k+1)\pi, (2l+1)\pi)^t$ with $k, l \in \mathbb{Z}$ appear to be essentially unoccupied by the quantum state even after 10^4 iterations of the quantum map. A closer inspection of the numerical data reveals that, although generally speaking $F^{\mathrm{H}}(x, p, nT - 0; 1)$ takes on nonzero values in all meshes, the meshes at $((2k+1)\pi, (2l+1)\pi)^t$ are characterized by much smaller values of $F^{\mathrm{H}}(x, p, nT - 0; 1)$ than elsewhere. This is a consequence of the choice of the initial state $|\psi_0\rangle = |0\rangle$. A different initial state, for example $|\psi_0\rangle = \hat{D}(\pi, \pi)\,|0\rangle$ (with $\hat{D}$ defined in equations (4.1) below), would lead to nonzero $F^{\mathrm{H}}(x, p, nT - 0; 1)$ in the meshes at $((2k+1)\pi, (2l+1)\pi)^t$.

[3] By equation (4.2), in place of $F^{\mathrm{H}}(x, p, t; 1)$ one may equivalently write $F^{\mathrm{H}}(\alpha, t; 1)$; this is used below in subsection 4.2.2, for example.

can be constructed. While $|m\rangle$ gives a HUSIMI distribution that is an annular structure centered around the origin — see figures A.2 and A.3 of the appendix — $\hat{D}(0,p_0)\,|m\rangle$ accordingly gives an annular HUSIMI distribution centered around $(0,p_0)^t$, where $p_0 \in \mathbb{R}$ can be chosen as desired. In the present context, either I choose $p_0 = 0$ in order to obtain an initial state centered in the central mesh of the classical stochastic web, or I choose $p_0 \neq 0$ in such a way that the initial state is centered around a point where two channels of the classical web intersect. Note that all states $\hat{D}(x,p)\,|0\rangle$, i.e. the ground state of the harmonic oscillator shifted to $(x,p)^t$ in the phase plane, are called *coherent states* — cf. subsection A.3.1 of the appendix.

For the following figures, depending on the value of $\hbar$, $m = m(\hbar)$ in the initial states $|m(\hbar)\rangle$ and $\hat{D}(0,p_0)\,|m(\hbar)\rangle$ is chosen in such a way that the energy of $|m(\hbar)\rangle$ is as close to 1/2 as possible:

$$E_0 \;=\; \hbar\left(m(\hbar) + \frac{1}{2}\right) \;\approx\; \frac{1}{2}. \tag{4.4}$$

Any other sufficiently large value of E_0 could have been chosen as well in (4.4). The point about this expression is that one wants to compare the dynamics — and in particular, in the next subsection, the behaviour of the energy as a function of time — corresponding to initial states with similar energies, although the values of $\hbar$ may be different.

This formalism is applied for figure 4.3, where for $\hbar = 0.01$ the initial condition $|\psi_0\rangle = \hat{D}\big(0,\pi\big)\,|50\rangle$ is used; for $T = \pi/2$, the point $(0,\pi)^t$ marks the intersection of the two stochastic channels right above the origin. In the figure, HUSIMI contour lines are drawn at 10%, 20%, ..., 90%, 99% of the respective maximum values of $F^{\mathrm{H}}(x,p,nT;1)$ for each state. Unless stated otherwise, this contour line scheme is applied in all HUSIMI contour plots throughout this study.

Figure 4.3 convincingly shows how the central portions of the stochastic web get filled by the phase space density evolving with time. Iterating for more than the 10^4 kicks shown in the figure, one would also see that not only the central portion of the channels of the web gets visited by the dynamics, but that later on the phase space density to a larger degree flows into the outer parts of the web, too. (Because of the declining numerical accuracy at large n for fixed $m_{\max}$, such figures are not shown here.) Note how closely the quantum web sticks to the square grid marking the skeleton of the classical web. In this sense, the part of phase space covered in figure 4.3 at $n = 10^4$ — plus its periodic continuation along the grid lines — marks the *skeleton of the quantum stochastic web.*

It is also interesting to see how in this particular case at least for some time the quantum dynamics quite closely mimics the classical evolution of an ensemble of phase space points: the figure for $n = 100$ shows how the quantum distribution is stretched along the skeleton line described by $p(x) = \pi + x$, while being

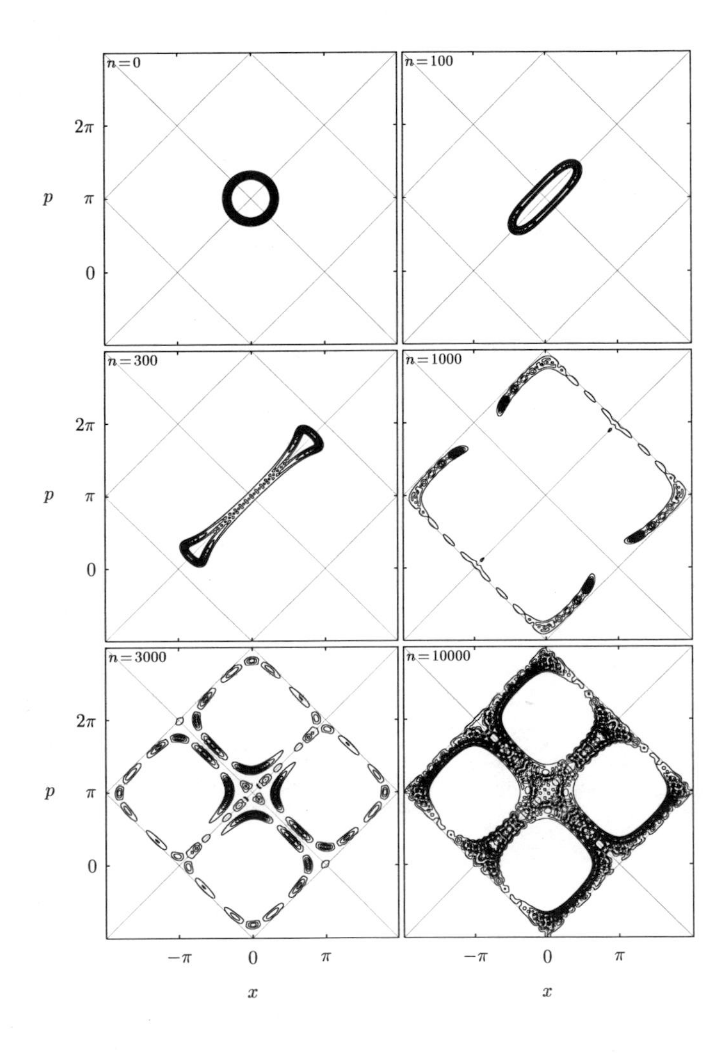

Figure 4.3: Contour plots of HUSIMI distributions of $|\psi_n\rangle$. The contour lines are drawn at 10%, 20%, ..., 90%, 99% of the maximum value of $F^{\mathrm{H}}(x, p, nT - 0; 1)$. Parameters: $T = \pi/2$, $\hbar = 0.01$, $V_0 = 1.0$. Initial state: $|\psi_0\rangle = \hat{D}(0, \pi)\,|50\rangle$.

compressed in the direction of $p(x) = \pi - x$. This is just the classical behaviour near the separatrices — near the stable and unstable manifolds — of the fixed point $(0, \pi)^t$ (with respect to the mapping M_4^4; cf. equation (1.29)). Similarly, the figures for $n = 300$, $n = 1000$ and $n = 3000$ demonstrate the dynamics in the neighbourhood of other classical separatrices: the quantum distribution roughly follows the unstable manifold of a fixed point until it comes close enough to the next fixed point, where it again changes direction, as determined by the respective separatrix. Sections C.1 and C.2 of the appendix contain several additional examples where this behaviour can clearly be recognized; see for example figures C.15, C.17, C.35 and C.36.

Similar HUSIMI contour plots, but for the resonance case $T = \pi/3$ ($q = 6$), are displayed in figures 4.4–4.6. This value of T classically produces hexagonal stochastic webs, one example of which is given in figure 1.3a. The classical skeleton is given by the straight level lines (1.48) of the Hamiltonian (1.47a), the contours of which are plotted in figure 1.10a. In the figures displaying hexagonal *quantum* stochastic webs, the classical skeleton is plotted using thin lines. According to the figures, the quantum skeleton again develops around these grid lines.

The 6-fold (respectively 2-fold: cf. the footnote on page 90) rotational symmetry of the quantum state shown in figure 4.4 becomes clearly visible, once the quantum map is iterated often enough: $n \gtrsim 1000$. The picture for $n = 1000$ also gives a good indication already of what the quantum web would look like for $n \to \infty$.

In figures 4.5 and 4.6, in addition to the usual level lines also the 2.5%-contours are plotted. In this way, for a suitably chosen initial state, the central portion of the channels of the web becomes densely filled by level lines after sufficiently large time, thereby exhibiting the stochastic region of the quantum stochastic web more clearly.

For all $n \leq 10^4$ shown, the dynamics in figure 4.5 is obviously still in a transient stage. While for growing n the HUSIMI distributions spread more and more in phase space, different phase space cells are visited one after the other in a way that mimics the corresponding classical dynamics in a periodic web: classically, for $q = 6$ the iteration of the web map (1.24) yields sequences of points that, qualitatively speaking, *rapidly* encircle the origin of phase space, completing approximately one rotation after every six iterations; superimposed to this is the much slower component of the dynamics in radial direction. This quantum-classical mimicry leads to the phase portraits in figure 4.5 being asymmetric with respect to 6-fold rotations around the origin of phase space. For values of n exceeding 10^4, the phase space distributions can be expected to become more symmetric — in a way that is similar to the development of the web displayed in the following figure.

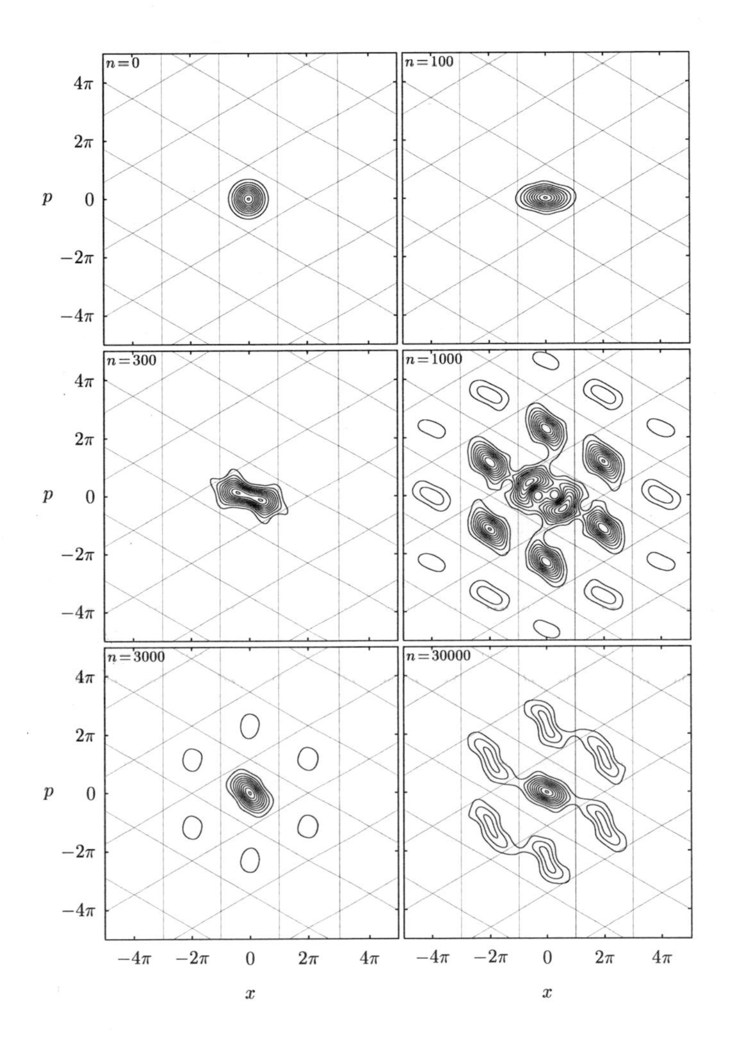

Figure 4.4: Contour plots of HUSIMI distributions of $|\psi_n\rangle$.
Parameters: $T = \pi/3$, $\hbar = 1.0$, $V_0 = 2.0$. Initial state: $|\psi_0\rangle = |0\rangle$.

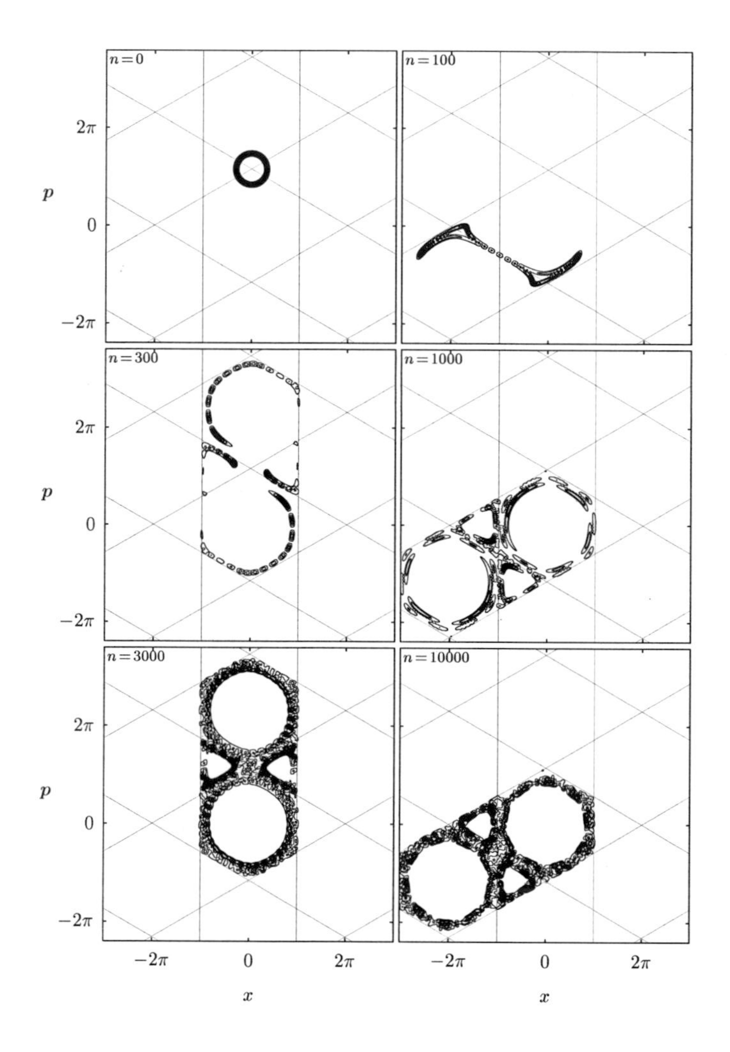

Figure 4.5: Contour plots of HUSIMI distributions of $|\psi_n\rangle$.
Parameters: $T = \pi/3$, $\hbar = 0.01$, $V_0 = 10.0$. Initial state: $|\psi_0\rangle = \hat{D}\big(0, 2\pi/\sqrt{3}\big)\,|50\rangle$.

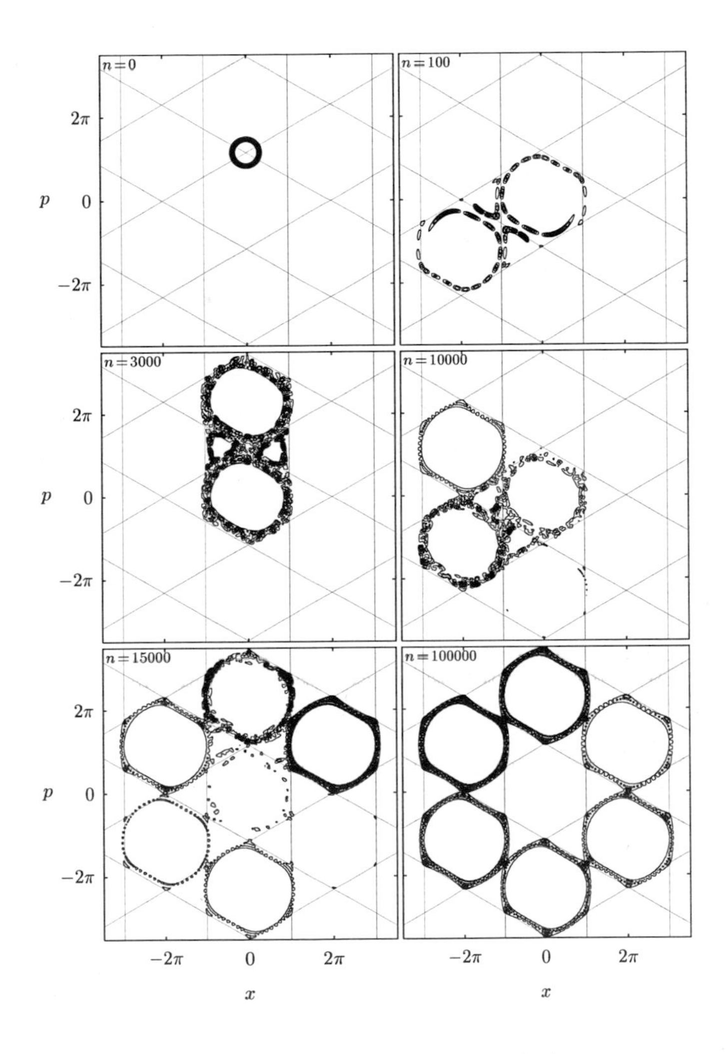

Figure 4.6: Contour plots of HUSIMI distributions of $|\psi_n\rangle$.
Parameters: $T = \pi/3$, $\hbar = 0.01$, $V_0 = 50.0$. Initial state: $|\psi_0\rangle = \hat{D}(0, 2\pi/\sqrt{3})\,|50\rangle$.

Finally, figure 4.6 is remarkable in that it shows the way in which a larger part of the quantum web becomes explored in the course of the dynamics. For $n < 10^4$, the phase portraits look similar to those of figure 4.5 and the phase space regions with essentially nonzero $F^{\mathrm{H}}(x, p, nT - 0; 1)$ are still approximately confined to only two meshes of the web at any given time. For larger times, though, $F^{\mathrm{H}}(x, p, nT - 0; 1)$ increasingly becomes distributed over more than just two meshes, and it is natural to assume that, for times larger than $n = 10^5$, this process continues, comprising even more — and finally all — meshes of the web.

When the quantum map is iterated as often as $n = 10^5$, the question of accuracy of the computed states needs to be addressed again. After so many iterations, the numerical $|\psi_n\rangle$ should not be expected to be exactly the state $\hat{U}^n |\psi_0\rangle$. But in the spirit of the Shadowing Lemma of dynamical systems theory [GH83], one may hope that the sequence $\{ |\psi_n\rangle \}$ the computer finds is nonetheless an approximation to some true quantum dynamics of the system, presumably with respect to a somewhat different initial state $|\tilde{\psi}_0\rangle$. Relying on this heuristic argument, one can more or less safely iterate as long as desired, provided the norms of the computed states do not deviate too much from 1.

Some more contour plots of quantum stochastic webs generated by the quantum kicked harmonic oscillator for $T = \pi/2$ ($q = 4$), $T = 2\pi/3$ ($q = 3$) and $T = 2\pi$ ($q = 1$) can be found in sections C.1, C.2 and C.3 of the appendix.

The portion of phase space significantly covered by the HUSIMI distribution can be visualized approximately by plotting the 1%-contour lines of $F^{\mathrm{H}}(x, p, nT - 0; 1)$. For $T = \pi/2$, $\hbar = 0.1$, $V_0 = 1.5$ and $p_0 = \pi$, this is done in figure 4.7. In this way, the borders of (the central) channels of the web are outlined more clearly, and the spreading of the state along these channels with increasing n becomes more evident. The following subsection addresses the question how this rather vague notion of spreading can be cast into a more exact, measurable form.

4.1.2 Diffusing Wave Packets

In subsection 1.2.4 I have discussed the diffusive energy growth in the channels of *classical* stochastic webs, i.e. in the cases of resonance given by equation (1.33). In the present subsection I study the *quantum* energy growth in these cases leading to the emergence of quantum stochastic webs, as shown in subsection 4.1.1 and in sections C.1 and C.2 of the appendix.

Figures 4.8a and 4.9a display the quantum energy expectation value

$$\langle E \rangle_n = \langle E \rangle \Big|_{t=nT-0} = \sum_{m=0}^{\infty} \left| a_m^{(n)} \right|^2 \hbar \left(m + \frac{1}{2} \right) \tag{4.5}$$

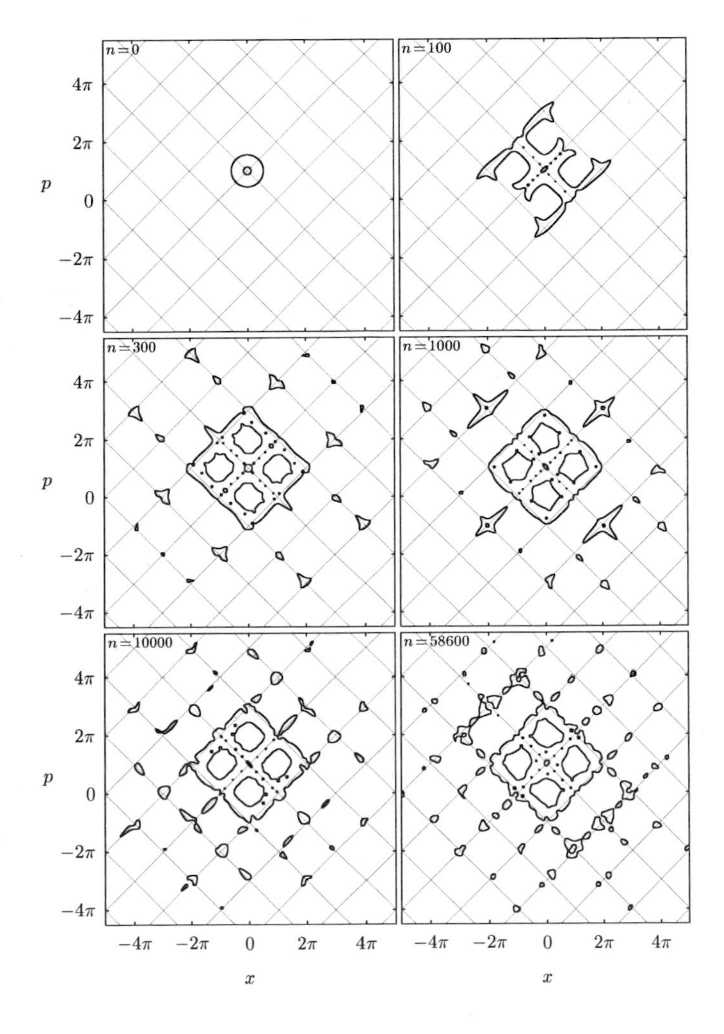

Figure 4.7: Contour plots of HUSIMI distributions of $|\psi_n\rangle$: 1%-level lines. Parameters: $T = \pi/2$, $\hbar = 0.1$, $V_0 = 1.5$. Initial state: $|\psi_0\rangle = \hat{D}(0, \pi)\,|5\rangle$.

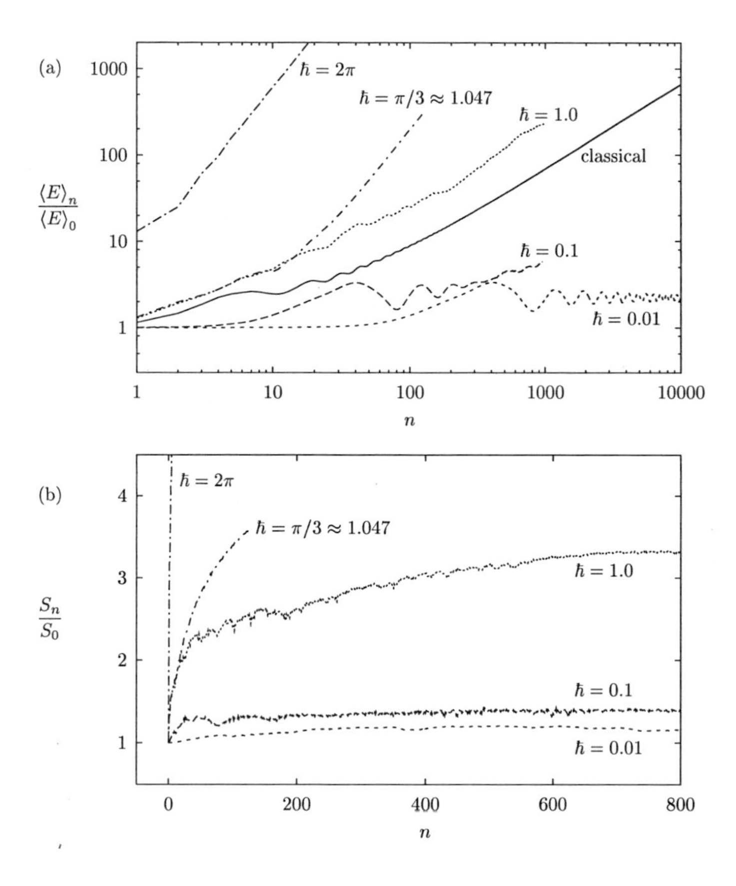

Figure 4.8: Energy and entropy growth for $T = \pi/2$, $V_0 = 2.0$, $p_0 = \pi$ and several values of $\hbar$. (a) Averaged classical energy $\langle E\rangle_n^{\text{cl}}$ and quantum energy expectation value $\langle E\rangle_n$ as a function of n. For $\langle E\rangle_n^{\text{cl}}$, a Gaussian ensemble of initial conditions centered around $(0, p_0)^t$ is used; the initial quantum states are $|\psi_0\rangle = \hat{D}(0, p_0)\,|m\rangle$ with $m = 0/0/0/5/50$ for $\hbar = 2\pi/\pi/3/1.0/0.1/0.01$. Note the logarithmic scaling of both axes. (b) Entropy of $|\psi_n\rangle$ as a function of n.

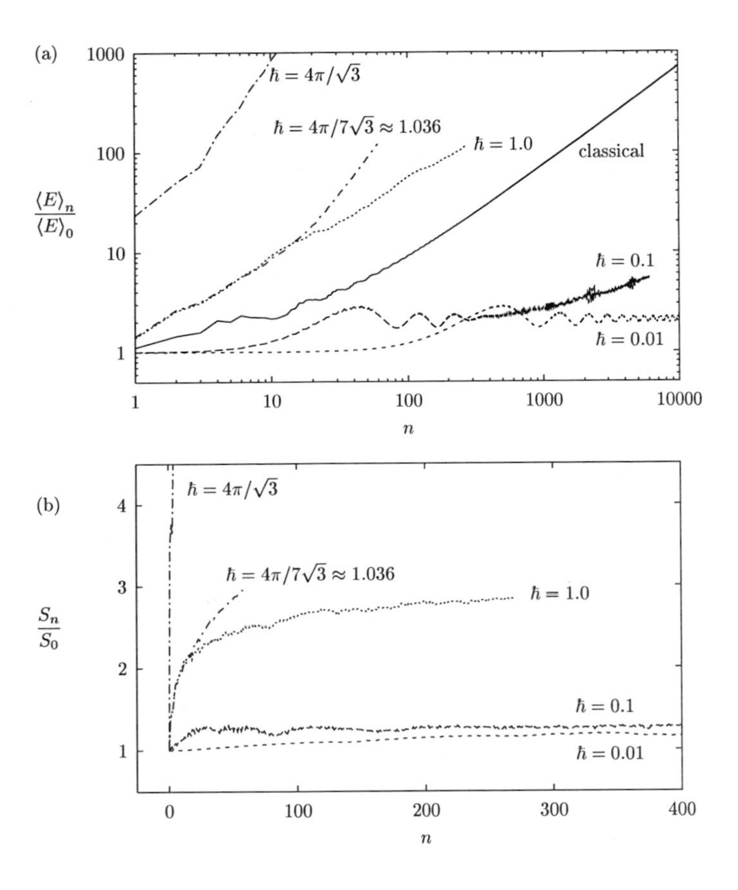

Figure 4.9: The same as figure 4.8, but for $T = 2\pi/3$, $p_0 = 2\pi/\sqrt{3}$.

as a function of discretized time for rectangular and hexagonal stochastic webs with $T = \pi/2$ and $T = 2\pi/3$. The results of numerical simulations for several values of $\hbar$ are shown, where the initial states are located at the unstable periodic points $(0, \pi)^t$ and $(0, 2\pi/\sqrt{3})^t$ of intersections of separatrices, as seems suitable for studying unbounded dynamics. The figures are for $V_0 = 2.0$ — other values of V_0 lead to similar results; but note that very large values of $V_0 \gtrsim 10.0$ can lead to a very fast spreading of the states and thus may necessitate to stop the algorithm after only a short period of time. For comparison, results for $\langle E \rangle_n^{\text{cl}}$, i.e. for the ensemble averaged classical energies (1.74), of classical computations with corresponding parameter values are also shown.

Within the accuracy of the computation, the figures indicate that for *generic* values of $\hbar$ — here: for $\hbar = 1.0/0.1$; nongenericity of $\hbar$ in the present context is defined in equations (4.7) below — the quantum energy grows in the same way as the classical energy average does: in the doubly logarithmic plots, the slopes of the respective graphs indicate an asymptotically *linear* dependence on time,

$$\langle E \rangle_n \sim n. \tag{4.6}$$

Needless to say, this observation can only be made with respect to the scaling region of these graphs, i.e. for large enough n, after the initial transient dynamics has been left behind. For $\langle E \rangle_n^{\text{cl}}$ and $\hbar = 1.0$ this is easily seen in the plots; for $\hbar = 0.1$ the dynamics develops into linear energy growth just around the time when the simulation has to be terminated for numerical reasons; and for $\hbar = 0.01$ the scaling region is not yet reached after even 10^4 kicks. It would be desirable to further confirm the result of linear scaling by using a much larger m_{max}. But this requires some considerable additional numerical effort, and no other result than linear energy growth should be expected to be found for these generic values of $\hbar$.

Many more results of this kind, for other values of generic $\hbar$ and V_0, have been obtained numerically, but are not shown here. Similarly, numerical results with respect to the third nontrivial type of resonance, given by $T = \pi/3$, are not shown here either, but have been obtained in large numbers; they lead to similar observations as described above. Summarizing I have the result that, generically, in quantum mechanics *diffusive energy growth* within stochastic webs is obtained, just as in the classical case.

The figures also show results for some *nongeneric* values of $\hbar$. In the present context, $\hbar$ is called nongeneric if there is an integer s such that

$$\hbar = \frac{2\pi}{s} \qquad \text{for} \quad T = \frac{\pi}{2} \tag{4.7a}$$

or

$$\hbar = \frac{4\pi}{\sqrt{3}s} \qquad \text{for} \quad T = \frac{2\pi}{3} \quad \text{or} \quad T = \frac{\pi}{3}. \tag{4.7b}$$

This definition of nongenericity of $\hbar$ is based on equations (4.44) and (4.49) below, which are obtained in a natural way in subsection 4.2.2 when analytically discussing dynamical consequences of the symmetries of quantum stochastic webs. The cases of nongeneric $\hbar$ as given by equations (4.7) are examples of *quantum resonances*; this is explained in subsection 4.2.2 as well.

With more than convincing numerical accuracy, the figures — with $s = 1$ and $s = 6$ in figure 4.8, and $s = 1$ and $s = 7$ in figure 4.9 — indicate that these nongeneric values of $\hbar$ asymptotically lead to faster, namely quadratic, energy growth,

$$\langle E \rangle_n \sim n^2, \tag{4.8}$$

as opposed to the linear n-dependence for generic $\hbar$-values. This *ballistic energy growth* result is remarkable, as for example the values of $\hbar = 1.0$ and $\hbar = \pi/3 \approx 1.047$ differ by less than five percent and still account for the much differing energy growth laws (4.6) and (4.8). Interestingly, for $n \lesssim 10$ the curves for $\hbar = 1.0$ and $\hbar = \pi/3$ agree within the numerical accuracy, due to the nearly identical values of $\hbar$. Only at later times the respective regimes of linear and quadratic energy growth take over, causing the slopes of the curves to take on differing values. The same observation holds for $\hbar = 1.0$ and $\hbar = 4\pi/7\sqrt{3} \approx 1.036$ in the case of $T = 2\pi/3$.

Both energy growth rate results, diffusive (4.6) and ballistic (4.8), are given a theoretical explanation in section 4.2. Note that, irrespective of the (non-) genericity of the value of $\hbar$, for a given resonant T always the same symmetric phase space patterns are obtained. In other words, depending on the value of $\hbar$, the same quantum stochastic web may be subject to diffusive or ballistic energy growth of the quantum state moving within the web.

Considering the diffusive energy growth of the quantum states $|\psi_n\rangle$ is one way of discussing the spreading of the states in the phase plane with time. Another way is to discuss their VON NEUMANN entropy

$$S_n = S(nT - 0) = -\sum_{m=0}^{\infty} \left|a_m^{(n)}\right|^2 \log \left|a_m^{(n)}\right|^2. \tag{4.9}$$

According to [MS99], S_n can be related to the classical KOLMOGOROV-SINAI entropy [BS93], and it can be interpreted as a measure for the degree to which the quantum state is spread within phase space. With this interpretation of S_n, but without going deeper into the theory of quantum entropies, figures 4.8b and 4.9b, showing S_n as a function of time, indicate that the corresponding sequences of quantum states continuously cover larger and larger portions of phase space — in agreement with the result obtained with respect to the energies of these dynamics.

4.2 An Analytical Explanation of Quantum Stochastic Webs

In subsection 4.1.2 unbounded diffusive motion has been shown to exist for the *quantum* kicked harmonic oscillator in the *classical* resonance cases (1.23) defined by $q \in \tilde{\mathcal{Q}}$ as given by equation (1.33). What is more, the quantum phase portraits in subsection 4.1.1 clearly demonstrate that the unbounded *quantum motion* takes place in just the channels of the *classical periodic stochastic webs* discussed in section 1.2. In other words, it has been found numerically that there exist quantum stochastic webs covering the whole phase plane and exhibiting the same symmetries as their classical counterparts described in chapter 1.

In this section 4.2 I review, along the lines of [BR95], an analytical explanation for these observations, using an argument that relies on exploiting the symmetries of the FLOQUET operator and on constructing groups of mutually commuting translation operators in the phase plane that also commute with the FLOQUET operator. A related, though slightly less transparent, line of reasoning may be found in [BRZ91]. The cases $q = 1$ and $q = 2$, belonging to $\mathcal{Q}$ — see equation (1.30) — but not to $\tilde{\mathcal{Q}}$, are discussed here, too, as well as the remaining resonances with $q \notin \mathcal{Q}$.

4.2.1 Translational Invariance of the Quantum Skeleton

One of the defining properties of classical periodic stochastic webs is their translational invariance. Therefore it is natural to look for a similar property in the quantum realm. To this end, the commutation properties of the FLOQUET operator $\hat{U}$, as given by equations (2.28, 2.32a, 2.35a), with respect to the displacement operator $\hat{D}(\cdot)$ (equations (4.1)) need to be considered.

For the first contribution to $\hat{U}$, the kick propagator $\hat{U}_{\text{kick}} = e^{-\frac{i}{\hbar}V_0 \cos \hat{x}}$,

$$\hat{D}(x', p')\, e^{-\frac{i}{\hbar}V_0 \cos \hat{x}} = e^{-\frac{i}{\hbar}V_0 \cos(\hat{x} - x')}\, \hat{D}(x', p') \tag{4.10}$$

is obtained, which is a direct consequence of the definition of the translation operator and can be seen by inspection of equations (A.39, A.40a), for example.

Next I consider the propagator $\hat{U}_{\text{free}} = \hat{U}_{\text{free}}(T)$ which describes the free harmonic oscillator dynamics. At this stage of the present discussion, T is not yet restricted to one of the resonance values specified in equation (1.23), but can take on any value. Since $\hat{U}_{\text{free}}$ brings about clockwise rotation in phase space through the angle T, corresponding to the free part $M_{\text{free}}(T)$ of the classical web map (1.20c), the operators $\hat{D}(x',p')$ and $\hat{U}_{\text{free}}$ can be interchanged by introducing a new set $\tilde{x}', \tilde{p}'$ of parameters which is obtained from x', p' by anti-clockwise rotation through T:

$$\hat{D}(x',p')\,\hat{U}_{\text{free}} = \hat{U}_{\text{free}}\,\hat{D}\left(M_{\text{free}}(-T)\begin{pmatrix}x'\\p'\end{pmatrix}\right) \tag{4.11a}$$

$$= \hat{U}_{\text{free}}\,\hat{D}(\tilde{x}',\tilde{p}') \tag{4.11b}$$

with the real parameters

$$\tilde{x}' = x'\cos T - p'\sin T \tag{4.12a}$$

$$\tilde{p}' = x'\sin T + p'\cos T \tag{4.12b}$$

— cf. equation (1.20c). Using the definition

$$\tilde{\alpha}(\tilde{x}',\tilde{p}') = \frac{1}{\sqrt{2\hbar}}\,(\tilde{x}' + i\tilde{p}')\,, \tag{4.13}$$

the new translation operator

$$\hat{D}(\tilde{x}',\tilde{p}') = e^{\frac{i}{\hbar}(\tilde{p}'\hat{x} - \tilde{x}'\hat{p})} \tag{4.14a}$$

can also be denoted as

$$\hat{D}(\tilde{\alpha}) = e^{\tilde{\alpha}\hat{a}^\dagger - \tilde{\alpha}^*\hat{a}}, \tag{4.14b}$$

and the new $\tilde{\alpha}$ in terms of the old α is

$$\tilde{\alpha} = \frac{1}{\sqrt{2\hbar}}\,(x' + ip')\,e^{iT} = \alpha\,e^{iT}. \tag{4.15}$$

Combination of equations (4.10) and (4.11b) with the splitting (2.28) then gives

$$\begin{aligned}\hat{D}(x',p')\,\hat{U} &= \hat{D}(x',p')\,\hat{U}_{\text{free}}\,\hat{U}_{\text{kick}}\\ &= \hat{U}_{\text{free}}\,\hat{D}(\tilde{x}',\tilde{p}')\,e^{-\frac{i}{\hbar}V_0\cos\hat{x}}\\ &= \hat{U}_{\text{free}}\,e^{-\frac{i}{\hbar}V_0\cos(\hat{x}-\tilde{x}')}\,\hat{D}(\tilde{x}',\tilde{p}').\end{aligned} \tag{4.16}$$

for a *single* iteration of $\hat{U}$. For $q \in \mathbb{N}$ iterations of $\hat{U}$, this implies

$$\begin{aligned} \hat{D}(x_0', p_0')\, \hat{U}^q &= \hat{U}_{\text{free}}\, e^{-\frac{i}{\hbar} V_0 \cos(\hat{x} - x_1')}\, \hat{D}(x_1', p_1')\, \hat{U}^{q-1} \\ &= \cdots \\ &= \prod_{j=1}^{q} \left\{ \hat{U}_{\text{free}}\, e^{-\frac{i}{\hbar} V_0 \cos(\hat{x} - x_j')} \right\} \hat{D}(x_q', p_q'), \end{aligned} \tag{4.17}$$

where — with equations (4.12) — the real parameters

$$x_j' = x_0' \cos jT - p_0' \sin jT \tag{4.18a}$$

$$\text{for} \quad 0 \leq j \leq q$$

$$p_j' = x_0' \sin jT + p_0' \cos jT \tag{4.18b}$$

have been introduced. Expressing x_j', p_j' in terms of α where this seems appropriate, and using equation (4.15), equation (4.17) can also be formulated as

$$\hat{D}(\alpha_0) \left\{ \hat{U}_{\text{free}}\, e^{-\frac{i}{\hbar} V_0 \cos \hat{x}} \right\}^q = \prod_{j=1}^{q} \left\{ \hat{U}_{\text{free}}\, e^{-\frac{i}{\hbar} V_0 \cos(\hat{x} - x_j')} \right\} \hat{D}\left(\alpha_0\, e^{iqT}\right) \tag{4.19}$$

with

$$\alpha_0 = \frac{1}{\sqrt{2\hbar}} (x_0' + i p_0'). \tag{4.20}$$

This expression can now be used to derive conditions that must be satisfied for a $\hat{D}(\alpha_0)$ commuting with some power q of the FLOQUET operator $\hat{U}$.

Concluding from equation (4.19), the operators $\hat{D}(\alpha_0)$ and $\hat{U}^q$ commute if and only if there are integers $k_1, \ldots, k_q, l \in \mathbb{Z}$ such that

$$qT = 2\pi l \tag{4.21a}$$

$$x_j' = 2\pi k_j \qquad \text{for} \quad 1 \leq j \leq q. \tag{4.21b}$$

These conditions need to be analyzed further in order to understand the consequences for the quantum phase portrait, but some consequences can be read off from equations (4.21) directly.

Equation (4.21a) imposes a restriction on the rotational part $\hat{U}_{\text{free}}$ of the FLOQUET operator and thus leads to *rotational symmetries*. For general values of l, equation (4.21a) is identical with the general classical resonance condition (1.22). For $l = 1$ it restricts the values of T to

$$T = \frac{2\pi}{q}. \tag{4.22}$$

This means that if the *classical* resonance condition (1.23) is satisfied then $\hat{D}(\alpha_0)$ and $\hat{U}^q$ do commute, provided certain additional conditions, resulting from equation (4.21b), are satisfied as well; these additional conditions are discussed below. In the present chapter, up to this point q was just the number of iterations of $\hat{U}$ being considered. Equations (4.21a) and (4.22) are noteworthy because they link this number of iterations with the parameter q of the classical resonance condition.

In the context of equation (4.19), a lucid interpretation of the resonance condition (1.23/4.22) can be given: it refers to symmetries that come about after the q successive applications of $\hat{U}_{\text{free}}$ in equation (4.19) have accounted for exactly one full rotation in phase space. For general values of l, T belongs to the more general category of resonances as defined by equation (1.22), corresponding to l rotations in phase space after q applications of $\hat{U}_{\text{free}}$. Below I show that it suffices to consider the simplest case $l = 1$ in order to discuss and explain the symmetries of the quantum stochastic webs that have been observed in the previous two sections.

Combination of equation (4.21a) with the definition (4.18) for $j = q$ shows that the resonance condition implies

$$x'_q = x'_0 \tag{4.23a}$$

$$p'_q = p'_0, \tag{4.23b}$$

which is another way of expressing that a full rotation in phase space is obtained after not more than q iterations of $\hat{U}$.

Equation (4.21b) determines the allowed translations and therefore concerns the *translational symmetries* of the phase portrait. Substituting equation (4.21b) into the definitions (4.18),

$$k_j = k_0 \cos\frac{2\pi jl}{q} - \frac{p'_0}{2\pi}\sin\frac{2\pi jl}{q} \tag{4.24a}$$

$$\text{for} \quad 0 \leq j \leq q$$

$$p'_j = 2\pi k_0 \sin\frac{2\pi jl}{q} + p'_0 \cos\frac{2\pi jl}{q} \tag{4.24b}$$

is obtained; here, $k_0 := k_q$ is used, which is consistent with equation (4.23a). Equation (4.24a) can be rewritten in the form

$$\frac{k_j + k_{q-j}}{2k_0} = \cos\frac{2\pi jl}{q} \tag{4.25a}$$

$$k_j - k_{q-j} = -\frac{p_0'}{\pi}\sin\frac{2\pi jl}{q}, \tag{4.25b}$$

which holds for $q \geq 3$ and $j = 1, 2, \ldots, j_{\max}$ with

$$j_{\max} = \begin{cases} \frac{q}{2} - 1 & \text{if } q \text{ is even} \\ \frac{q-1}{2} & \text{if } q \text{ is odd.} \end{cases} \tag{4.26}$$

For even q, an additional nontrivial relation follows from equation (4.24a):

$$k_{\frac{q}{2}} = k_0\,(-1)^l. \tag{4.27}$$

It can be concluded from (4.25a) that, since the k_j are integers, in order to obtain solutions of equation (4.24a) $\cos 2\pi jl/q$ must take on rational values for all $1 \leq j \leq j_{\max}$.

Note that — with any $z \in \mathbb{R}$, and in fact with any $j_{\max} \in \mathbb{N}$ — the following two assertions are equivalent:

$$\cos jz \in \mathbb{Q} \quad \text{for } 1 \leq j \leq j_{\max} \tag{4.28a}$$

$$\cos z \in \mathbb{Q}. \tag{4.28b}$$

This is a useful observation, since it allows to discuss the rationality of $\cos 2\pi jl/q$ for all j by considering just the case of $j = l = 1$. The equivalence of (4.28a) and (4.28b) can be shown in the following way. Clearly, (4.28b) is implied by (4.28a) with $j = 1$. Conversely, assuming that $\cos z$ is rational, the same is true for $\cos 2z = 2\cos^2 z - 1$. Rationality of $\cos jz$ for all $j \geq 3$ then follows by induction using

$$\cos jz = 2\cos(j-1)z\cos z - \cos(j-2)z. \tag{4.29}$$

Finally, for general values of $q \in \mathbb{N}$ one has the result that

$$\cos\frac{2\pi}{q} \in \mathbb{Q} \quad \Leftrightarrow \quad q \in \mathcal{Q}. \tag{4.30}$$

The "⇐" part of (4.30) is trivial; for a proof of the "⇒" part see [CR62].

For simplicity, from here on I consider the case of $l = 1$ only. Higher order symmetries that are associated with $l \neq 1$ are thus excluded from the following considerations. On the other hand, the choice $l = 1$ is sufficient to discuss and explain the symmetries of the quantum stochastic webs that have been observed in the previous two subsections.

Combining the above arguments (4.25–4.30), it is now shown that if T satisfies the *classical* resonance condition (1.23) with $q \in \mathcal{Q}$, then integers k_j can in fact be found which satisfy equation (4.25a), and therefore equation (4.24a) as well, provided the k_0, p_0' are chosen suitably. This being granted, the result

$$\left[\hat{D}(x_0', p_0'),\, \hat{U}^q\right] \;=\; 0 \tag{4.31}$$

is established. This means that there exists a complete set of common eigenstates of $\hat{D}(x_0', p_0')$ and $\hat{U}^q$, and consequently these eigenstates of $\hat{U}^q$ are invariant with respect to the translations defined by x_0', p_0'. Furthermore, the eigenstates of $\hat{U}$ are also periodic in phase space and invariant with respect to rotations through $T = 2\pi/q$. All eigenstates are extended in phase space. Below, this scenario is analyzed in some more detail for each of the elements q of $\mathcal{Q}$.

In this way the classical resonance condition with $q \in \mathcal{Q}$ is proven to be a necessary condition for the existence of periodic quantum stochastic webs; in this sense it plays the same role both in classical *and* quantum mechanics. This observation is supported by the fact that the results described above are obtained irrespective of the values of V_0 and $\hbar$. As in the classical case, the overall structure of the stochastic web is entirely and solely determined by the parameter T.

In order to check that the k_j determined in accordance with equation (4.25a) also satisfy (4.24a) it remains to check that equation (4.25b) is obeyed, too. In particular, it must be confirmed that the k_j determined by equation (4.24a) are indeed integers, as required by equation (4.21b). This cannot be discussed in general terms, but needs to be checked for each individual $q \in \mathcal{Q}$.

Evaluating equation (4.24a) it turns out — not surprisingly — that the cases of $q = 1, 2$, belonging to $\mathcal{Q}$ but not to $\tilde{\mathcal{Q}}$, are in a characteristic way different from the cases of $q = 3, 4, 6$ belonging to $\tilde{\mathcal{Q}}$. In the following equations, both k_0 and l_0 are integers that can be chosen as desired.

- $\boldsymbol{q = 1, 2}$:

  The symmetric quantum phase space structures obtained for $q = 1$ and $q = 2$ are identical. The restrictions discussed above imply that the parameters x_0', p_0' describing the translational symmetries have to be chosen according to

$$x_0' = 2\pi k_0, \quad k_0 \in \mathbb{Z} \tag{4.32a}$$

$$p_0' \in \mathbb{R}. \tag{4.32b}$$

The phase portrait is periodic in x-direction with period 2π; translations by p_0' in p-direction are admissible with any p_0'. This is exactly the symmetry pattern visible in the contour plot 1.9 of the classical time averaged Hamiltonians $\mathcal{H}_1$, $\mathcal{H}_2$. And while it is not confirmed — due to the limited number of iterations — by the quantum phase portraits C.38–C.40 for $q = 1$ (see section C.3 of the appendix), these figures at least do not stand in contradiction to the symmetries described here.

- **$q = 3, 6$:**

  For these values of q, translational invariance is obtained with respect to the translations given by

$$x_0' = 2\pi k_0 \tag{4.33a}$$

$$p_0' = \frac{2\pi}{\sqrt{3}}(k_0 + 2l_0), \qquad k_0, l_0 \in \mathbb{Z} \tag{4.33b}$$

  in agreement with both the skeleton of the classical stochastic webs displayed in figure 1.10a and the quantum phase portraits 4.4–4.6 and C.18–C.37 (in the appendix) for $q = 3$ and $q = 6$.

- **$q = 4$:**

  In this case, the translations

$$x_0' = 2\pi k_0, \quad k_0 \in \mathbb{Z} \tag{4.34a}$$

$$p_0' = 2\pi l_0, \quad l_0 \in \mathbb{Z} \tag{4.34b}$$

  are obtained, reproducing the classical square grid of figure 1.10b and of the quantum stochastic webs for $q = 4$ in figures 4.2, 4.3, 4.7 and C.1–C.17.

In this way, the symmetries of the quantum stochastic webs that have been obtained in subsection 4.1.1 using numerical means are explained analytically by exploiting the translation invariance of the FLOQUET operator. This explanation of the infinitely extended eigenstates of the system works for all $q \in \mathcal{Q}$ and for all values of V_0 and $\hbar$.

Note that this analytical explanation of the quantum skeletons nicely parallels the analytical explanation of the skeletons of classical stochastic webs (see subsection 1.2.2) in the following sense: the rotational and translational symmetries of the quantum webs arise from the *combination* of $\hat{U}_{\text{kick}}$ and $\hat{U}_{\text{free}}$. None of these operators alone gives rise to the symmetries of the webs. The same is true for M_{kick} and M_{free} with respect to the classical webs — cf. the footnote on page 28. In summary, both the classical and quantum stochastic webs are the result of the combination of both indispensable contributions H_{kick} and H_{free} to the Hamiltonian. In particular, the symmetries of the quantum webs do *not* reflect any symmetries of the elements of the basis $\{|m\rangle\}$ used for expanding the quantum states in the web.

4.2.2 Energy Growth within Quantum Stochastic Webs

The analysis described in the previous section can be extended in order to give a qualitative explanation of the energy growth that has been observed in subsection 4.1.2 as a result of the unbounded quantum dynamics in the channels of the quantum stochastic web. In this way, contact is made with the classical counterpart not only with respect to the symmetries of the phase portrait — thereby taking a *static* point of view — but also with respect to the most important *dynamical* property of the system.

The existence of a complete set of extended states implies that, with respect to almost all initial states $|\psi(0)\rangle$, unbounded growth of the energy expectation value as a function of time is to be expected.

In order to check the explicit time dependence of this energy growth the commutation properties of different translation operators must be considered. Using the BCH formula

$$e^{\hat{A}}e^{\hat{B}} = e^{\hat{A}+\hat{B}}\,e^{\frac{1}{2}[\hat{A},\hat{B}]}, \tag{4.35}$$

which holds for operators $\hat{A}$ and $\hat{B}$ satisfying $[\hat{A},[\hat{A},\hat{B}]] = [\hat{B},[\hat{A},\hat{B}]] = 0$ [Per93], it is easily confirmed that

$$\left[\hat{D}(x_1',p_1'),\hat{D}(x_0',p_0')\right] = 2i\,\sin\frac{x_1'p_0'-x_0'p_1'}{2\hbar}\,\hat{D}(x_0'+x_1',p_0'+p_1'). \tag{4.36}$$

This equation is now used to identify those translations defined by equations (4.32–4.34), respectively, that are independent of each other in the sense that they form *commutative* groups of translations, with all elements satisfying

$$\sin\frac{x_1'p_0'-x_0'p_1'}{2\hbar} = 0 \qquad \text{for all} \quad x_0',p_0',x_1',p_1'. \tag{4.37}$$

For these groups of commuting translation operators the obvious multiplication rule is used:

$$\hat{D}(x_1', p_1')\,\hat{D}(x_0', p_0') \;=\; \hat{D}(x_0' + x_1', p_0' + p_1'). \tag{4.38}$$

- **$q = 4$:**

  For this value of q it follows from equations (4.34) and (4.37) that $[\hat{D}(x_1', p_1'), \hat{D}(x_0', p_0')] = 0$ is equivalent to

  $$\sin\left(\frac{2\pi^2}{\hbar}(k_1 l_0 - k_0 l_1)\right) \;=\; 0. \tag{4.39}$$

  There are two ways for this equality to hold for all k_0, l_0, k_1, l_1:

  - First there is the case of

    $$k_1 l_0 \;=\; k_0 l_1. \tag{4.40}$$

    The translation operators of this type can be organized as a family — indexed by $r \in \mathbb{Q} \cup \{\infty\}$ — of commutative *one*-parameter groups of translations. For $r \in \mathbb{Q}$, these groups are given by

    $$\left\{\hat{D}(2\pi k, 2\pi r k)\;\middle|\; k \in \mathbb{Z}\right\}. \tag{4.41}$$

    All translations in such a group shift along the same direction in the (x, p)-plane, defined by the *rational* gradient

    $$r \;=\; \frac{l_0}{k_0} \;=\; \frac{l_1}{k_1}. \tag{4.42}$$

    For $r = \infty$, the group is

    $$\left\{\hat{D}(0, 2\pi k)\;\middle|\; k \in \mathbb{Z}\right\} \tag{4.43}$$

    and consists of vertical translations.

  - Alternatively, equation (4.39) is satisfied for all k_0, l_0, k_1, l_1 if there is an $s \in \mathbb{N}$ such that

    $$\hbar \;=\; \frac{2\pi}{s}, \tag{4.44}$$

    i.e. if 2π is an integer multiple of $\hbar$. Then the set

    $$\left\{\hat{D}(2\pi k, 2\pi l)\;\middle|\; k, l \in \mathbb{Z}\right\} \tag{4.45}$$

of all translations satisfying equations (4.34) is in fact a commutative *two*-parameter group.

If $\hbar$ is given by equation (4.44), then the commuting translation operators in any case form the two-parameter group (4.45), regardless of equation (4.42) being satisfied in addition or not.

- **$q = 3, 6$:**

For these two values of q, commuting translation operators are obtained from equations (4.33) and (4.37) if and only if

$$\sin\left(\frac{4\pi^2}{\sqrt{3}\hbar}(k_1 l_0 - k_0 l_1)\right) = 0, \tag{4.46}$$

which can be satisfied in two ways:

- Either k_0, l_0, k_1, l_1 satisfy equation (4.40), making this case very similar to the corresponding case for $q = 4$. Again, the family of Abelian one-parameter groups obtained in this way is indexed by the parameter $r \in \mathbb{Q} \cup \{\infty\}$. For $r \in \mathbb{Q}$, one has the groups

$$\left\{\hat{D}\left(2\pi k, \frac{2\pi}{\sqrt{3}}(2r+1)k\right) \,\middle|\, k \in \mathbb{Z}\right\}, \tag{4.47}$$

and for $r = \infty$, the group of vertical translations is

$$\left\{\hat{D}\left(0, \frac{4\pi}{\sqrt{3}}k\right) \,\middle|\, k \in \mathbb{Z}\right\}. \tag{4.48}$$

- Equation (4.46) is also satisfied for all k_0, l_0, k_1, l_1 if there is an $s \in \mathbb{N}$ such that

$$\hbar = \frac{4\pi}{\sqrt{3}s}. \tag{4.49}$$

This being granted, with equations (4.33) the commutative two-parameter group

$$\left\{\hat{D}\left(2\pi k, \frac{2\pi}{\sqrt{3}}(k+2l)\right) \,\middle|\, k, l \in \mathbb{Z}\right\} \tag{4.50}$$

is obtained, which is similar to the group (4.45) for $q = 4$.

- **$q = 1, 2$:**

While the above shows that the cases of $q = 3$, 4 and 6 are essentially equivalent with respect to commutation of translation operators, the situation is substantially different for $q = 1$ or $q = 2$, because here p_0' and p_1' can take on any real value. I do not discuss these cases any further.

Equations (4.44) and (4.49) represent a new kind of *quantum resonance with respect to $\hbar$*. It has no classical counterpart and is thus entirely different from the resonance condition (1.23/4.22) that concerns the parameter T and plays quite the same role both classically and quantum mechanically, as discussed earlier. The consequences of the quantum resonances (4.44, 4.49) have been studied numerically in subsection 4.1.2.

In this way, for $q \in \tilde{\mathcal{Q}}$ the existence of the commutative groups (4.41, 4.45, 4.47, 4.50) of translation operators, each commuting with the q-th power of the FLOQUET operator, has been established. This is the setting of BLOCH's theorem [Mad78]. The associated energy growth can now be estimated qualitatively as follows.

The first step is to express the energy expectation value in terms of the HUSIMI distribution function $F^{\mathrm{H}}(x,p,t;1)$ corresponding to the state $|\psi(t)\rangle$. With equation (2.31) and the overcompleteness relation (A.52) of the coherent states I obtain for all t:

$$\begin{aligned}
\langle E\rangle_t &= \big\langle \psi(t)\big|\hat{H}_{\text{free}}\big|\psi(t)\big\rangle \\
&= 2\hbar^2 \iint_{\mathbb{R}^2} \mathrm{d}^2\alpha \left(|\alpha|^2 + \frac{1}{2}\right) F^{\mathrm{H}}(\alpha,t;1) && \text{(4.51a)} \\
&= \frac{1}{2}\int_{-\infty}^{\infty}\mathrm{d}x \int_{-\infty}^{\infty}\mathrm{d}p \left(x^2 + p^2 + \hbar\right) F^{\mathrm{H}}(x,p,t;1). && \text{(4.51b)}
\end{aligned}$$

Integrals of this kind can be approximately evaluated by considering the phase space region $\mathcal{A}(t)$ significantly occupied by the HUSIMI distribution at time t. From the normalization property (A.53) of $F^{\mathrm{H}}(x,p,t;1)$,

$$\int_{-\infty}^{\infty}\mathrm{d}x \int_{-\infty}^{\infty}\mathrm{d}p\, F^{\mathrm{H}}(x,p,t;1) = 1, \tag{4.52}$$

which is easily confirmed using the overcompleteness relation (A.52),

$$\iint_{\mathcal{A}(t)} \mathrm{d}x\,\mathrm{d}p\, F^{\mathrm{H}}(x,p,t;1) \approx 1 \tag{4.53}$$

is obtained, making it possible to write

$$|\mathcal{A}(t)| \approx \frac{1}{\left\langle F^{\mathrm{H}}\right\rangle_t}, \tag{4.54}$$

with $\left\langle F^{\mathrm{H}}\right\rangle_t$ being the average value of $F^{\mathrm{H}}(x,p,t;1)$ in $\mathcal{A}(t)$. For sufficiently large t, and on the assumption that F^{H} covers $\mathcal{A}(t)$ more or less uniformly,[4] $\langle E\rangle_t$ can now be approximated as

$$\langle E\rangle_t \approx \frac{1}{2}\left\langle F^{\mathrm{H}}\right\rangle_t \iint\limits_{\mathcal{A}(t)} \mathrm{d}x\,\mathrm{d}p\left(x^2+p^2+\hbar\right) \tag{4.55a}$$

$$\approx \frac{\left\langle F^{\mathrm{H}}\right\rangle_t |\mathcal{A}(t)|^2}{4\pi}$$

$$\approx \frac{1}{4\pi\left\langle F^{\mathrm{H}}\right\rangle_t}, \tag{4.55b}$$

where in the intermediate step it has also been assumed that on the average $\mathcal{A}(t)$ grows isotropically in the phase plane — a behaviour that is typical for the dynamics with $q \in \tilde{\mathcal{Q}}$: cf. figures 4.2, 4.4 and 4.6 (this behaviour is in contrast to the nonisotropic growth of $\mathcal{A}(t)$ for $q = 1, 2$, examples of which are shown in figures C.38–C.40 in appendix C).

From here on the discussion has to distinguish between the cases of the one-parameter groups (4.41, 4.47) and the two-parameter groups (4.45, 4.50) of commuting translation operators. In the first case, there exists a complete set of common eigenstates $\left|\tilde{\phi}_m(\tilde{k},t\right\rangle$ of $\hat{U}^q$ and, for example, $\hat{D}(2\pi, 2\pi r)$ or $\hat{D}\left(2\pi, 2\pi(2r+1)/\sqrt{3}\right)$; the indices $m \in \mathbb{N}$, $\tilde{k} \in \mathbb{R}$ label the eigenstates of $\hat{U}^q$ and $\hat{D}(\cdot,\cdot)$, respectively. These states share the most important properties of the quasienergy states (2.20). In particular, they can be written as

$$\left|\tilde{\phi}_m(\tilde{k},t)\right\rangle = e^{-\frac{i}{\hbar}F_m(\tilde{k})t}\left|\tilde{u}_m(\tilde{k},t)\right\rangle \tag{4.56a}$$

[4] As the contour plots in subsection 4.1.1 and in appendix C show, this assumption is somewhat ill-justified, but at least within the family F^f of phase space distributions discussed in appendix A, F^{H} is the smoothest (cf. equation (A.92)) and thus comes closest to satisfying this assumption. What is more, for the estimate (4.55a) it is also used that the HUSIMI distribution — by equations (A.86) — is non-negative. Table A.1 shows that this advantageous feature is a unique property of F^{H} (and thus of F^{AN}) within the family F^f

These observations, together with the fact that the conclusions obtained on the basis of the assumptions leading to the approximation (4.55a) obviously agree with the numerical results of section 4.1.2 (this is discussed on pages 117ff below), further explain the importance of the HUSIMI distribution for the analysis of the quantum dynamics of classically chaotic systems.

with the reduced states $\big|\tilde{u}_m(\tilde{k},t)\big\rangle$ satisfying

$$\big|\tilde{u}_m(\tilde{k},t+qT)\big\rangle \;=\; \big|\tilde{u}_m(\tilde{k},t)\big\rangle, \tag{4.56b}$$

analogous to the $|u_E(t)\rangle$ of equation (2.23). The phase in the exponential in the full states (4.56a) is essentially determined by $F_m(\tilde{k})$ which parallels the quasienergy E in the states (2.20). Due to the completeness of $\big\{\big|\tilde{\phi}_m(\tilde{k},0)\big\rangle \,\big|\, m \in \mathbb{N},\ \tilde{k} \in \mathbb{R}\big\}$, any initial state $|\psi(0)\rangle$ can be expanded as

$$|\psi(0)\rangle \;=\; \sum_m \int \mathrm{d}\tilde{k}\; A_m(\tilde{k})\; \big|\tilde{\phi}_m(\tilde{k},0)\big\rangle, \tag{4.57}$$

with suitable coefficients $A_m(\tilde{k}) \in \mathbb{C}$, and by equation (4.56a) evolves according to

$$|\psi(t)\rangle \;=\; \sum_m \int \mathrm{d}\tilde{k}\; A_m(\tilde{k})\; e^{-\frac{i}{\hbar}F_m(\tilde{k})t}\; \big|\tilde{u}_m(\tilde{k},t)\big\rangle. \tag{4.58}$$

The HUSIMI distribution at stroboscopic times nT is then obtained as[5]

$$\begin{aligned} F^{\mathrm{H}}(x,p,nT;1) \;&=\; \frac{1}{2\pi\hbar}\,\Big|\,\langle\alpha(x,p)\,|\psi(nT)\rangle\,\Big|^2 \\ &=\; \frac{1}{2\pi\hbar}\,\left|\sum_m \int \mathrm{d}\tilde{k}\; A_m(\tilde{k})\; e^{-\frac{i}{\hbar}F_m(\tilde{k})nT}\; \big\langle\alpha(x,p)\big|\tilde{u}_m(\tilde{k},0)\big\rangle\right|^2. \end{aligned} \tag{4.59}$$

For sufficiently large n, the exponential in equation (4.59) becomes a rapidly oscillating function of $\tilde{k}$, such that the integral can be evaluated using a stationary phase argument [JJ72]. Assuming without loss of generality that, for each m, $F_m(\tilde{k})$ has a single critical point at $\tilde{k}_{0,m}$, I have:

$$F^{\mathrm{H}}(x,p,nT;1) \;\approx\; \frac{1}{nT}\left|\sum_m A_m(\tilde{k}_{0,m})\;\big\langle\alpha(x,p)\big|\tilde{u}_m(\tilde{k}_{0,m},0)\big\rangle \times \frac{e^{-\frac{i}{\hbar}F_m(\tilde{k}_{0,m})nT \,+\, \frac{i\pi}{4}\nu_m}}{\sqrt{F''_m(\tilde{k}_{0,m})}}\right|^2, \tag{4.60}$$

where $\nu_m = \pm 1$ for $F''_m(\tilde{k}_{0,m}) \lessgtr 0$. Initial states $|\psi(0)\rangle$ with $A_m(\tilde{k}_{0,m}) = 0$ can be excluded on generic grounds, but even in such cases comparable results may

[5] Due to the approximative nature of the derivations in this subsection it is not necessary here to consider, more exactly, the stroboscopic times $nT - 0$ just before the n-th kick, rather than nT.

be found [Olv74]. Similarly, the method can be generalized to cover situations with $F''_m(\tilde{k}_{0,m}) = 0$ as well. The important point about equation (4.60) is that it indicates that the HUSIMI distribution at a given phase space point $(x, p)^t$ asymptotically decays like

$$F^{\mathrm{H}}(x, p, nT; 1) \sim \frac{1}{n} \tag{4.61}$$

with time nT. This in turn, together with equation (4.55b), gives

$$\langle E \rangle_n \sim n, \tag{4.62}$$

that has already been found numerically — cf. equation (4.6). In this way, asymptotically unbounded energy growth has been proven, and this energy growth follows a linear, *diffusive* pattern indeed. This is the generic result, which holds for most values of $\hbar$ and is demonstrated for example in figures 4.8a and 4.9a in subsection 4.1.2. The other, exceptional cases are discussed in the following.

For the *two*-parameter groups of commuting translation operators in the cases of the $\hbar$-resonances (4.44) and (4.49), the above reasoning can be repeated in a similar, but not identical, fashion. The differences in some details account for a result that is remarkably different from equation (4.62).

With the two-parameter groups (4.45, 4.50), there is a complete set of common eigenstates $|\tilde{\phi}_m(\tilde{k}, \tilde{l}, t)\rangle$ of $\hat{U}^q$ and, for example, $\hat{D}(2\pi, 0)$ and $\hat{D}(0, 2\pi)$ (or $\hat{D}\left(0, 4\pi/\sqrt{3}\right)$ and $\hat{D}\left(2\pi, 2\pi/\sqrt{3}\right)$); corresponding to the two group parameters there are now two indices $\tilde{k}, \tilde{l} \in \mathbb{R}$ in addition to m. Parallel to equations (4.56) I now have

$$\big|\tilde{\phi}_m(\tilde{k}, \tilde{l}, t)\big\rangle = e^{-\frac{i}{\hbar} F_m(\tilde{k}, \tilde{l}) t} \big|\tilde{u}_m(\tilde{k}, \tilde{l}, t)\big\rangle, \tag{4.63a}$$

with the reduced states $|\tilde{u}_m(\tilde{k}, \tilde{l}, t)\rangle$ satisfying

$$\big|\tilde{u}_m(\tilde{k}, \tilde{l}, t + qT)\big\rangle = \big|\tilde{u}_m(\tilde{k}, \tilde{l}, t)\big\rangle. \tag{4.63b}$$

The initial state

$$|\psi(0)\rangle = \sum_m \int \mathrm{d}\tilde{k} \int \mathrm{d}\tilde{l}\; A_m(\tilde{k}, \tilde{l})\; \big|\tilde{\phi}_m(\tilde{k}, \tilde{l}, 0)\big\rangle, \tag{4.64}$$

with coefficients $A_m(\tilde{k}, \tilde{l}) \in \mathbb{C}$, becomes

$$|\psi(t)\rangle = \sum_m \int \mathrm{d}\tilde{k} \int \mathrm{d}\tilde{l}\; A_m(\tilde{k}, \tilde{l})\; e^{-\frac{i}{\hbar} F_m(\tilde{k}, \tilde{l}) t} \big|\tilde{u}_m(\tilde{k}, \tilde{l}, t)\big\rangle \tag{4.65}$$

after time t and gives the HUSIMI distribution

$$F^{\mathrm{H}}(x,p,nT;1) = \frac{1}{2\pi\hbar}\left|\sum_m \int \mathrm{d}\tilde{k} \int \mathrm{d}\tilde{l}\, A_m(\tilde{k},\tilde{l})\, e^{-\frac{i}{\hbar}F_m(\tilde{k},\tilde{l})nT} \times \langle \alpha(x,p)|\tilde{u}_m(\tilde{k},\tilde{l},0)\rangle \right|^2. \quad (4.66)$$

As in equation (4.60), this integral is evaluated using a stationary phase argument. For each m, it is assumed without loss of generality that $F_m(\tilde{k},\tilde{l})$ has a single critical point at $(\tilde{k}_{0,m},\tilde{l}_{0,m})$ and that the Hessian at this point is either negative or positive definite, such that $\nu_m = \pm 1$ for negative/positive definite $(\mathrm{Hess}\, F_m)(\tilde{k}_{0,m},\tilde{l}_{0,m})$ can be defined. Then I obtain for large enough n:

$$F^{\mathrm{H}}(x,p,nT;1) \approx \frac{2\pi}{n^2 T}\left|\sum_m A_m(\tilde{k}_{0,m},\tilde{l}_{0,m})\, \langle \alpha(x,p)|\tilde{u}_m(\tilde{k}_{0,m},\tilde{l}_{0,m},0)\rangle \times \frac{e^{-\frac{i}{\hbar}F_m(\tilde{k}_{0,m},\tilde{l}_{0,m})nT + \frac{i\pi}{2}\nu_m}}{\sqrt{\det(\mathrm{Hess}\, F_m)(\tilde{k}_{0,m},\tilde{l}_{0,m})}}\right|^2. \quad (4.67)$$

In this way, as a result of the *two*-dimensional integration, the HUSIMI distribution scales like

$$F^{\mathrm{H}}(x,p,nT;1) \sim \frac{1}{n^2}, \quad (4.68)$$

as opposed to the previous result (4.61). This, with equation (4.55b), implies

$$\langle E\rangle_n \sim n^2, \quad (4.69)$$

thereby explaining the asymptotically quadratic, *ballistic* energy growth in the resonance cases (4.44) and (4.49) that is demonstrated in figures 4.8a and 4.9a and in equation (4.8).

The above explanation for the energy growth relies on the possibility to expand the states as in equations (4.57) and (4.64), and on the applicability of the stationary phase approximation. In particular the expansions (4.57, 4.64) are somewhat questionable; their — at least approximate — validity depends on details of the spectral properties of the respective operators. Nevertheless, the arguments based on these assumptions yield suggestive results, in accordance with the numerical findings in section 4.1.2. These points are discussed in some more detail in [GB93, BR95].

This chapter provides a description of the typical quantum dynamics of the kicked harmonic oscillator in the cases of resonance with $q \in \mathcal{Q}$. It is natural to ask in which way the quantum dynamics in the complementary cases of nonresonance — where there are no phase space structures characterized by combined translational and rotational symmetries — differs from the scenario of the present chapter. It might be conjectured that in the absence of resonance, without the condition (1.23/4.22) enforcing the existence of infinitely extended quantum states, the typical quantum dynamics is characterized by some kind of localization phenomenon. That conjecture can be considered to be motivated by the multitude of localization results that have been obtained for the quantum kicked rotor, for which the absence of a classical resonance condition like (1.23) is an essential feature. This question is addressed in the following chapter 5, where the localization approach to the dynamics is taken.

Chapter 5

Anderson Localization

It is nice to know that the computer understands the problem. But I would like to understand it, too.

EUGENE P. WIGNER
Physics Today **46** (1993) 38

The kicked rotor plays a key role in the field of quantum chaos. On the one hand the 1979 paper by CASATI, CHIRIKOV, IZRAELEV and FORD [CCIF79] marks the starting point of the systematic analysis of the quantum dynamics of classically chaotic dynamical systems; in this study characteristic differences between the classical and quantum behaviours of the same model systems were found for the first time. In particular, the *numerical* investigation of the quantum kicked rotor for certain parameter values clearly yielded localized quantum states, whereas the classical counterpart of the system exhibited unbounded diffusive dynamics with respect to the energy in sharp contrast to the quantum result that established an upper limit for the energy.

On the other hand, the *theoretical* explanation of this observation was given in 1982 by FISHMAN, GREMPEL and PRANGE [FGP82, GFP82, PGF83], who showed that the quantum kicked rotor can be mapped onto a well-known model of solid state physics. This model, the ANDERSON model [And58, And61, And78], is characterized by pronounced localization phenomena of the quantum states, provided certain conditions are met that I discuss below. Localization in the ANDERSON model thus carries over to the quantum kicked rotor and provides the desired explanation.

In the present chapter I analyze the *nonresonant* (with respect to T) *quantum kicked harmonic oscillator* along lines similar to those sketched above with respect to the quantum kicked rotor. In this way, by making contact between the quantum kicked harmonic oscillator and the ANDERSON model, I provide a novel theoretical explanation for the suppressed diffusion that is numerically found to be typical for the quantum dynamics in the absence of stochastic webs as discussed in the previous chapter.

Section 5.1 is dedicated to the exposition of the theory of ANDERSON localization in its well-established area of application, the quantum kicked rotor. In this context I also briefly summarize the most important properties of the ANDERSON model of disordered solids, as far as these properties are needed for the present discussion. In section 5.2 I present numerical evidence for localization phenomena in the quantum kicked harmonic oscillator for nonresonant values of T, i.e. in the absence of quantum stochastic webs as discussed in the previous chapter. In section 5.3 I then show analytically that the nonresonant quantum kicked oscillator exhibits ANDERSON localization.

5.1 Anderson Localization in the Quantum Kicked Rotor

Following a short exposition of the theory of the kicked rotor in classical and quantum mechanics in subsection 5.1.1 and a survey of ANDERSON's theory of localization in subsection 5.1.2, I outline in subsection 5.1.3 how localization phenomena in the quantum kicked rotor can be explained by combining these two theories.

5.1.1 The Kicked Rotor

The kicked rotor is one of the most frequently studied model systems in dynamical systems theory. It emerges in many physical systems (see for example [Zas85, MRB$^+$95]), and despite being very simple it has successfully been used for modelling the onset of chaos, retaining many of the typical and complex features of the underlying physical system.

After suitable scaling, leading to dimensionless variables, the kicked rotor can be defined by the Hamiltonian

$$H_{\mathrm{kr}}(\vartheta, I, t) = \frac{1}{2}I^2 + V(\vartheta) \sum_{n=-\infty}^{\infty} \delta(t-n) \tag{5.1}$$

with the angular displacement ϑ and the angular momentum I conjugate to ϑ. In the present subsection I choose the kick potential in the conventional way,

$$V(\vartheta) = -V_0 \cos\vartheta, \tag{5.2}$$

such that H_{kr} models, for example, a mathematical pendulum that is driven by "impulsively acting gravity" [CCIF79, LL92]. The first summand $I^2/2$ of H_{kr} describes free rotation; the second specifies the periodic "gravitational" kicks the strength of which depends on the kick function. See figure 5.1 for a schematic illustration. In subsection 5.1.3 other kick functions are considered as well.

Note that while the Hamiltonian (5.1) with the harmonic forcing (5.2) is similar to the Hamiltonian (1.17) of the kicked harmonic oscillator there are two essential differences: the phase space of the rotor is a cylinder — as opposed to the phase plane of the oscillator — and there is no harmonic potential term like $x^2/2$ in H_{kr}. Notice further that the rotor depends on the single parameter V_0 only that controls the amplitude of the kicks. This is in contrast to the oscillator, where a second parameter (for example the period T of the kicks) cannot be eliminated by scaling. As I outline below these seemingly minor differences account for remarkably different dynamics of the two model systems, both in classical and quantum mechanics.

Defining ϑ_n, I_n as the values of ϑ, I immediately before the n-th kick,

$$\vartheta_n := \lim_{t \nearrow n} \vartheta(t) \tag{5.3a}$$

$$I_n := \lim_{t \nearrow n} I(t), \tag{5.3b}$$

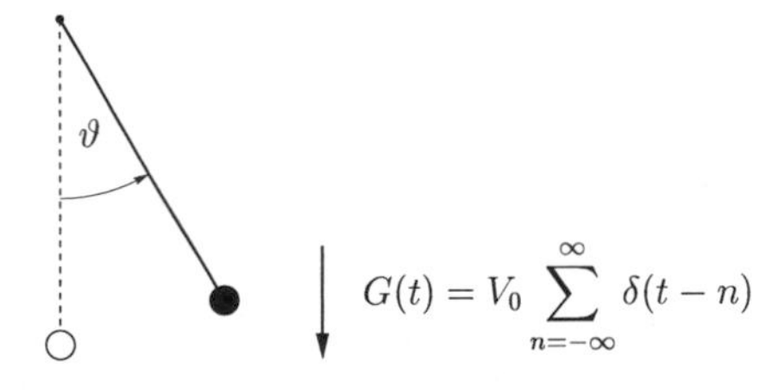

Figure 5.1: Schematic sketch of the kicked rotor as a mathematical pendulum driven by a periodic impulsive gravitational force. Note that by scaling the mass and the length of the pendulum are set to unity. $G(t)$ is the time-dependent gravitational force.

one obtains from $H_{\rm kr}$ the discrete time dynamics

$$\begin{aligned} \vartheta_{n+1} &= \vartheta_n + I_{n+1} && \pmod{2\pi} \\ I_{n+1} &= I_n - V'(\vartheta_n) &&= I_n - V_0 \sin\vartheta_n. \end{aligned} \tag{5.4}$$

This is the CHIRIKOV-TAYLOR *map* [Chi79] or *standard map*.[1] The behaviour of this discrete dynamical system has been studied extensively in the literature [Sch89, LL92] (the second reference contains a long list of references on the standard map).

The bifurcation scenario of this map for increasing values of V_0 is a typical KAM scenario in the sense that, as V_0 is increased, more and more invariant tori, guaranteed to exist by the KAM theorem [GH83], are destroyed. Some phase portraits — periodic with period 2π both in the ϑ- and I-directions — of this transition to chaos are shown in figure 5.2. For $V_0 = 0.2$ (figure 5.2a) the phase portrait is dominated by the invariant lines of regular dynamics. For the intermediate parameter value $V_0 = 1.0$ (figure 5.2b) the regime of *weak chaos* has been reached where invariant lines, POINCARÉ-BIRKHOFF island chains and chaotic regions coexist. For large enough V_0 unbounded motion in the direction of I becomes possible: at a critical value $V_{0,\rm c} \approx 0.9716$ only a single global torus — the "golden" KAM torus — persists, and for $V_0 > V_{0,\rm c}$ it is destroyed, giving way for global diffusion in phase space (figure 5.2c, for $V_0 = 5.0$).

Energy diffusion of the rotor in the classical (ϑ, I)-phase space can be described by considering the rotational energy[2] before the n-th kick:

$$E_n := \frac{1}{2} I_n^2. \tag{5.5}$$

The standard map (5.4) makes E_n evolve according to

$$E_{n+1} = E_n - V_0 I_n \sin\vartheta_n + \frac{V_0^2}{2}\sin^2\vartheta_n, \tag{5.6}$$

which by averaging over an ensemble of orbits and employing the random phase approximation (cf. subsection 1.2.4) becomes the diffusion law

$$\langle E\rangle_n^{\rm cl} \approx \langle E\rangle_0^{\rm cl} + D_{\rm kr}(V_0)\, n \tag{5.7a}$$

[1] The sign of the force term in the standard map (5.4) or in the potential (5.2) is a matter of convention. Changing this sign is equivalent to shifting ϑ by π.

[2] Due to the boundedness of the rotor's phase space in the direction of ϑ, energy diffusion occurs only in the (angular) momentum coordinate I here, as opposed to the case of the kicked harmonic oscillator, where both the position and momentum variables x, p are unbounded and subject to diffusion — cf. equations (1.74, 1.77).

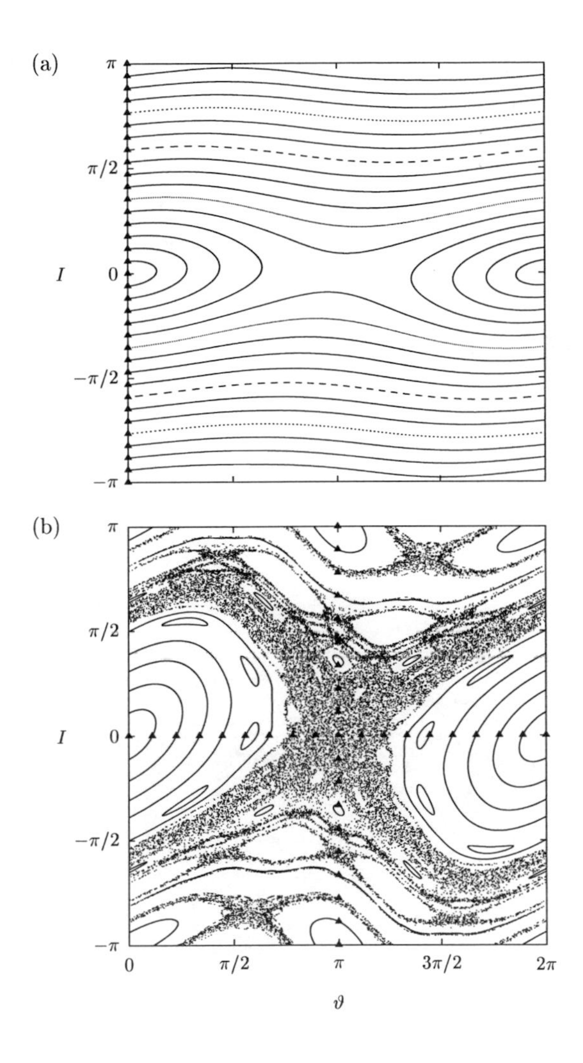

Figure 5.2: Phase portraits of the standard map for (a) $V_0 = 0.2$; (b) $V_0 = 1.0$, just beyond the critical value $V_{0,c} \approx 0.9716$; (c) $V_0 = 5.0$. The initial values (marked by ▲) are iterated 2000 times each. (Continued on page 126.)

with the ("quasilinear") diffusion coefficient

$$D_{\mathrm{kr}}(V_0) \approx \frac{V_0^2}{4}, \tag{5.7b}$$

such that normally diffusive dynamics is to be expected. This averaging is justified for large enough V_0, when unhindered diffusion through phase space is possible. Corrections to formula (5.7b) resulting from accelerator modes and angular correlations are discussed in [LL92], for example.

In analogy with the quantum map of the kicked harmonic oscillator introduced in section 2.1, the quantum dynamics of the kicked rotor is given by

$$|\psi_{n+1}\rangle = \hat{U}_{\mathrm{kr,free}}\hat{U}_{\mathrm{kr,kick}}|\psi_n\rangle, \quad n \in \mathbb{Z}, \tag{5.8}$$

with the quantum state $|\psi_n\rangle$ just before the n-th kick,

$$|\psi_n\rangle := \lim_{t \nearrow n} |\psi(t)\rangle, \tag{5.9}$$

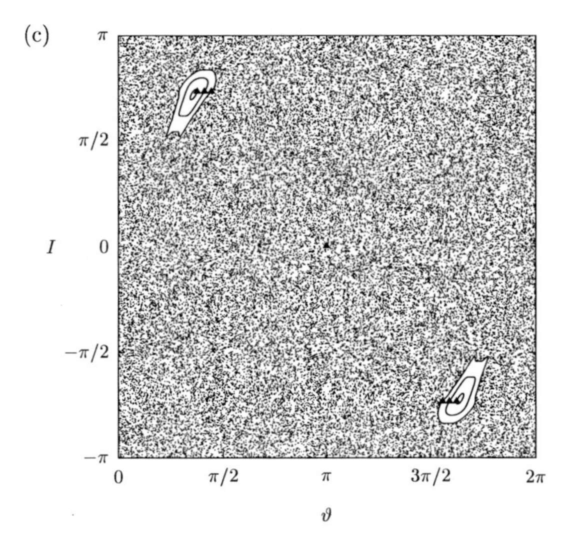

Figure 5.2: (continued) (c) $V_0 = 5.0$. The initial value at $(\vartheta, I)^t = (3.1415, 0.0)^t$ is iterated 50000 times.

and the time evolution operators

$$\hat{U}_{\text{kr,kick}} = e^{-\frac{i}{\hbar}V(\hat{\vartheta})} = e^{\frac{i}{\hbar}V_0 \cos\vartheta} \tag{5.10a}$$

$$\hat{U}_{\text{kr,free}} = e^{-\frac{i}{2\hbar}\hat{I}^2} = e^{\frac{i\hbar}{2}\frac{\partial^2}{\partial\vartheta^2}} \tag{5.10b}$$

for the kick and the unperturbed dynamics, respectively, where $\hat{\vartheta}$ and $\hat{I}$ are the angle and angular momentum operators, and $V(\hat{\vartheta})$ is the potential energy operator. Unlike the classical system that contains the single parameter V_0 only, the quantum system depends on *both* the parameters V_0 and $\hbar$. The dependence on $\hbar$ is essential for the proof of quantum localization in subsection 5.1.3 below.

The quantum map (5.8) can be iterated comparatively easily since the unperturbed dynamics of the rotor is *free rotation* and the propagator $\hat{U}_{\text{kr,free}}$ becomes a mere multiplication operator in the angular momentum representation: for $|\psi_n\rangle$ expanded according to

$$\langle\vartheta\,|\psi_n\rangle = \frac{1}{\sqrt{2\pi}}\sum_{m=-\infty}^{\infty}\Psi_{nm}e^{im\vartheta}, \tag{5.11}$$

using the FOURIER coefficients $\Psi_{nm} = {}_{\text{r}}\langle m\,|\psi_n\rangle$ with the eigenstates of angular momentum

$$\langle\vartheta\,|m\rangle_{\text{r}} = \frac{1}{\sqrt{2\pi}}e^{im\vartheta}, \quad m \in \mathbb{Z}, \tag{5.12}$$

with respect to the eigenvalues $\hbar m$, one obtains for the free rotation part of the dynamics:

$$\left\langle\vartheta\middle|\hat{U}_{\text{kr,free}}\middle|\psi_n\right\rangle = \frac{1}{\sqrt{2\pi}}\sum_{m=-\infty}^{\infty}\tilde{\Psi}_{nm}e^{im\vartheta} \tag{5.13a}$$

with

$$\tilde{\Psi}_{nm} := e^{-\frac{i\hbar m^2}{2}}\Psi_{nm}. \tag{5.13b}$$

Switching between the angle and angular momentum representations can be accomplished with little numerical effort by *fast* FOURIER *transformation* (cf. the footnote on page 70). This makes the kicked rotor a numerically much more accessible model than the kicked harmonic oscillator, where the quantum map typically has to be iterated by multiplying with huge matrices, as described in chapter 3. In [CCIF79] another numerical method for iterating the quantum map of the cosine-kicked rotor is described, which is even more efficient than the FFT-based method, but has the disadvantage of being less general because it is tailored to the special kick potential (5.2).

Results of numerical experiments for both the classical and quantum rotors are shown in figure 5.3, where classical normal diffusion in the case $V_0 = 4.0$ is contrasted with quantum mechanically suppressed diffusion. The classical diffusion coefficient is found numerically as $D_{\mathrm{kr}}(4.0) \approx 3.07$, a somewhat smaller value than given by the large V_0 approximation (5.7b), due to the small value of V_0 (cf. the discussion, at the end of subsection 1.2.4, of a similar situation with respect to the kicked harmonic oscillator). On the other hand, the quantum energy expectation value as a function of discretized time n,

$$\langle E \rangle_n := \left\langle \psi_n \middle| \frac{1}{2}\hat{I}^2 \middle| \psi_n \right\rangle = \frac{\hbar^2}{2} \sum_{m=-\infty}^{\infty} |\Psi_{nm}|^2 m^2, \tag{5.14}$$

exhibits a notably distinct behaviour. Up to a *quantum break time* n^*, $\langle E \rangle_n$ follows the classical curve (5.7a), but is suppressed for larger times; $\langle E \rangle_n$ even appears to be bounded for all n. (In the present example one has $n^* \approx 5$ for $\hbar = 0.1$ and $n^* \approx 10$ for $\hbar = 0.2$; but note that these values of n^* are subject not only to the value of $\hbar$, but also depend on V_0 and the initial state $|\psi_0\rangle$.)

It is this quantum mechanically suppressed energy diffusion, or *quantum localization*, of the kicked rotor that is to be explained in the following two subsections.

5.1.2 Anderson Localization on One-dimensional Lattices

An important subject in solid state physics is the investigation of electronic motion in disordered solids at low temperature. ANDERSON initiated a particular type of research in this field, focusing on one-dimensional lattices as model systems for the crystal lattices of solids [And58, And61, And78]. (A more recent review may be found in [LR85].) Here, I discuss only those aspects of the theory that are relevant for the understanding of *localization phenomena* on such lattices that can be taken as simple model systems for *disordered* solids.

The concept of *transfer matrices* [Pen94] is an essential ingredient of the theory. As an introductory example for the use of these matrices I consider a simplified discussion of a stationary quantum wave function within the one-dimensional potential

$$V_{\mathrm{A}}(x) = \sum_{m=-\infty}^{\infty} w_m v(x - md), \tag{5.15}$$

consisting of a bi-infinite sequence of attractive potentials at the equally spaced lattice sites md, with the lattice constant d. For simplicity, the individual potentials are all assumed to be of the same shape given by $v(x)$, but the weight factors $w_m \in \mathbb{R}$ may vary. The single site potential $v(x)$ is assumed to be suffi-

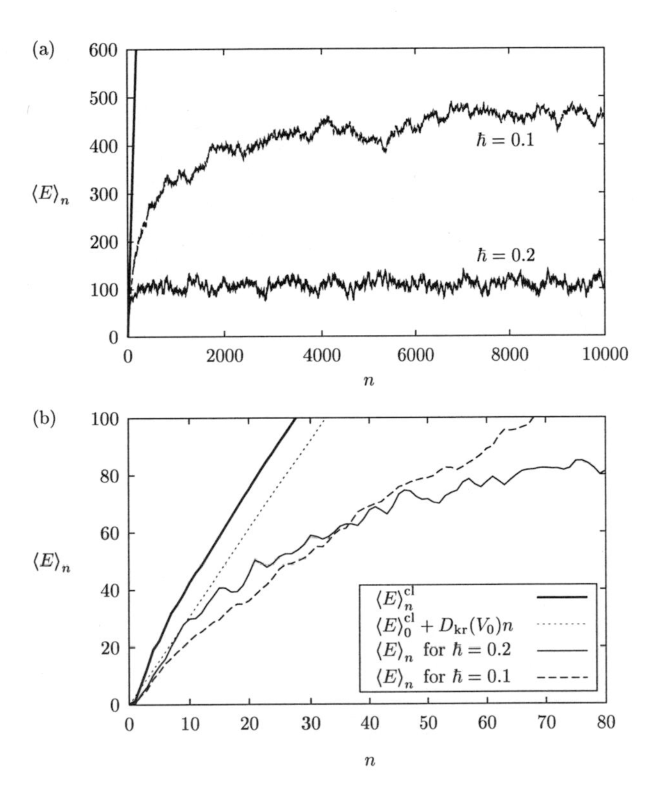

Figure 5.3: Averaged energies $\langle E \rangle_n^{\mathrm{cl}}$ and $\langle E \rangle_n$ of the kicked rotor with $V_0 = 4.0$ versus discretized time n. The classical value of $\langle E \rangle_n^{\mathrm{cl}}$ for a Gaussian ensemble of initial conditions centered at $(\pi, 0)^t$ is drawn as a fat line; the dotted line is the graph of the diffusion formula (5.7a) with the numerically determined value of $D_{\mathrm{kr}}(4.0) \approx 3.07$. The thin and the dashed lines show the time evolution of the energy expectation value $\langle E \rangle_n$ of the quantum kicked rotor for $\hbar = 0.2$ and $\hbar = 0.1$, respectively, where coherent states $\hat{D}(\pi, 0) |0\rangle$ (cf. equation (4.1b) and page 92) have been used as initial conditions. Figure (b) is a magnification of figure (a).

ciently localized around $x = 0$, such that the overlap of potentials belonging to different sites is small and $V_A(x)$ may be considered zero in the inter-site regions. In other words, an electron moving within such a lattice essentially interacts at most with one lattice atom at any given time, and it propagates freely while not being scattered on-site. A lattice of this type may serve as a model system for weakly bound electrons moving within a disordered solid in the framework of a single particle approximation; the lack of order due to impurities (e.g. different types of atoms [Fis96]) or other perturbations of the crystal lattice is described by the varying w_m, leading to reflection and transmission properties that vary from site to site.[3] See figure 5.4 for a schematic sketch of such an ANDERSON lattice.[4]

In the zero potential region between the $(m-1)$-st and m-th site the quantum wave function can be written in the form

$$\langle x \,|\psi\rangle \;=\; A_m e^{ikx} + B_m e^{-ikx} \quad \text{for} \quad md + \delta x < x < (m+1)d - \delta x, \tag{5.16}$$

[3]The model can also be used to describe light propagation in a randomly stratified transparent medium, where the medium is found to reflect the light perfectly. In this way an optical realization of ANDERSON localization is obtained [BK97].

[4]It is also possible to choose $v(x)$ as a narrow rectangular potential barrier, thus automatically guaranteeing good localization of $v(x)$ [PGF85]. But this choice requires a different physical interpretation and is not used in the present context.

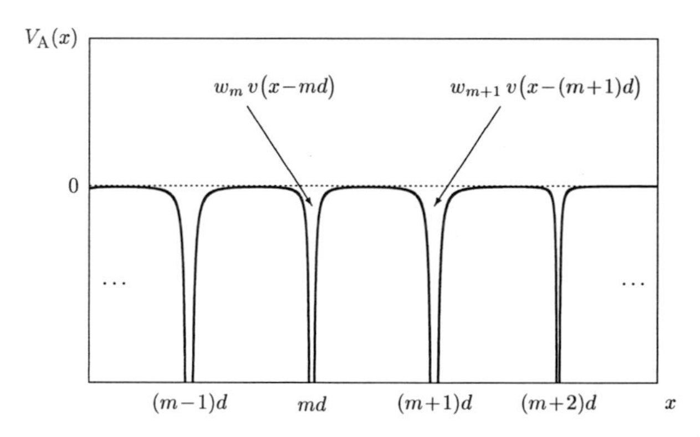

Figure 5.4: Schematic sketch of a one-dimensional ANDERSON lattice, consisting of a bi-infinite sequence of equidistant attractive sharply localized potentials. The individual on-site potentials are of the same shape given by $v(x)$, but the weight factors w_m may vary.

with coefficients $A_m, B_m \in \mathbb{C}$, the wave number $k \in \mathbb{R}$ and some suitable $\delta x \in \mathbb{R}$. As the model does not depend on time explicitly it suffices to consider stationary states. The values of the coefficients in neighbouring flat regions can be related by the $(2,2)$ transfer matrix $\mathcal{T}_m$, which in this context is defined implicitly by

$$\begin{pmatrix} A_{m+1} \\ B_{m+1} \end{pmatrix} = \mathcal{T}_m \begin{pmatrix} A_m \\ B_m \end{pmatrix}. \tag{5.17}$$

In this way the wave function in any of the inter-site regions of the lattice, given by $(A_m, B_m)^t$, can be constructed systematically by specifying a suitable "initial condition" $(A_0, B_0)^t$ and iterating the discrete "dynamical system" (5.17). The particular properties of $V_{\mathrm{A}}(x)$ enter into the matrix elements of $\mathcal{T}_m$ via the reflection and transmission coefficients $T_m^{(\varsigma)}, R_m^{(\varsigma)} \in \mathbb{C}$, $\varsigma = \mathrm{l,r}$, at the m-th site:

$$\mathcal{T}_m = \frac{1}{T_m^{(\mathrm{r})}} \begin{pmatrix} T_m^{(\mathrm{l})} T_m^{(\mathrm{r})} - R_m^{(\mathrm{l})} R_m^{(\mathrm{r})} & R_m^{(\mathrm{r})} \\ -R_m^{(\mathrm{l})} & 1 \end{pmatrix}. \tag{5.18}$$

The phases of these coefficients and the transmission probability $T_m^2 := \left|T_m^{(\varsigma)}\right|^2 = 1 - \left|R_m^{(\varsigma)}\right|^2$, $\varsigma = \mathrm{l,r}$, are subject of the shape of the m-th potential $w_m v(x - md)$ and depend on its width and depth, for example. Due to the time-reversal symmetry of the system defined by the potential (5.15), $\mathcal{T}_m$ can be simplified to give

$$\mathcal{T}_m = \frac{1}{T_m} \begin{pmatrix} e^{i\phi_m^{(1)}} & \sqrt{1 - T_m^2}\, e^{i\phi_m^{(2)}} \\ \sqrt{1 - T_m^2}\, e^{-i\phi_m^{(2)}} & e^{-i\phi_m^{(1)}} \end{pmatrix}, \tag{5.19}$$

where the only remaining parameters are $T_m \in \mathbb{R}$ and two phases $\phi_m^{(1)}, \phi_m^{(2)}$.

For a *periodic* potential $V_{\mathrm{A}}(x)$ with $w_m = w \ \forall m$ the conditions for complete transmission through a lattice site are identical for all sites and can be satisfied by adjusting the energy of the electron. This is a consequence of BLOCH's theorem which asserts the existence of delocalized motion [Mad78].

For *aperiodic* potentials, on the other hand, which — as discussed above — provide a more appropriate model for disordered solids than periodic potentials, the situation is different and more conveniently discussed in terms of another class of systems, giving rise to another type of transfer matrices. Here, the role of the time-independent SCHRÖDINGER equation is taken by the *discrete* SCHRÖDINGER *equation* [And58, And78]:

$$\epsilon_m u_m + \sum_{\substack{m'=-\infty \\ m' \neq m}}^{\infty} W_{mm'}\, u_{m'} = \eta\, u_m. \tag{5.20}$$

It describes a particle with energy $\eta \in \mathbb{R}$, moving on a one-dimensional lattice of equidistant sites, and is to be solved for $u_m \in \mathbb{C}$, the probability amplitude of finding the particle at the m-th site. At each site a random potential $\epsilon_m \in \mathbb{R}$ acts as a *diagonal energy*; the way in which this potential term becomes "randomized" is discussed extensively below. The off-diagonal matrix elements $W_{mm'} \in \mathbb{C}$ describe the coupling of the sites labelled by m and m', respectively. Often the $W_{mm'}$ are referred to as *interaction energies* (cf. [Zim79]), although in general they cannot be restricted to take on real values only.

The discrete SCHRÖDINGER equation (5.20) can be derived from its continuous counterpart, the time-independent SCHRÖDINGER equation for a state $|\psi\rangle$ with energy η,

$$\hat{H}_{\mathrm{A}} |\psi\rangle = \eta |\psi\rangle \tag{5.21a}$$

with the Hamiltonian

$$\hat{H}_{\mathrm{A}} = \frac{1}{2m_{\mathrm{e}}}\hat{p}^2 + V_{\mathrm{A}}(\hat{x}) \tag{5.21b}$$

and the electron mass m_{e}, in the following way. The potential (5.15) again has been assumed to be composed of well localized, nonoverlapping individual potentials $v(x)$. This assumption is often called the *tight binding approximation* (see for example [Stö99]) — although some authors also refer to tight binding with respect to a special case of the model that I discuss below on page 134.

Let the eigenstates of $v(x)$ be denoted by $|\psi^{(\alpha)}\rangle$, with α labelling the different eigenstates. While a general state needs to be constructed by superposition of all of these $|\psi^{(\alpha)}\rangle$, for the model system to be discussed here it suffices to consider a single eigenstate, $|\psi^{(0)}\rangle$, taken to be normalized. It may be looked upon as that single atom state interacting strongest with the passing electron. Obviously, the restriction to a single eigenstate is a further assumption on the system, but allowed in the present context, since the resulting model system is still powerful enough to explain some of the key features of electronic states in disordered crystalline lattices. The localization property of $v(x)$ carries over to $|\psi^{(0)}\rangle$, such that the following LCAO ansatz (linear combination of atomic orbitals, [Zim79]) for an eigenstate $|\psi\rangle$ of the complete lattice can be made,

$$|\psi\rangle = \sum_{m'=-\infty}^{\infty} u_{m'} \hat{T}(m'd) |\psi^{(0)}\rangle , \tag{5.22}$$

characterizing $|\psi\rangle$ as a linear combination of the eigenstates $|\psi^{(0)}\rangle$ shifted to each lattice site by means of the translation operator $\hat{T}(x')$:

$$|x + x'\rangle = \hat{T}(x') |x\rangle = e^{-\frac{i}{\hbar}x'\hat{p}} |x\rangle , \quad x' \in \mathbb{R}; \tag{5.23}$$

$\hat{T}(\cdot)$ is a special case of the more general translation operator $\hat{D}(\cdot,\cdot)$ of equation (4.1b): $\hat{T}(x') = \hat{D}(x', 0)$.

Inserting the expansion (5.22) into the SCHRÖDINGER equation (5.21), and making use of the orthonormality relation

$$\left\langle \psi^{(0)} \right| \left(\hat{T}(md)\right)^{\dagger} \hat{T}(m'd) \left| \psi^{(0)} \right\rangle = \delta_{mm'}, \tag{5.24}$$

which expresses the vanishing of the overlap between eigenstates belonging to different sites, I obtain the discrete SCHRÖDINGER equation (5.20) once the matrix elements

$$W_{mm'} := \left\langle \psi^{(0)} \right| \left(\hat{T}(md)\right)^{\dagger} \hat{H}_{\mathrm{A}}\, \hat{T}(m'd) \left| \psi^{(0)} \right\rangle \tag{5.25}$$

and the diagonal energies

$$\epsilon_m := W_{mm} \tag{5.26}$$

have been defined.

Essentially, the expectation values ϵ_m, taken with respect to $\hat{T}(md) \left| \psi^{(0)} \right\rangle$, depend on the weight factors w_m of V_{A} only,

$$\epsilon_m = \frac{1}{2m_{\mathrm{e}}} \left\langle \psi^{(0)} \,\middle|\, \hat{p}^2 \,\middle|\, \psi^{(0)} \right\rangle + w_m \left\langle \psi^{(0)} \,\middle|\, \hat{v} \,\middle|\, \psi^{(0)} \right\rangle, \tag{5.27}$$

as the two expectation values in equation (5.27), taken with respect to $\left| \psi^{(0)} \right\rangle$, do not depend on m any more. This observation is useful when discussing the way in which disorder or "randomness" is introduced into the theory.

Up to this point, the dynamics on the lattice as given by equation (5.20) is completely deterministic, and its parameters are all fixed by specifying V_{A}. On the other hand, since a priori the eigenstates $\left| \psi^{(0)} \right\rangle$ are unknown and the calculation of the $W_{mm'}$ and ϵ_m might be a difficult task, it is much simpler to *choose* these quantities in an appropriate way, making the model as simple as possible, while still retaining its essential characteristics. In the following I discuss the most important of the possible choices, which is known as *diagonal* or *site disorder*. Here, the matrix elements $W_{mm'}$ are assumed to be "constant", i.e. they are not treated as varying much with m, m'. Frequently, most of the $W_{mm'}$ are taken to be zero, and just a few of them take on a very limited number of nontrivial values. A typical example for such a choice is given below in equation (5.30). The disorder part of the case of site disorder is constituted by assuming the diagonal energies to vary with m, in such a way that the ϵ_m are statistically distributed according to a given distribution and nothing more is known about them. It is a nice feature of this particular model that, as indicated by equation (5.27), the assumptions concerning the distribution of the ϵ_m carry over to the w_m, and vice versa. In this way it is possible to adjust the disorder properties of the model by specifying the weight factors of the "random potential" V_A in the very beginning of the discussion.[5]

Often, the model potential V_A is defined in such a way that the resulting discrete SCHRÖDINGER equation is at least approximately translation invariant with the lattice constant d, i.e. the matrix elements $W_{mm'}$ depend on the distance of the two respective sites m and m' only. In these cases the simplified discrete SCHRÖDINGER equation

$$\epsilon_m u_m + \sum_{\substack{m'=-\infty \\ m' \neq m}}^{\infty} W_{m'-m}\, u_{m'} \;=\; \eta\, u_m \tag{5.28}$$

can be used, where the *hopping matrix elements*

$$W_{m'} \;:=\; W_{m,m+m'} \tag{5.29}$$

specify the interaction of the particle at a given site with its m'-th nearest neighbour site. An especially simple example of such a model is defined by a potential V_A for which all weight factors w_m are identical; but in the present context this example expresses an oversimplification that rules out ANDERSON localization and therefore is not considered in the following.

Depending on V_A, the model may be simplified even further. In order to study ANDERSON localization it suffices to consider the special case where only the interaction with nearest neighbours is taken into account and assumed to be symmetric,

$$W_{m'} \;=\; 0 \quad \text{for} \quad |m'| > 1 \tag{5.30a}$$

$$W_{-1} \;=\; W_1. \tag{5.30b}$$

The restriction (5.30a) is also frequently referred to as the tight binding approximation [Fis93]. (Cf. the discussion of tight binding on page 132.) The resulting *tight binding equation* is

$$\epsilon_m u_m + W_1\,(u_{m-1} + u_{m+1}) \;=\; \eta\, u_m. \tag{5.31}$$

[5]Another way to introduce disorder into the theory is to consider *bond disorder*. In this case, the $W_{mm'}$ are assumed to be statistically distributed, and the ϵ_m are constant.

By scaling the energies ϵ_m and η, W_1 can be set to unity without loss of generality. What is left is a model system that contains just a single external parameter $\eta - \epsilon_m$, fluctuating with m in a random fashion but in accordance with some given distribution of values.

Again, this time as a result of the tight binding approximation (5.30a), the problem can be formulated using a transfer matrix approach by setting

$$\vec{u}_m := \begin{pmatrix} u_m \\ u_{m-1} \end{pmatrix}. \tag{5.32}$$

Defining the transfer matrix as

$$\mathcal{T}_m = \begin{pmatrix} \eta - \epsilon_m & -1 \\ 1 & 0 \end{pmatrix}, \tag{5.33}$$

the tight binding equation (5.31) becomes

$$\vec{u}_{m+1} = \mathcal{T}_m \vec{u}_m. \tag{5.34}$$

Every point (here: corresponding to the index $m+1$) to the right of some initial point (with index 0) on the lattice can be reached by repeated application of the transfer matrix:

$$\vec{u}_{m+1} = \mathcal{T}_m \mathcal{T}_{m-1} \cdots \mathcal{T}_1 \mathcal{T}_0 \vec{u}_0, \quad m \geq 0. \tag{5.35}$$

Similarly, with

$$\vec{v}_m := \begin{pmatrix} u_{m-1} \\ u_m \end{pmatrix} \tag{5.36}$$

every point (with index $m-2$) to the left of another initial point (with index -1) is obtained by

$$\begin{aligned} \vec{v}_{m-1} &= \mathcal{T}_m \vec{v}_m \\ &= \mathcal{T}_m \mathcal{T}_{m+1} \cdots \mathcal{T}_{-1} \mathcal{T}_0 \vec{v}_0, \quad m \leq 0. \end{aligned} \tag{5.37}$$

In such a setting FURSTENBERG's theorem [Fur63] can be applied, which deals with the more general class of unimodular random matrices. Obviously the $\mathcal{T}_m$ are unimodular for all m,

$$\det(\mathcal{T}_m) = 1, \tag{5.38}$$

and $\eta - \epsilon_m$ is a random variable for all η if and only if ϵ_m is random. The theorem then ensures that with probability one the limit

$$\gamma := \lim_{m\to\infty} \frac{\log \|\vec{u}_m\|}{m} = \lim_{m\to\infty} \frac{\log \|\vec{v}_m\|}{m} \tag{5.39}$$

for the "LIAPUNOV exponent γ of the transfer matrix" [Fis93] exists and that γ is positive.[6] (See appendix B for a more detailed exposition of FURSTENBERG's theorem.) This means that in general — i.e. for generic values of the energy η — $\|\vec{u}_m\|$ and $\|\vec{v}_m\|$ essentially grow exponentially with m. For boundary conditions I choose the values of the components of two vectors $\vec{u}_{m_1}$, $\vec{v}_{m_2}$ describing distant points on the lattice, i.e. with $m_1 \ll m_2$. While by the above theorem both sequences $\{\|\vec{u}_m\|\}$ and $\{\|\vec{v}_m\|\}$ *generically* grow exponentially — thus yielding no physical (normalizable or at least improper) states — it is still possible to adjust the energy in such a way that the wave functions iterated from both ends match somewhere in between; this was first observed in [Bor63] and rigorously proven in [DLS85a, DLS85b] to be the generic case. As a result, one obtains almost surely a discrete point spectrum of energy eigenvalues with eigenstates which are exponentially localized around a site $m_c \in \mathbb{Z}$. They can be written as

$$u_m = A\, f_m\, e^{-\gamma|m-m_c|} \tag{5.40}$$

with a normalization constant $A \in \mathbb{R}$, and with coefficients $f_m \in \mathbb{C}$ the absolute values of which do not exceed unity and typically oscillate rapidly with m. Note that in this way the LIAPUNOV exponent γ is identified as the inverse of the localization length of the state described by (5.40). See figure 5.5 for a sketch of the envelope $|u_m|_{\max} = Ae^{-\gamma|m-m_c|}$ of an ANDERSON-localized wave function.

A necessary condition for this explanation to work is that the diagonal energies ϵ_m are randomly distributed under variation of m. The definition of a tight binding model thus has to be completed by specifying the distribution function $p(\epsilon_m)$ for the values of ϵ_m. Actually, in order to ensure that FURSTENBERG's theorem is applicable, $p(\epsilon_m)$ must be "sufficiently well-behaved" in a way that is described in some more detail in appendix B. (In [FMSS85, CKM87] ANDERSON localization is even proved under assumptions which are somewhat weaker than those of FURSTENBERG's theorem, as discussed in appendix B.) Several different choices for $p(\epsilon_m)$ satisfying these requirements have been applied successfully; choosing a Lorentzian distribution function — in which case the model is often referred to as the ANDERSON-LLOYD model —,

$$p(\epsilon_m) = \frac{\delta}{\pi(\delta^2 + \epsilon_m^2)}, \quad \delta \in \mathbb{R}, \tag{5.41}$$

[6]Note that FURSTENBERG's theorem can also be utilized to prove chaoticity of the *classical* kicked rotor for sufficiently large V_0, by reformulating the linearized standard mapping (5.4) in the form of equation (5.35) and thus establishing positivity of the larger LIAPUNOV exponent of the system [Haa01].

has the advantage that in this case an explicit expression for the localization length $1/\gamma$ can be derived analytically:

$$\cosh\gamma = \frac{1}{4}\left(\sqrt{(2+\eta)^2+\delta^2}+\sqrt{(2-\eta)^2+\delta^2}\right); \tag{5.42}$$

i.e. it depends on the energy η and the half width 2δ of the Lorentzian only. A proof of this statement is reviewed in [Haa01].

To summarize, a discrete SCHRÖDINGER equation describing electronic motion within the framework of a one-dimensional lattice model of a disordered solid has been derived, and I have demonstrated that typical solutions exhibit exponential localization.

5.1.3 Mapping of the Quantum Kicked Rotor onto the Anderson Model

Combining the results of the two preceding subsections, I now present the argument showing that the quantum kicked rotor exhibits ANDERSON localization, by deriving a mapping of the rotor model onto the ANDERSON model [FGP82, GFP82, PGF83].

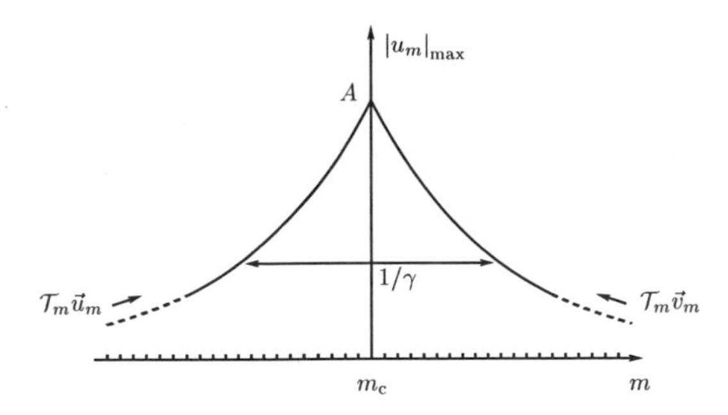

Figure 5.5: Schematic sketch of the envelope $|u_m|_{\max}$ of the absolute value of an ANDERSON-localized state on a one-dimensional lattice. The way in which the transfer matrix $\mathcal{T}_m$ acts on the lattice is also indicated. The "wave function" u_m typically oscillates rapidly between the envelopes given by $\pm|u_m|_{\max}$.

I begin this discussion of the quantum map (5.8) with a general kick potential $V(\vartheta)$,

$$|\psi_{n+1}\rangle = e^{-\frac{i}{2\hbar}\hat{I}^2} e^{-\frac{i}{\hbar}V(\hat{\vartheta})} |\psi_n\rangle, \tag{5.43}$$

and specialize to particular potentials — the cosine potential (5.2) being one of two to be considered — later in the course of the discussion. Using the results of subsection 2.1.1, $|\psi_n\rangle$ can be expanded in terms of the quasienergy states of the kicked rotor $|\phi_E(t)\rangle$ with *constant* coefficients A_E:

$$|\psi_n\rangle = \sum_E A_E |\phi_E(n)\rangle. \tag{5.44}$$

Since the $A_E \in \mathbb{C}$ are time-independent it suffices to consider in the following the time evolution of a single quasienergy state rather than the general $|\psi_n\rangle$ of equation (5.44).

For the investigation of the kicked rotor the states $|u_E(t)\rangle$ are much better suited than the full quasienergy states $|\phi_E(t)\rangle$, since by the periodicity (2.24) the time argument can be dropped altogether. This allows to map the dynamical system (5.43) onto a static problem. In accordance with the discrete nature of the kick, I define the (reduced) quasienergy states immediately before and after the kicks as

$$\big|u_E^{\mp}\big\rangle := \lim_{\varepsilon \searrow 0} \big|u_E(n \mp \varepsilon)\big\rangle. \tag{5.45}$$

In terms of the new states the quantum map can now be formulated as

$$\big|u_E^{+}\big\rangle = e^{-\frac{i}{\hbar}V(\hat{\vartheta})} \big|u_E^{-}\big\rangle \tag{5.46a}$$

$$\big|u_E^{-}\big\rangle = e^{\frac{i}{\hbar}\left(E - \frac{1}{2}\hat{I}^2\right)} \big|u_E^{+}\big\rangle. \tag{5.46b}$$

It turns out that the kick part (5.46a) of the dynamics is most conveniently described using the averaged state

$$\big|\overline{u}_E\big\rangle := \frac{1}{2}\Big(\big|u_E^{-}\big\rangle + \big|u_E^{+}\big\rangle\Big) \tag{5.47}$$

and exchanging the potential $V(\hat{\vartheta})$ for the expression

$$W(\hat{\vartheta}) := -\tan\frac{V(\hat{\vartheta})}{2\hbar}. \tag{5.48}$$

This definition of $W(\hat{\vartheta})$ can be used to rewrite the kick propagator as

$$e^{-\frac{i}{\hbar}V(\hat{\vartheta})} = \frac{1 + iW(\hat{\vartheta})}{1 - iW(\hat{\vartheta})}, \tag{5.49}$$

such that I obtain by substitution into equation (5.46a):

$$\left|u_E^+\right\rangle = \Big(1 + iW(\hat{\vartheta})\Big)\left|\overline{u}_E\right\rangle \tag{5.50a}$$

$$\left|u_E^-\right\rangle = \Big(1 - iW(\hat{\vartheta})\Big)\left|\overline{u}_E\right\rangle. \tag{5.50b}$$

Inserting these expressions into the equation (5.46b) that describes the free part of the dynamics, the $\left|u_E^{\mp}\right\rangle$ are eliminated from the equation and the dynamics is entirely formulated in terms of the averaged quasienergy state $\left|\overline{u}_E\right\rangle$:

$$\Big(1 - iW(\hat{\vartheta})\Big)\left|\overline{u}_E\right\rangle = e^{\frac{i}{\hbar}\left(E - \frac{1}{2}\hat{I}^2\right)}\Big(1 + iW(\hat{\vartheta})\Big)\left|\overline{u}_E\right\rangle. \tag{5.51}$$

In order to facilitate the evaluation of the action of the angular momentum operator $\hat{I}$, this equation is projected onto the eigenstates $\left|m\right\rangle_{\mathrm{r}}$ of $\hat{I}$ (see equation (5.12)). To this end I define

$$\overline{u}_{E,m} := {}_{\mathrm{r}}\!\left\langle m\middle|\overline{u}_E\right\rangle, \quad m \in \mathbb{Z} \tag{5.52}$$

and obtain

$$\tan\left(\frac{1}{2\hbar}\left(E - \frac{\hbar^2 m^2}{2}\right)\right)\overline{u}_{E,m} + {}_{\mathrm{r}}\!\left\langle m\middle|\hat{W}\middle|\overline{u}_E\right\rangle = 0. \tag{5.53}$$

It remains to evaluate the matrix element ${}_{\mathrm{r}}\!\left\langle m\middle|\hat{W}\middle|\overline{u}_E\right\rangle$. By virtue of the representation (5.12) of $\left|m\right\rangle_{\mathrm{r}}$ it is easily seen that the matrix element ${}_{\mathrm{r}}\!\left\langle m\middle|\hat{W}\middle|m'\right\rangle_{\mathrm{r}}$ satisfies

$${}_{\mathrm{r}}\!\left\langle m\middle|\hat{W}\middle|m'\right\rangle_{\mathrm{r}} = {}_{\mathrm{r}}\!\left\langle m - m'\middle|\hat{W}\middle|0\right\rangle_{\mathrm{r}}, \quad m' \in \mathbb{Z}, \tag{5.54}$$

i.e. the matrix element depends on the *difference* of the quantum numbers m, m' only. Therefore it makes sense to use just a single index m' and define the "interaction energy" as[7]

$$W_{m'} := {}_{\mathrm{r}}\!\left\langle m'\middle|\hat{W}\middle|0\right\rangle_{\mathrm{r}}, \tag{5.55}$$

[7] In general, $W_{m'}$ does not take on real values only. However, for kick potentials $V(\vartheta)$ that are even with respect to $\vartheta = \pi$, $W_{m'}$ is real, thus justifying the characterization as an energy. The potential (5.2) falls into this category.

in terms of which the matrix element ${}_{\mathrm{r}}\langle m|\hat{W}|\overline{u}_E\rangle$ reads:

$$ {}_{\mathrm{r}}\langle m|\hat{W}|\overline{u}_E\rangle \;=\; \sum_{m'=-\infty}^{\infty} W_{m'}\,\overline{u}_{E,m-m'}\;. \tag{5.56}$$

In the end one thus arrives at an equation analogous to the discrete SCHRÖDINGER equation (5.20) with $\eta = 0$,

$$ \epsilon_m\overline{u}_{E,m} + \sum_{m'=-\infty}^{\infty} W_{m'}\,\overline{u}_{E,m-m'} \;=\; 0, \tag{5.57}$$

with the diagonal energies $\epsilon_m \in \mathbb{R}$ given by

$$ \epsilon_m(E,\hbar) \;:=\; \tan\left(\frac{1}{2\hbar}\left(E - \frac{\hbar^2 m^2}{2}\right)\right)\,; \tag{5.58}$$

the hopping matrix elements $W_{m'}$ describe the coupling of a "site of the lattice" $\{\overline{u}_{E,m}\}$ with its m'-th nearest neighbour, due to the kick. By the property (5.54) this coupling depends on the distance between the two sites only, which means that the system is translation invariant.

Note that in general equation (5.57) is *not* a tight binding equation (5.31), since the interaction is not restricted to just the respective nearest neighbours of each site. In fact, the cosine potential (5.2) leads to

$$ W_{m'} \;=\; \frac{1}{2\pi}\int_0^{2\pi} \mathrm{d}\vartheta\,\tan\frac{V_0\cos\vartheta}{2\hbar}\;\cos m'\vartheta, \tag{5.59}$$

which implies that $W_{m'} = 0$ for even m', but also that, generally speaking, *all* $W_{m'}$ for odd m' are possibly nonzero. Nevertheless, for this cosine potential $W_{m'}$ decays rapidly — obviously exponentially — with m', as figure 5.6 shows, where $|W_{m'}|$ (obtained by numerical evaluation of the integral in equation (5.59), since a closed formula for this integral is lacking) as a function of m' is plotted for several parameter combinations. The figure indicates that the corresponding discrete SCHRÖDINGER equation (5.57), while not exactly describing a tight binding model, is a good approximation to a tight binding equation for suitable combinations of the parameters V_0, $\hbar$. Namely, equation (5.57) becomes a true tight binding equation in the limit $V_0/\hbar \to 0$, i.e. for weak perturbations and/or in the fully quantum mechanical regime specified by large values of $\hbar$. The tight binding system

$$ \epsilon_m\overline{u}_{E,m} + W_1\left(\overline{u}_{E,m-1} + \overline{u}_{E,m+1}\right) \;=\; 0 \tag{5.60}$$

obtained in this limit can be described — in analogy to the transfer matrix (5.33) in ANDERSON's theory — by the transfer matrix[8]

$$\mathcal{T}_m = \begin{pmatrix} -\dfrac{\epsilon_m}{W_1} & -1 \\ 1 & 0 \end{pmatrix}, \tag{5.61}$$

acting on the vectors

$$\vec{u}_{E,m} := \begin{pmatrix} \overline{u}_{E,m} \\ \overline{u}_{E,m-1} \end{pmatrix} \tag{5.62a}$$

$$\vec{v}_{E,m} := \begin{pmatrix} \overline{u}_{E,m-1} \\ \overline{u}_{E,m} \end{pmatrix} \tag{5.62b}$$

[8] Without loss of generality W_1 can be assumed to be nonzero — this assumption is identical with assuming nontriviality of the tight-binding equation (5.60) — such that dividing by W_1 is not a problem: for the construction of the appropriate tight binding system it suffices to consider the first nonvanishing matrix element $W_{m'_1}$. A sample tight binding equation with $m'_1 = 2$ is given below in equation (5.101). (But note that that equation is for the kicked harmonic oscillator rather than the kicked rotor; this accounts for some differences in details which are discussed there.)

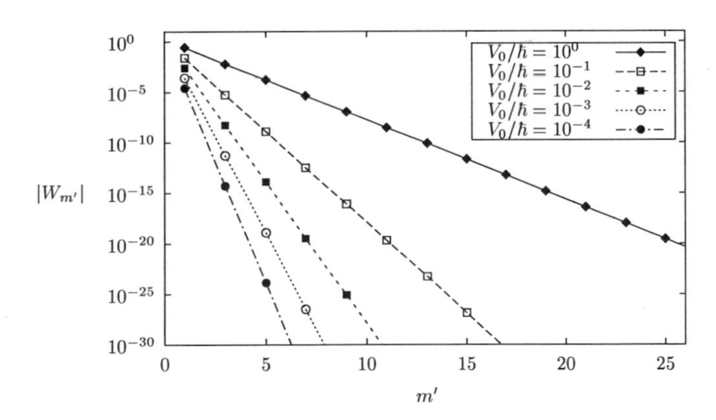

Figure 5.6: Absolute values of the hopping matrix elements $W_{m'}$ (5.59) of the cosine-kicked rotor versus site index m', for odd m'. Note that $W_{m'} = 0$ for even m', and $W_{m'} = W_{-m'}$.

via

$$\vec{u}_{E,m+1} = \mathcal{T}_m \, \vec{u}_{E,m} \tag{5.63a}$$

$$\vec{v}_{E,m-1} = \mathcal{T}_m \, \vec{v}_{E,m} \tag{5.63b}$$

— cf. the vectors (5.32, 5.36) and the “equations of motion” (5.34, 5.37) of the ANDERSON model in subsection 5.1.2.

By choosing a different kick potential $V(\vartheta)$ the transition to a true tight binding model may be achieved. The choice of the potential affects the hopping matrix elements only and leaves the diagonal energies unchanged. For the alternate potential

$$\tilde{V}(\vartheta) := -2\hbar \arctan\left(2\tilde{\lambda}\cos\vartheta - \tilde{\eta}\right) \tag{5.64}$$

with arbitrary real constants $\tilde{\eta}$ and $\tilde{\lambda} \neq 0$, the corresponding alternate hopping matrix elements are

$$\tilde{W}_{m'} = \begin{cases} \tilde{\lambda} & |m'| = 1 \\ -\tilde{\eta} \quad \text{for} & m' = 0 \\ 0 & |m'| > 1, \end{cases} \tag{5.65}$$

such that with this potential one gets a tight binding model without having to rely on an approximation:

$$\epsilon_m \overline{u}_{E,m} + \tilde{\lambda}\left(\overline{u}_{E,m-1} + \overline{u}_{E,m+1}\right) = \tilde{\eta}\,\overline{u}_{E,m}. \tag{5.66}$$

For all practical purposes the diagonal energies ϵ_m of (5.58) can be taken to be randomly distributed as in the ANDERSON-LLOYD model, which can be seen as follows. In the absence of the *quantum resonances* [Fis93] defined by

$$\hbar_{\text{res}} = \frac{P}{Q}\pi \quad \text{with} \quad P, Q \in \mathbb{N}, \tag{5.67}$$

the argument

$$\varphi_m = \frac{1}{2\hbar}\left(E - \frac{\hbar^2 m^2}{2}\right) \pmod{\pi} \tag{5.68}$$

of the tangent in equation (5.58) is pseudo- or quasi-random[9] and uniformly distributed in the interval $[-\pi/2, \pi/2]$. This implies that the corresponding distribution function $p(\epsilon_m)$ for the diagonal energies satisfies

$$p(\epsilon_m)\,\mathrm{d}\epsilon_m \;=\; \frac{1}{\pi}\,\mathrm{d}\varphi_m \;=\; \frac{1}{\pi}\,\arctan'\epsilon_m\,\mathrm{d}\epsilon_m, \tag{5.69}$$

and therefore $p(\epsilon_m)$ turns out to be a Lorentzian (with $\delta = 1$ in terms of equation (5.41)):

$$p(\epsilon_m) \;=\; \frac{1}{\pi(1+\epsilon_m^2)}. \tag{5.70}$$

Figure 5.7a confirms this generic distribution of the ϵ_m in the case of $E = 1.0$ and $\hbar = 1.0$. Figure 5.7b, on the other hand, shows the distribution of the ϵ_m for the just slightly detuned value $\hbar = 7\pi/22 \approx 0.9996$; in this particular example of quantum resonance, due to the implicit modulo operation in equation (5.58) ϵ_m takes on only 18 different values, as opposed to the continuous range of values obtained for $\hbar = 1.0$. Thus it may be claimed that for resonant values of $\hbar$ the resulting $p(\epsilon_m)$ is not well-behaved enough (e.g. not smooth enough) for FURSTENBERG's theorem to apply. This interpretation is in agreement with the well-known fact that in the resonant case the quantum states typically (but not always) are delocalized [Fis93]. For nonresonant $\hbar$, and thus for a Lorentzian quasi-random distribution of the diagonal energies, it can be shown (see equation (B.6) in appendix B) that FURSTENBERG's integrability condition (B.2) is satisfied.

For the above reasoning the presence of the additional factor $1/W_1$ in the matrix element $(\mathcal{T}_m)_{11}$ of equation (5.61) (as compared with equation (5.33)) is irrelevant: the condition (B.2) is satisfied for the matrix (5.61) if and only if it is satisfied for the matrix (5.33).

Note that the quantum resonances (5.67) are an entirely quantum mechanical phenomenon that does not have a classical counterpart. In this respect the quantum resonances of the kicked rotor are similar to the quantum resonances (4.7, 4.44, 4.49) of the resonant (with respect to T) kicked harmonic oscillator discussed in subsections 4.1.2 and 4.2.2. The resonances (1.23/4.22) with respect to T, on the other hand, can also be viewed as a truly quantum mechanical phenomenon, but nevertheless they play an important role classically, too.

Putting all the above pieces together, a mapping from the quantum kicked rotor in the case of quantum nonresonance onto the ANDERSON-LLOYD model

[9] In the case of quantum-nonresonance in the sense of equation (5.67), despite the deterministic quadratic dependence on m, φ_m effectively becomes randomly distributed in the interval $[-\pi/2, \pi/2]$ by the implicit modulo operation, due to the periodicity of the tangent in equation (5.58).

— defined by equations (5.33, 5.34, 5.41) — has been established, and the results on ANDERSON localization reviewed in the previous subsection carry over to the rotor. Using the same genericity assumptions as in subsection 5.1.2 (see pages 135f), the averaged time-independent quasienergy states $|\overline{u}_E\rangle$ are found to be exponentially localized with respect to rotor (angular momentum) eigenstates; the localization length $1/\gamma$ can be derived along the same lines as equation (5.42) for $\eta = 0$, $\delta = 1$:

$$\cosh\gamma \;=\; \sqrt{1 + \frac{1}{4W_1^2}}. \tag{5.71}$$

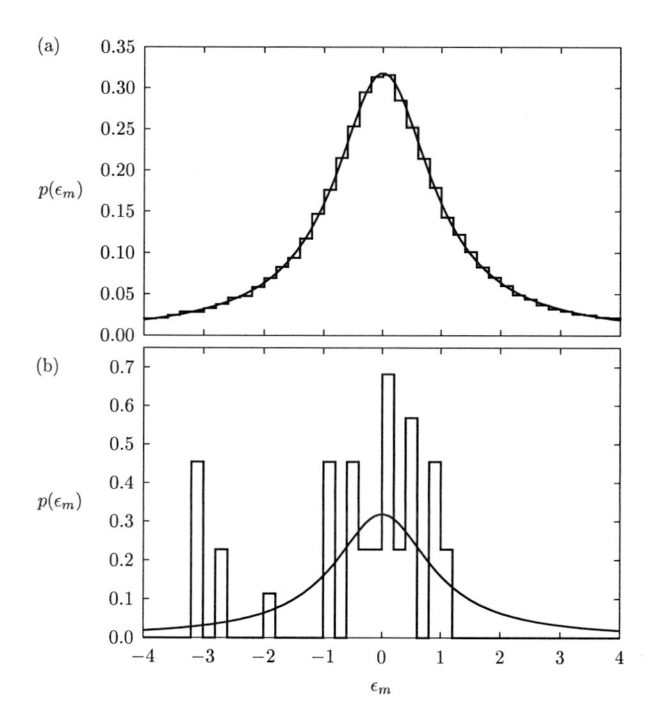

Figure 5.7: Histograms for the distribution of the values $\epsilon_m(E,\hbar)$ of the diagonal energy of the kicked rotor with $1 \leq m \leq 10^5$ and $E = 1.0$, compared with the graph of the Lorentzian (5.70): (a) for the nonresonant value $\hbar = 1.0$; (b) for the quantum resonance given by $\hbar = 7\pi/22 \approx 0.9996$.

This in turn means that every generic sequence of states $|\psi_n\rangle$, $n \in \mathbb{Z}$, generated by the quantum map (5.43) consists of localized states. In other words: the quantum kicked rotor is ANDERSON-localized. Consequences of this type of localization of quantum states are discussed below in sections 5.2 and 5.3 with respect to similarly localized states of the quantum kicked harmonic oscillator.

It is interesting to note the different roles the parameters V_0 and $\hbar$ play with respect to the details of the localization mechanism. Regardless of the value of V_0, the (resonant or nonresonant) value of $\hbar$ alone decides on the existence of localization. V_0, on the other hand, divided by $\hbar$ controls the localization length via equations (5.59) and (5.71).

While by the above arguments it has been shown that there exists a very close relationship between the quantum kicked rotor and the ANDERSON model, it has to be stressed that until now no mathematically *rigorous* proof has been found for exact equivalence of both model systems [CIS98], one of the reasons being that a tight binding equation for the rotor is obtained as an — albeit good — approximation only; in addition, the consequences of the quantum resonances (5.67) for such a potential proof are not yet completely understood. The ongoing effort to establish a complete analogy between the quantum kicked rotor and an ANDERSON-like model is documented in [AZ96, CIS98, AZ98], for example. The lack of mathematically rigorous equivalence may also be expressed by noting that the sample kick potential (5.64), which leads to the tight binding equation (5.66), differs from the original kick potential (5.2). Nevertheless, this explanation of localization in the quantum kicked rotor is generally accepted, and all available numerical evidence supports the equivalence of the two model systems, as far as localization is concerned [Haa01].

In this way the numerical findings of subsection 5.1.1, especially the results on saturation of energy growth of the quantum kicked rotor displayed in figure 5.3, find a satisfactory explanation.

5.2 Localized Wave Packets in the Quantum Kicked Harmonic Oscillator

Subsequent to the introduction into phenomenology and theory of ANDERSON localization and the kicked rotor in the preceding section, I now return to the quantum kicked harmonic oscillator. Complementary to the discussion in chapter 4, in the following sections 5.2 and 5.3 I consider the case of *non-resonant values of T*. Using the numerical techniques described in chapter 3 I demonstrate in section 5.2 that quantum mechanical suppression of classical diffusion is abundant in this model system as well. These findings are then explained analytically in section 5.3.

5.2.1 Numerical Indications of Localization

In chapter 4 I have shown how the quantum manifestations of classical stochastic webs develop under iteration of the quantum map of the kicked harmonic oscillator. Now I turn to the complementary case of nonresonance, in which classically no stochastic web is present and typically the system is characterized by diffusive classical dynamics if V_0 is sufficiently large.

Figure 5.8 shows the time evolution of an initial coherent state, centered at $(0,0)^t$, under the dynamics of the quantum map (2.37) for the nonresonant $T = 1.0$. This figure is to be compared with figure 1.4 showing a phase portrait of the corresponding classical dynamics. In agreement with the classical case there is no spreading of the quantum wave packet; it is just slightly deformed under the dynamics of the quantum map. This observation is not too surprising, as the given initial state $|0\rangle$ is concentrated in the phase space region that is surrounded by the yet unbroken outer invariant line visible in figure 1.4.

This changes in figure 5.9, which shows the same as figure 5.8, but for a larger value of V_0: $V_0 = 3.0$. Classically, for this V_0 the last remaining invariant line has already broken up and given way for global diffusive dynamics of any ensemble of initial conditions. But the quantum dynamics in figure 5.9 shows no sign of any corresponding spreading of the initial wave packet. It is distorted slightly more than in the previous figure, due to the larger V_0, but obviously it does not flow apart. Even after 10^5 iterations of the quantum map the wave packet is well localized. There is no indication of further spreading beyond what is already seen, for example, after 100 kicks.

Figure 5.10 supports this observation. The figure displays the time evolution of the quantum states for a smaller value of $\hbar$: $\hbar = 0.01$; accordingly, $|\psi_0\rangle = |50\rangle$ has been chosen for the initial state. — Cf. equation (4.4) in subsection 4.1.1 for an explanation of this choice for the initial state. — Again obviously localized quantum dynamics is found, even though for this figure the kick strength has been chosen very large, $V_0 = 50.0$, such that in classical mechanics strongly diffusive dynamics would have been expected.

Some further remarks on the choice of the initial state: While the preceding figures for $\hbar = 1.0$ had $|\psi_0\rangle = |0\rangle$, in figure 5.10, for the lower value of $\hbar$, I have $|\psi_0\rangle = |50\rangle$. As discussed earlier, this choice has the advantage that energies of the respective initial states are the same, roughly, for all three figures: $\hbar/2|_{\hbar=1.0} = 0.5 \approx \hbar(50 + 1/2)\big|_{\hbar=0.01} = 0.505$. In addition, in this way all three initial states cover approximately the same area in phase space (near the classical path of the harmonic oscillator with $E = 0.5$.), thus facilitating the comparison of the corresponding phase space distributions. What is more, the alternative possibility of choosing $|\psi_0\rangle = |0\rangle$ for all values of $\hbar$ would have the disadvantage of already starting with states that are more localized for smaller values of $\hbar$.

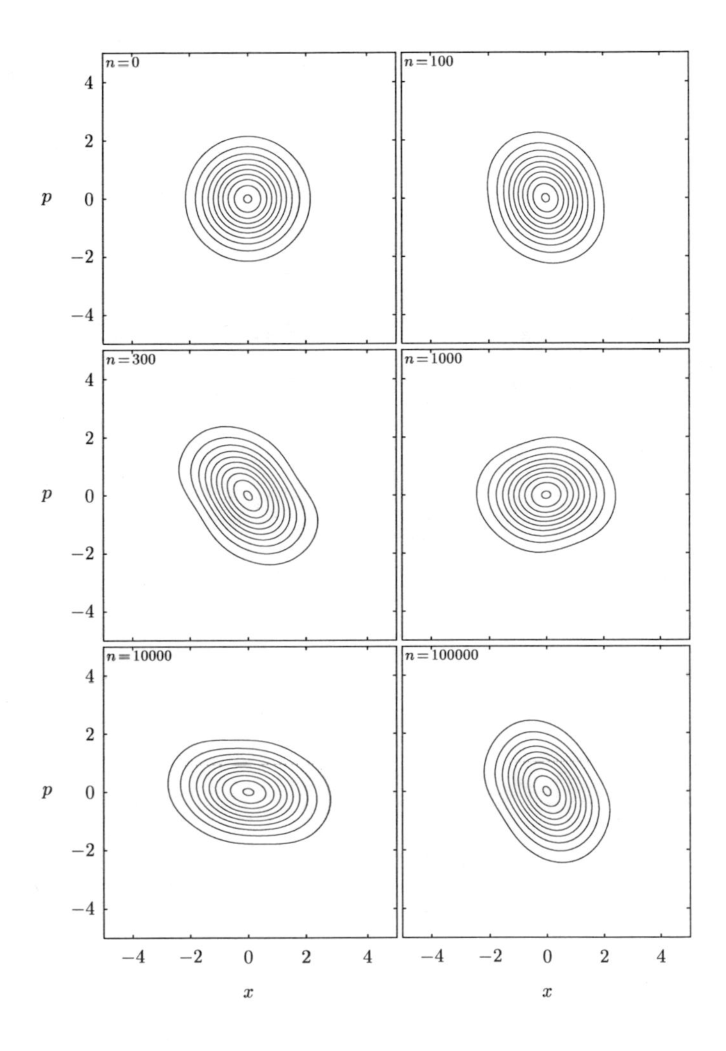

Figure 5.8: Contour plots of HUSIMI distributions of $|\psi_n\rangle$ in a case of nonresonance with respect to T. For each state, the contour lines are drawn at 10%, 20%, ..., 90%, 99% of the maximum value of $F^{\mathrm{H}}(x, p, nT - 0; 1)$, respectively. Parameters: $T = 1.0$, $\hbar = 1.0$, $V_0 = 1.0$. Initial state: $|\psi_0\rangle = |0\rangle$.

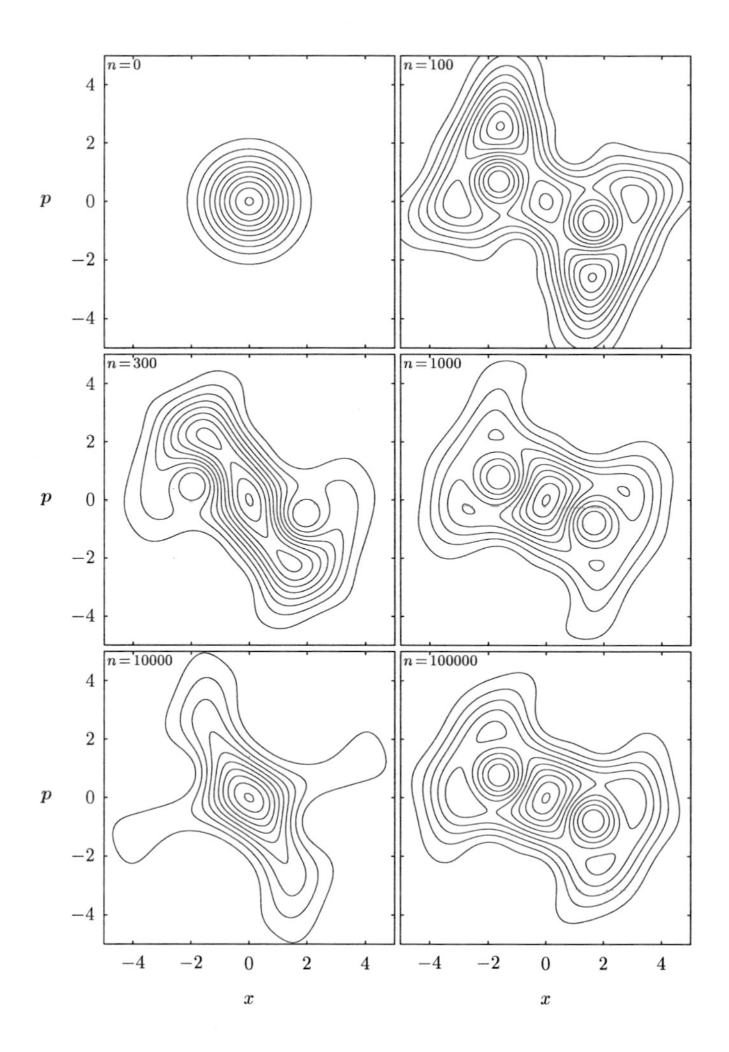

Figure 5.9: Contour plots of HUSIMI distributions of $|\psi_n\rangle$.
Parameters: $T = 1.0$, $\hbar = 1.0$, $V_0 = 3.0$. Initial state: $|\psi_0\rangle = |0\rangle$.

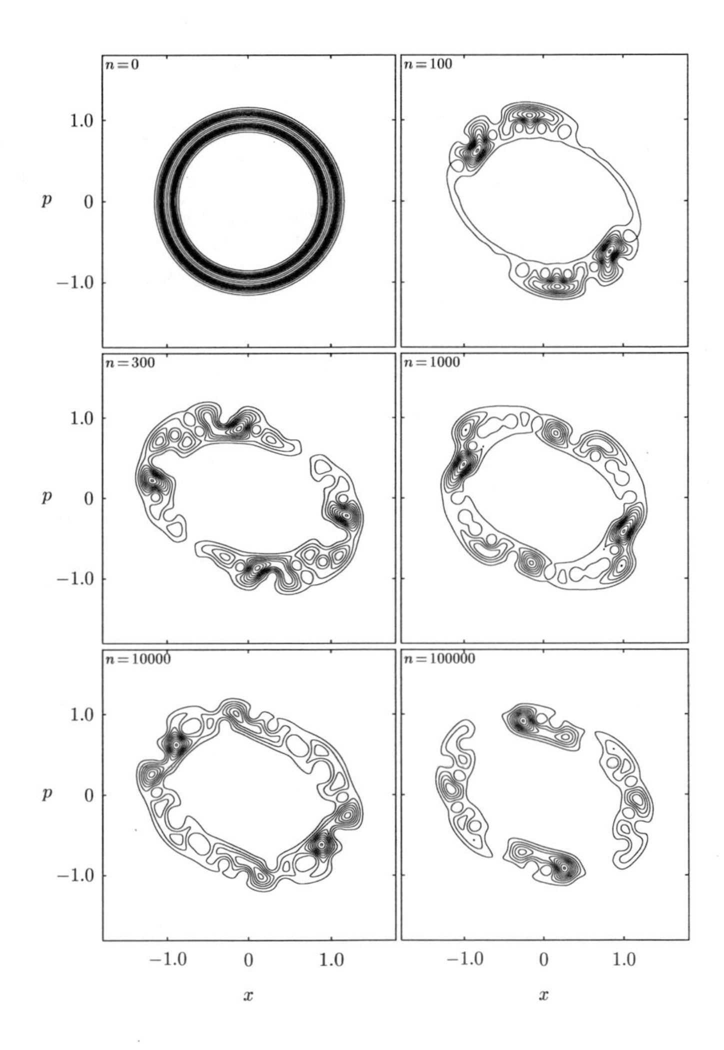

Figure 5.10: Contour plots of HUSIMI distributions of $|\psi_n\rangle$.
Parameters: $T = 1.0$, $\hbar = 0.01$, $V_0 = 50.0$. Initial state: $|\psi_0\rangle = |50\rangle$.

The numerical evidence collected by iterating the quantum map for many different combinations of parameters and initial states — in addition to the figures shown here, a small selection of these simulations is documented in section C.4 of the appendix — indicates that regarding localization the dynamics is "robust" with respect to the initial state: having iterated just long enough, for a given parameter combination always the same type of quantum dynamics evolves, regardless of the exact choice of $|\psi_0\rangle$ — as long as $|\psi_0\rangle$ is concentrated in a phase space region that is characterized by localized dynamics. (But also take into account the considerations in subsection 5.2.2 concerning the choice of the initial position of $|\psi_0\rangle$ in quantum phase space.) In this sense the notion of arbitrariness that entered the discussion via the ad hoc choice of initial conditions is rendered irrelevant.

The above observations on the lack of quantum phase space diffusion are fairly convincing, as the quantum map has been iterated a *very* large number of times, up to and beyond $n = 10^5$, and the available numerical indications show that the computations are not too erroneous: typically, the error of the norm of the numerically computed states does not exceed 10^{-10}, and often it is even much smaller. Observations of the same kind can be made for "any" other combination of parameters, as long as T takes a nonresonant value. The situation becomes completely different in the resonance case, of course, as discussed in chapter 4. Some more examples of HUSIMI contour plots demonstrating localization in quantum phase space can be found in section C.4 of the appendix.

Another manifestation of this localization is obtained by plotting the expansion coefficients $a_m^{(n)}$ — cf. the expansion (2.40) — of wave packets that have been generated by iterating the quantum map (2.37) n times, with n sufficiently large. Figure 5.11 shows typical states in this eigenrepresentation $\{|m\rangle\}$ of the harmonic oscillator. The exponential decay of the absolute values of the expansion coefficients to the right — and in figures 5.11a and 5.11b to varying degree to the left as well — is clearly visible. The nonexponential parts on the right hand sides of figures 5.11b and 5.11c are noise-floors, due to the unavoidable numerical errors in the computation; they do not interfere with the above interpretation of these graphs.

In order to get a less qualitative picture of the situation it is useful — although less intuitive than looking at quantum phase space pictures — to consider the behaviour of the energy expectation value (4.5) of the system as a function of time. Figures 5.12a,b/5.13a,b/5.14a,b, (corresponding, for $\hbar = 1.0$, to figures 5.8, C.41 and 5.9, for example) show the time evolution of quantum energy expectation values $\langle E\rangle_n$ of the kicked harmonic oscillator for $V_0 = 1.0/2.0/3.0$ and $T = 1.0$, i.e. in cases of classical nonresonance, for several values of $\hbar$. For the computation of the corresponding ensemble averaged classical energy $\langle E\rangle_n^{\mathrm{cl}}$ (cf. equation (1.74)) an ensemble of initial values has been used that is described by the same Gaussian in phase space as the HUSIMI distribution for the coherent state $|0\rangle$.

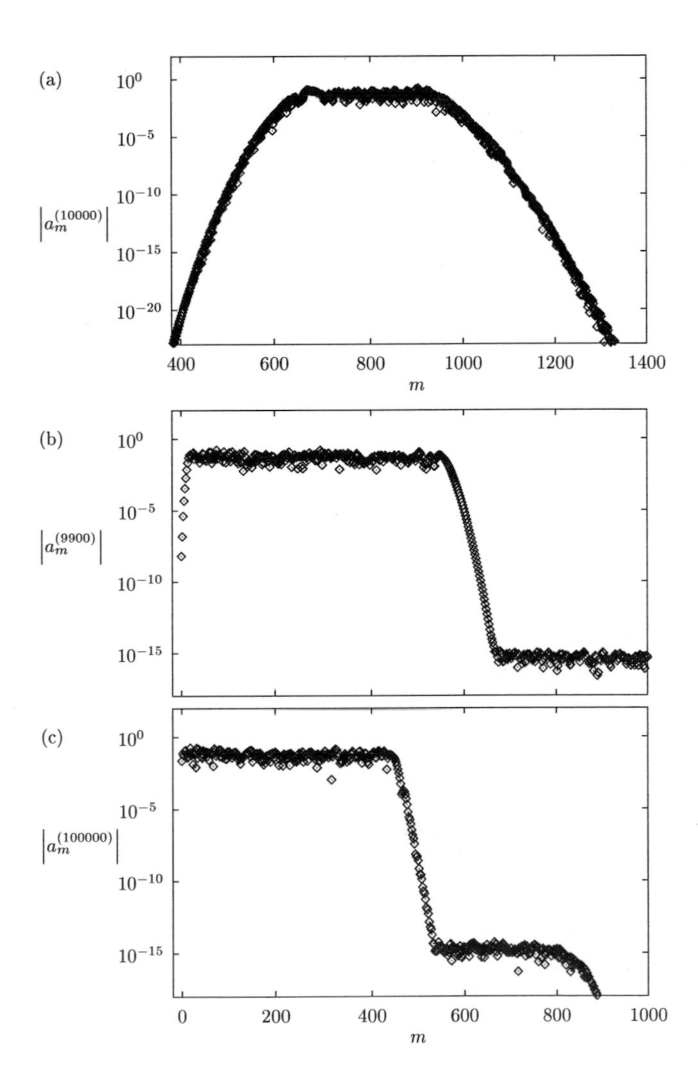

Figure 5.11: Three typical localized states $|\psi_n\rangle$ in the eigenrepresentation of the harmonic oscillator. (a) $T = 1.0$, $V_0 = 1.0$, $\hbar = 0.1$, $|\psi_0\rangle = \hat{D}(0, 12.5)\,|5\rangle$, $n = 10^4$; (b) $T = 1.0$, $V_0 = 1.0$, $\hbar = 0.003$, $|\psi_0\rangle = |166\rangle$, n=9900; (c) $T = 1.0$, $V_0 = 2.0$, $\hbar = 0.01$, $|\psi_0\rangle = |50\rangle$, $n = 10^5$.

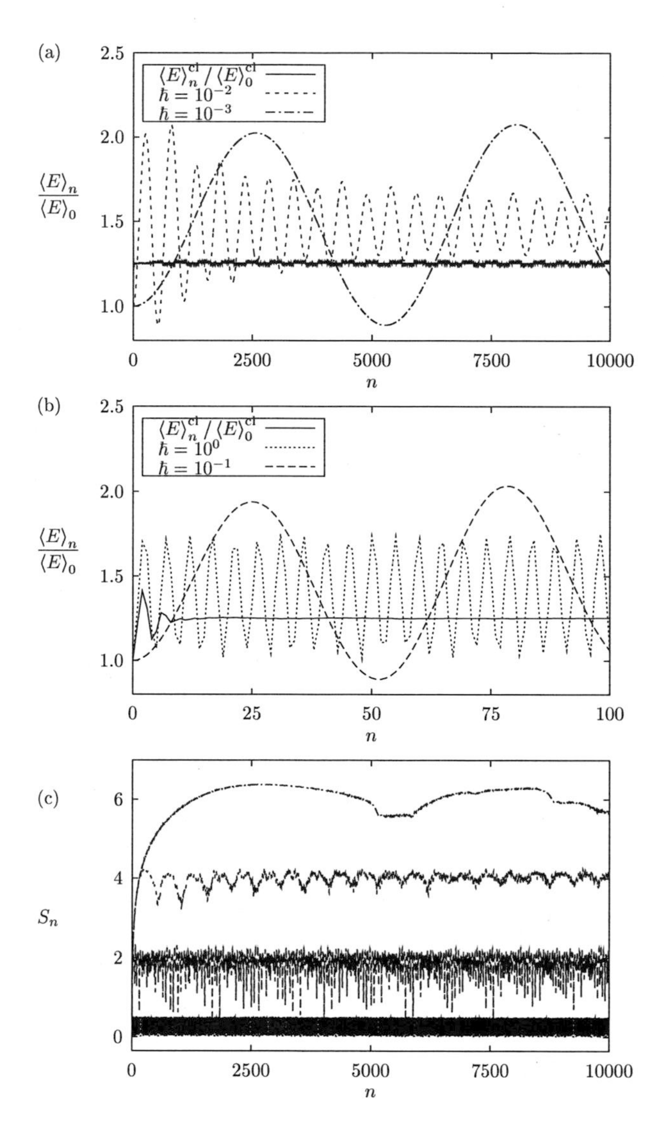

Figure 5.12: Suppressed energy growth in a case of nonresonance. Figures (a) and (b) show the energy expectation value $\langle E\rangle_n$ and the classical averaged energy $\langle E\rangle_n^{\rm cl}$ of the kicked harmonic oscillator with $T = 1.0$ and $V_0 = 1.0$ versus the number n of kicks. (a) $|\psi_0\rangle = |50/500\rangle$ for $\hbar = 10^{-2}/10^{-3}$, respectively; for details on these initial conditions see text. (b) $|\psi_0\rangle = |0/5\rangle$ for $\hbar = 10^{0}/10^{-1}$, respectively. (c) The entropy S_n versus n for the values of $\hbar$ and the initial conditions used in (a) and (b).

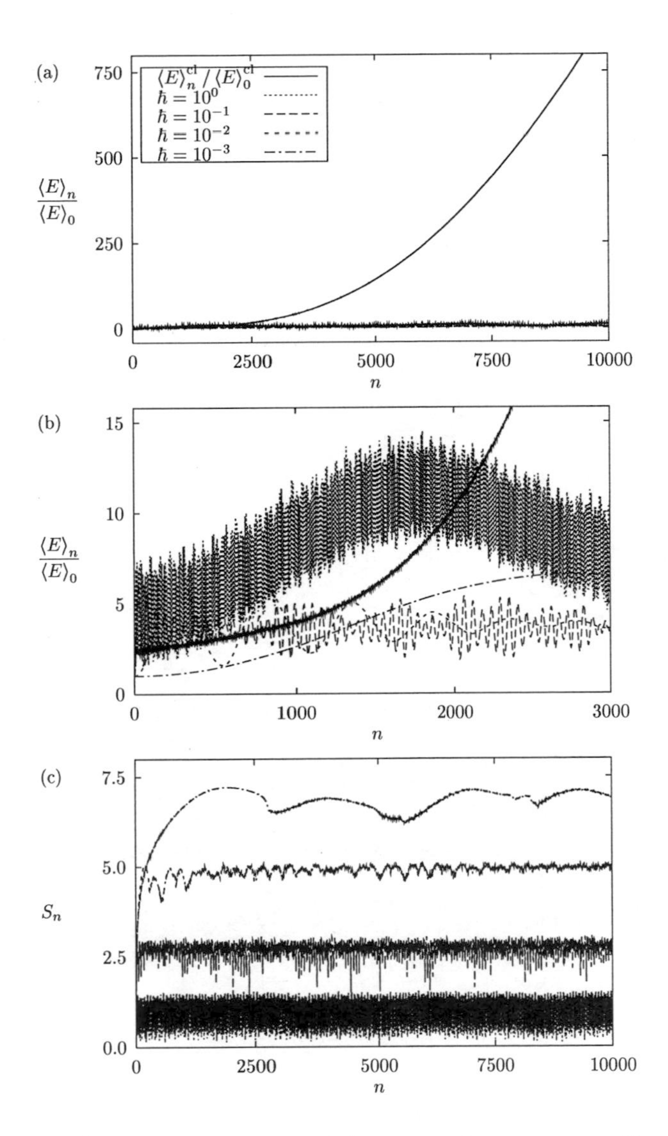

Figure 5.13: Suppressed energy growth in a case of nonresonance. Figures (a) and (b) show $\langle E\rangle_n$ and $\langle E\rangle_n^{\rm cl}$ versus n for $T = 1.0$ and $V_0 = 2.0$. $|\psi_0\rangle = |0/5/50/500\rangle$ for $\hbar = 10^0/10^{-1}/10^{-2}/10^{-3}$, respectively. (b) is a magnification of (a). (c) S_n versus n for the values of $\hbar$ and the initial conditions used in (a) and (b).

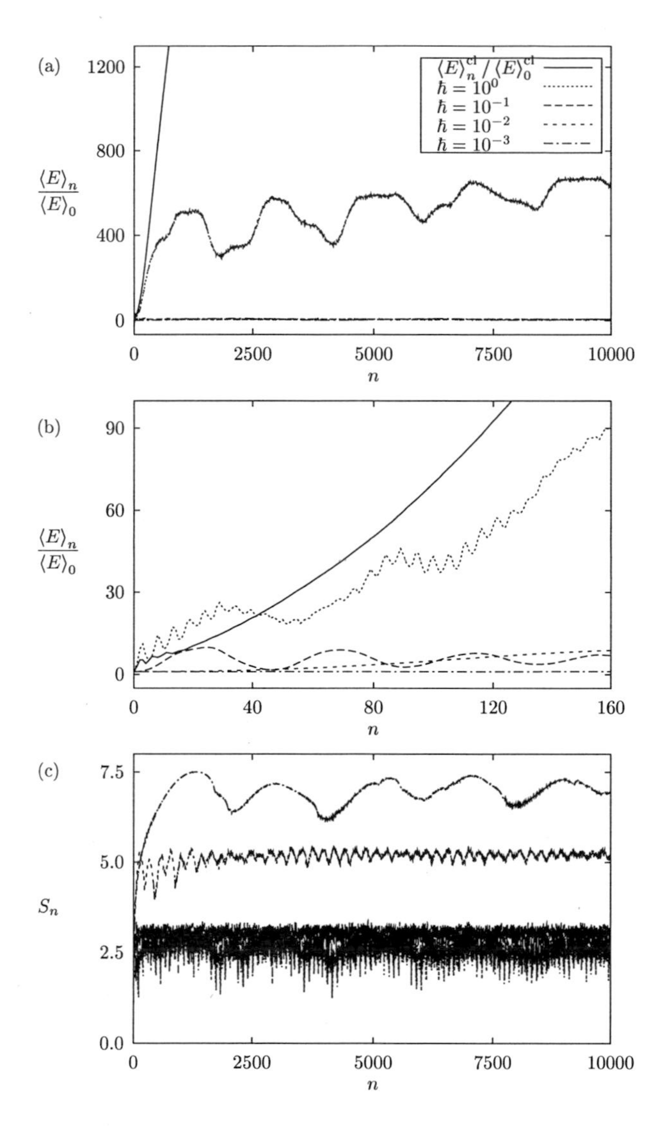

Figure 5.14: The same as figure 5.13, but for $V_0 = 3.0$.

In line with the above remarks concerning the case of $V_0 = 1.0$, figure 5.12 shows no unexpected deviation of the quantum energy expectation values from the classical energy averages (disregarding the typical quantum mechanical oscillations of $\langle E \rangle_n$ around the classical value of $\langle E \rangle_n^{\mathrm{cl}}$); in both the quantum and the classical cases the dynamics is localized and therefore the energy is bounded. As indicated by the previous figures, the situation is different for larger values of V_0, as displayed in figures 5.13 and 5.14: in contrast to the classical energies which grow without upper bound, the quantum curves (for all values of $\hbar$ considered) obviously are characterized by bounded energy growth. This is a clear indication to quantum suppressed energy growth, i.e. *quantum localization*, similar to the quantum localization demonstrated in figure 5.3 with respect to the quantum kicked rotor.

Figures 5.12c/5.13c/5.14c show another manifestation of quantum localization: in these figures the entropy (4.9) of the iterated quantum states is plotted, also exhibiting saturation after a limited number of iterations. The entropy is a good means for demonstrating localization, as by definition it measures the degree to which a state spreads within the HILBERT space of kicked harmonic oscillator states, and thereby in phase space — cf. subsection 4.1.2.

Using a different kick function (making the system much more tractable numerically but not allowing for stochastic webs to develop), a similarly slowed down energy growth in a related version of the quantum kicked harmonic oscillator has been observed numerically in [SHM00], but the authors could not provide an analytical explanation for their observation.

5.2.2 Dependence on Initial Conditions

Two different mechanisms of localization of a quantum state have to be distinguished. First, there are the quantum manifestations of classical properties leading to localization, such as invariant lines in classical phase space not yet having broken up und thus being obstacles to global dynamics in phase space. This is the dominant feature of the quantum dynamics displayed in figures 5.8 and 5.12; from a classical point of view it is to be expected and therefore less interesting. Second, there are the genuinely quantum effects that are leading to localization as visible in the remaining figures in subsection 5.2.1 (and in section C.4 of the appendix) and that are to be explained in section 5.3. Quite naturally, these quantum effects are in the focus of attention in the present chapter.

The quantum dynamics for $V_0 = 1.0$, while at first sight dominated by classical localization only, is also affected by the quantum localization mechanism in a crucial way. This can be seen by starting the dynamics with different initial conditions. For figures 5.8 and 5.12, the $|\psi_0\rangle = |m\rangle$ have been chosen to be localized *within* the closed invariant lines, leading to the classical localization result. For figure 5.15, on the other hand, the same initial states have been chosen (here for $\hbar = 1.0$ and $\hbar = 0.1$ only), but shifted to the point $(0, 12.5)^t$, well beyond the outermost invariant line: $|\psi_0\rangle = \hat{D}(0, 12.5)\,|m\rangle$. (Note that the topmost classical initial point in figure 1.4 is $(0, 12.5)^t$, too, making comparison between the classical and quantum cases easier.) The resulting dynamics obviously differs from the case with initial conditions near $(0, 0)^t$: the quantum dynamics remains localized, while the classical dynamics becomes diffusive for large enough n.

To summarize, depending on the initial condition, one may have quantum localization for classical or quantum reasons, or for a combination of both.

The localization phenomena that have been demonstrated in section 5.2 using numerical means are given a theoretical foundation in the following section 5.3.

5.3 The Nonresonant Quantum Kicked Harmonic Oscillator is Anderson-localized

Along the lines of subsection 5.1.3 I now show in subsection 5.3.1 that the quantum kicked harmonic oscillator can be mapped onto a model system that is very similar to the ANDERSON model. In subsections 5.3.2 and 5.3.3 I argue that although the oscillator case is more intricate than the rotor case, the former nevertheless exhibits ANDERSON localization in a certain well-defined way. In subsection 5.3.4 I then discuss some consequences of these results and mark a striking difference to the localization result discussed in subsection 5.1.3 with respect to the quantum kicked rotor.

5.3.1 Mapping of the Quantum Kicked Harmonic Oscillator onto an Anderson-like Model

Although classically the kicked rotor (5.1) and the kicked harmonic oscillator (1.17) are fundamentally different, for example in that the kicked rotor is characterized by *one* frequency only (namely the frequency of the kick) whereas the kicked oscillator has *two* frequencies (the second being the frequency of the unperturbed harmonic dynamics), the quantum localization phenomena in both models can be explained by the same mechanism. This mechanism is described by the theory of ANDERSON localization as outlined in section 5.1. I now show that this theory — with some alterations due to the differing eigenstates of the two model systems — can also be applied to the quantum kicked harmonic oscillator.

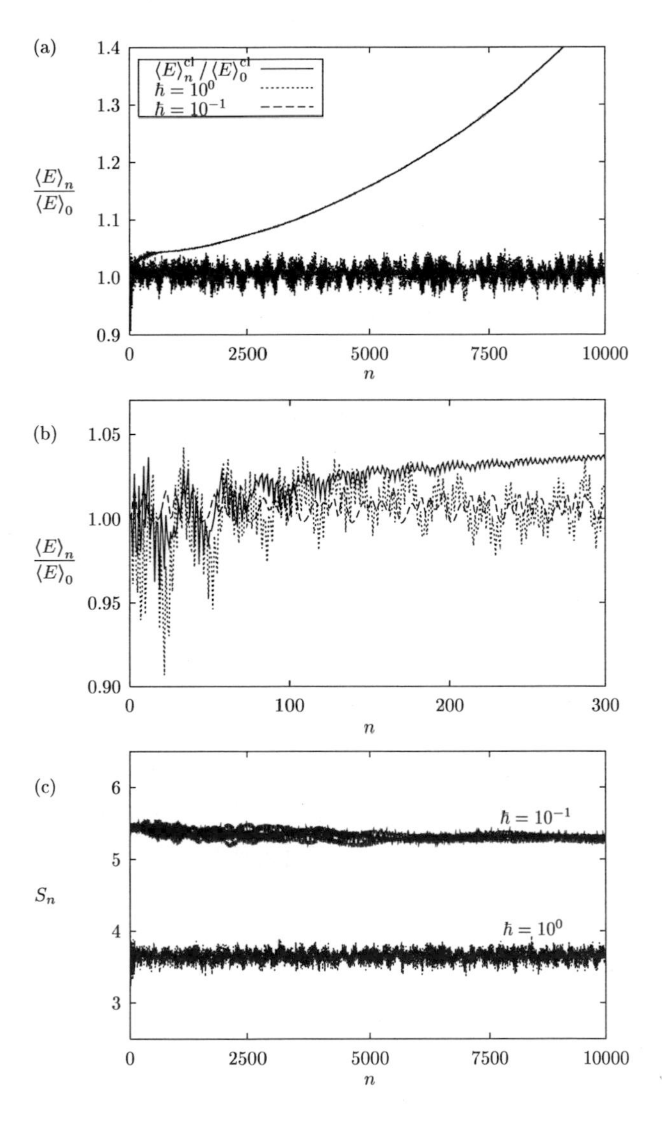

Figure 5.15: Dependence of the details of quantum localization on the initial conditions. $T = 1.0$ and $V_0 = 1.0$ as in figure 5.12, but different initial conditions are considered here: (a) $|\psi_0\rangle = \hat{D}(0, 12.5)\,|0/5\rangle$ for $\hbar = 10^0/10^{-1}$, respectively. (b) is a magnification of (a). (c) S_n versus n.

As in equation (5.43) I begin with the quantum map (2.37) for the kicked harmonic oscillator,

$$|\psi_{n+1}\rangle = e^{-iT\left(\hat{a}^\dagger\hat{a}+\frac{1}{2}\right)} e^{-\frac{i}{\hbar}V(\hat{x})} |\psi_n\rangle , \tag{5.72}$$

with the kick potential $V(x)$ not yet specified. This equation of motion is to be reformulated in terms of the (reduced) quasienergy states

$$\left|u_E^{\mp}\right\rangle := \lim_{\varepsilon\searrow 0} \left|u_E(nT\mp\varepsilon)\right\rangle \tag{5.73a}$$

$$\left|\overline{u}_E\right\rangle := \frac{1}{2}\Big(\left|u_E^-\right\rangle + \left|u_E^+\right\rangle\Big), \tag{5.73b}$$

which are similarly defined as their rotor counterparts in equations (5.45) and (5.47). Using the quasienergy states $\left|u_E^{\mp}\right\rangle$ before and after the kick, the quantum map (5.72) becomes

$$\left|u_E^+\right\rangle = e^{-\frac{i}{\hbar}V(\hat{x})} \left|u_E^-\right\rangle \tag{5.74a}$$

$$\left|u_E^-\right\rangle = e^{\frac{i}{\hbar}T\left(E-\hbar\left(\hat{a}^\dagger\hat{a}+\frac{1}{2}\right)\right)} \left|u_E^+\right\rangle . \tag{5.74b}$$

With the operator

$$W(\hat{x}) := -\tan\frac{V(\hat{x})}{2\hbar} \tag{5.75}$$

I transform the kick equation (5.74a) into

$$\left|u_E^+\right\rangle = \Big(1+iW(\hat{x})\Big)\left|\overline{u}_E\right\rangle \tag{5.76a}$$

$$\left|u_E^-\right\rangle = \Big(1-iW(\hat{x})\Big)\left|\overline{u}_E\right\rangle, \tag{5.76b}$$

which in terms of the averaged quasienergy state $\left|\overline{u}_E\right\rangle$ gives

$$\Big(1-iW(\hat{x})\Big)\left|\overline{u}_E\right\rangle = e^{\frac{i}{\hbar}T\left(E-\hbar\left(\hat{a}^\dagger\hat{a}+\frac{1}{2}\right)\right)}\Big(1+iW(\hat{x})\Big)\left|\overline{u}_E\right\rangle \tag{5.77}$$

by combination with the equation (5.74b) describing the free part of the dynamics. As in the case of the kicked rotor the original dynamical system given by equation (5.72) is thus mapped onto a static problem by exploiting the FLOQUET character of the system.

Projecting equation (5.77) onto the eigenstates $|m\rangle$ of the free harmonic oscillator (cf. equation (2.38)), I get

$$\overline{u}_{E,m} - i\big\langle m\big|\hat{W}\big|\overline{u}_E\big\rangle \;=\; e^{\frac{i}{\hbar}T\left(E-\hbar\left(m+\frac{1}{2}\right)\right)}\left(\overline{u}_{E,m} + i\big\langle m\big|\hat{W}\big|\overline{u}_E\big\rangle\right), \qquad (5.78)$$

where $\overline{u}_{E,m}$ is defined as in equation (5.52), but here with respect to oscillator eigenstates rather than rotor eigenstates:

$$\overline{u}_{E,m} \;:=\; \langle m\,|\overline{u}_E\rangle\,, \qquad m \in \mathbb{N}_0\,. \qquad (5.79)$$

Collecting terms I then arrive at the equation

$$\epsilon_m \overline{u}_{E,m} + \big\langle m\big|\hat{W}\big|\overline{u}_E\big\rangle \;=\; 0 \qquad (5.80)$$

with the diagonal energies

$$\epsilon_m(E,\hbar,T) \;:=\; \tan\left(\frac{T}{2\hbar}\left(E-\hbar\left(m+\frac{1}{2}\right)\right)\right). \qquad (5.81)$$

Completeness of $\{|m\rangle\}$ implies that $\big\langle m\big|\hat{W}\big|\overline{u}_E\big\rangle$ can be expanded according to

$$\big\langle m\big|\hat{W}\big|\overline{u}_E\big\rangle \;=\; \sum_{m'=0}^{\infty} W_{m,m'}\,\overline{u}_{E,m'}\,, \qquad (5.82)$$

where the hopping matrix elements $W_{m,m'} \in \mathbb{R}$,

$$W_{m,m'} \;:=\; \big\langle m\big|\hat{W}\big|m'\big\rangle\,, \qquad m,m' \in \mathbb{N}_0\,, \qquad (5.83)$$

have been introduced. Summarizing, the discrete SCHRÖDINGER equation that describes the quantum dynamics of the kicked harmonic oscillator has been derived:

$$\epsilon_m \overline{u}_{E,m} + \sum_{m'=0}^{\infty} W_{m,m'}\,\overline{u}_{E,m'} \;=\; 0. \qquad (5.84)$$

The argument list $(E,\hbar,T)$ of the ϵ_m is dropped for convenience, as usual. But note that here — in contrast to the diagonal energies $\epsilon_m(E,\hbar)$ of the rotor (5.58) — the ϵ_m depend on the kick period T as well. In the context of localization, this is a very important observation, as will be seen shortly when the issue of resonances is discussed. While the above derivation at first sight seems to be entirely analogous to the situation in subsection 5.1.3, there are several important differences that I discuss in the following.

The first remark concerns the diagonal energies. As their rotor counterparts, the $\epsilon_m(E, \hbar, T)$ can be viewed as being generated by a (quasi-) random number generator that follows a Lorentzian distribution (5.70), provided a condition of nonresonance is satisfied. Because of the different expressions (5.58) and (5.81) defining the respective diagonal energies, the (non-) resonances in question differ, too. For the quantum kicked oscillator, the *quantum resonances* inhibiting quasi-randomness of the ϵ_m are given by

$$T_{\text{res}} = \frac{P}{Q}\pi, \quad P, Q \in \mathbb{N}, \tag{5.85}$$

as can be concluded from equation (5.81), thus *coinciding with the classical resonance condition* (1.22) on the kick period. Figure 5.16 verifies this observation: for the nonresonant value $T = 1.0$ in figure 5.16a, a Lorentzian distribution of the values of ϵ_m is obtained, while the resonant values $T = 7\pi/22$ and $T = \pi/2$ in figures 5.16b and 5.16c yield no continuous distribution for ϵ_m at all, although $T = 7\pi/22 \approx 0.9996$ is very close to the nonresonant value $T = 1.0$ of figure 5.16a. Thus for *non*resonant values of T in the sense of equation (5.85) one can expect localization as in the ANDERSON-LLOYD model, as long as the other conditions discussed below are met. In this sense, quantum localization is complementary to the quantum webs discussed in chapter 4, which develop in the resonance case as defined by equation (5.85).

The fact that the resonance condition (1.22/4.21a) — and thereby the important special case (1.23/4.22) as well —, which is essential for the formation of *classical* and *quantum stochastic webs*, naturally arises as equation (5.85) in this quantum theory and characterizes those situations where no *quantum localization* can be expected, is a strong argument in favour of this formulation of the theory of localization in the kicked harmonic oscillator: in this way proper quantum-classical correspondence is automatically guaranteed. What is more, in this way also the quantum theories for the cases of resonance and nonresonance nicely fit together in a complementary fashion. The resonances with respect to T are true *quantum resonances* in the sense that they are obtained here as a consequence of a genuinely quantum mechanical theory. On the other hand, as discussed above, these resonances play an important role classically as well, such that the term "quantum resonance" might be questioned. This situation is to be compared with the quantum resonances (5.67) of the kicked rotor and (4.7, 4.44, 4.49) of the resonant kicked harmonic oscillator which have no classical counterpart.

The second important difference to the case of the rotor is that for the oscillator the evaluation of the hopping matrix elements is not as simple as in subsection 5.1.3, since the oscillator eigenfunctions $\langle x | m \rangle$ (2.39) are algebraically

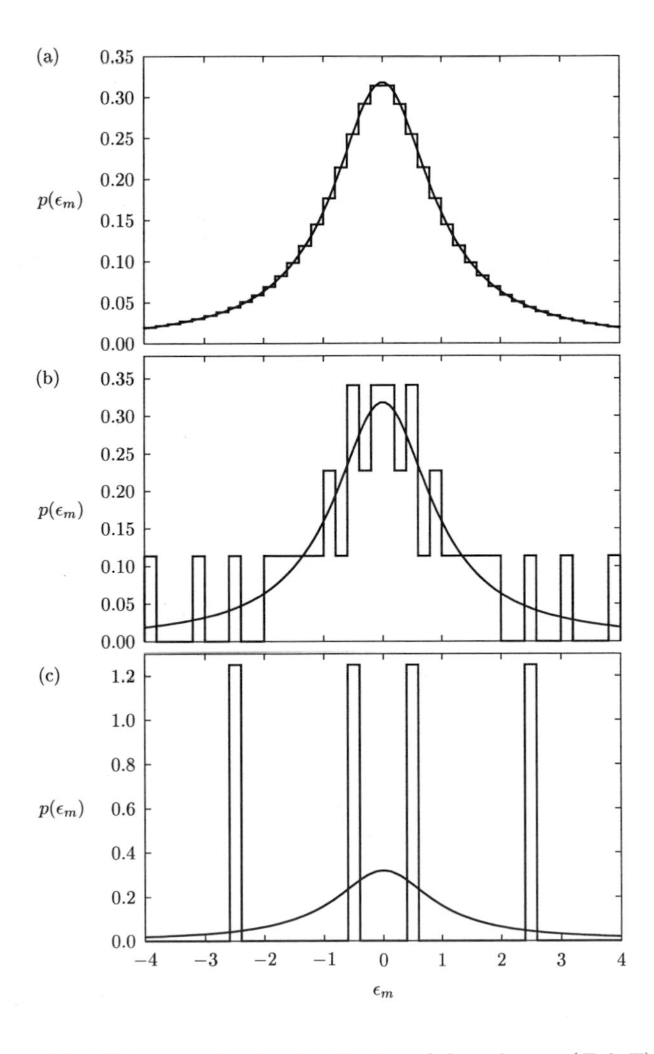

Figure 5.16: Histograms for the distribution of the values $\epsilon_m(E, \hbar, T)$ of the diagonal energy of the kicked harmonic oscillator with $1 \leq m \leq 10^5$, $E = 1.0$ and $\hbar = 1.0$, compared with the graph of the Lorentzian (5.70): (a) for the nonresonant value $T = 1.0$; (b) for the resonance given by $T = 7\pi/22 \approx 0.9996$; (c) for the resonance given by $T = \pi/2$ $(q = 4)$.

much more complicated than the simple exponentials of the rotor eigenfunctions (5.12). Explicitly, the matrix elements (5.83) for the cosine kick potential (1.18) are given by[10]

$$W_{m,m'} = -\frac{1}{\sqrt{\pi 2^{m+m'} m! m'!}} \int_{-\infty}^{\infty} \mathrm{d}x \, \tan \frac{V_0 \cos \sqrt{\hbar} x}{2\hbar} e^{-x^2} \mathrm{H}_m(x) \mathrm{H}_{m'}(x). \quad (5.86)$$

The indices m, m' run over the non-negative integers only, as opposed to all the integers in the case of the rotor. This leads to an asymmetric situation, since the lattice on which the discrete SCHRÖDINGER equation is defined is infinite on the right hand side only, rather than being bi-infinite. More asymmetry is added by the fact that the interaction with the left and right neighbours is not the same in general:

$$W_{m,m-m'} \neq W_{m,m+m'} , \quad (5.87)$$

while for the rotor $W_{m'} = W_{-m'}$ holds.

It is obvious that, because of the dependence on the HERMITE polynomials, the formula (5.86) cannot be simplified to yield an expression that depends on a single index only as the corresponding rotor matrix element (5.54): the matrix elements $W_{m,m'}$ have to be viewed as functions of both the position m and the difference $m - m'$. This means that the interaction between the sites of the ANDERSON lattice for the oscillator is *not* translation invariant; in other words, the strength of the interaction between two sites does not depend on their distance on the lattice alone. For the proof of ANDERSON localization this is not an obstacle, as long as for all m the absolute values $|W_{m,m'}|$ of the matrix elements decay rapidly enough with the distance $|m - m'|$ from the site m, in much the same way as the absolute values of the rotor matrix elements do.

Analytically not much can be said in general about the values $W_{m,m'}$ takes on. The only straightforward property to notice is that every second of them vanishes,

$$W_{m,m'} = 0 \quad \text{for } m + m' \text{ odd}, \quad (5.88)$$

[10]For $V_0/\hbar \geq \pi$ the integral in (5.86) is not RIEMANN integrable, because in this case the argument of the tangent can take on values equal to half-integer multiples of π, making the integrand singular. However, with probability one (with respect to the values of V_0 and $\hbar$), the $W_{m,m'}$ still take on well-defined values if the integral is evaluated as the corresponding CAUCHY principal value of (5.86). Therefore this issue does not challenge the proper physical interpretation of the discrete SCHRÖDINGER equation (5.84).

since the HERMITE polynomials H_m in equation (5.86) are odd if and only if m is odd, and the kick potential $V(x) = V_0 \cos x$ is an even function of x. (A similar selection rule, but with respect to the matrix elements (2.51) of the FLOQUET operator, has been discussed in subsection 2.1.3; in both cases the selection rules stem from the same symmetry of the kick potential.) Therefore, without loss of generality, it suffices to consider $W_{m,m+2m''}$ with $m'' \in \mathbb{Z}$ (and $m'' \geq -m/2$). In the special case of small values of $V_0/\hbar$, an estimate for $|W_{m,m+2m''}|$ can be obtained: using just the lowest order term of the TAYLOR series of the tangent I can write

$$W_{m,m+2m''} \approx \frac{-V_0}{2^{m+m''}\hbar\sqrt{\pi m!(m+2m'')!}} \int_0^\infty \mathrm{d}x \, \cos\sqrt{\hbar}x \; e^{-x^2} \, \mathrm{H}_m(x) \, \mathrm{H}_{m+2m''}(x). \tag{5.89}$$

According to [GR00] the integral can be evaluated in the following way: (For notational convenience, the following formulae up to (5.93) are meant for positive values of m'' only; similar formulae hold for negative m'' and large enough m.)

$$\int_0^\infty \mathrm{d}x \, \cos\sqrt{\hbar}x \; e^{-x^2} \mathrm{H}_m(x) \, \mathrm{H}_{m+2m''}(x) = (-1)^{m''} 2^{m-1} \sqrt{\pi} \, m! \, \hbar^m \, e^{-\frac{\hbar}{4}} \, \mathrm{L}_m^{(2m'')}\!\left(\frac{\hbar}{2}\right), \tag{5.90}$$

and the generalized LAGUERRE polynomials $\mathrm{L}_m^{(\alpha)}$ satisfy the inequality [AS72]

$$\left|\mathrm{L}_m^{(\alpha)}(x)\right| \leq \frac{\Gamma(m+\alpha+1)}{m! \, \Gamma(\alpha+1)} \, e^{\frac{x}{2}}, \tag{5.91}$$

such that I get the estimate

$$|W_{m,m+2m''}| \lesssim \frac{V_0}{4} \left(\frac{\hbar}{2}\right)^{m''-1} \sqrt{\frac{\binom{m+2m''}{m}}{(2m'')!}}. \tag{5.92}$$

By STIRLING's formula [AS72], in the limit of large m'' this expression becomes

$$|W_{m,m+2m''}| \lesssim \frac{V_0}{2\hbar} \frac{1}{\sqrt{m! \, e^m}} \left(\frac{27 e^2 \hbar^2}{64 m''}\right)^{\frac{m''}{2}}, \tag{5.93}$$

which shows that the absolute values of the matrix elements decay more than exponentially fast with the distance $2m''$; in addition, as in the case of the rotor the amplitudes of all $|W_{m,m+2m''}|$ are controlled by the factor $V_0/\hbar$ and can thus be made as small as desired by taking the limit $V_0/\hbar \to 0$. Summarizing, the inequality (5.93) shows that for weak perturbations and for not too large values of $\hbar$ the discrete SCHRÖDINGER equation (5.84) with the cosine kick potential (1.18) indeed approaches a tight binding equation.

The convergence $|W_{m,m+2m''}| \to 0$ for larger values of $|m''|$ is checked in figures 5.17 and 5.18; since the integral in equation (5.86) cannot be solved analytically in general, the matrix elements are evaluated numerically and plotted for several parameter combinations. Due to the oscillatory integrand, it is difficult to obtain exact results by numerical computation for some of these parameter combinations. This difficulty has an impact especially on the results for larger values of V_0 and becomes worse in the case of small values of $\hbar$, as can be seen in the figures: the data for $V_0 = 1$ and $V_0 = 0.1$ in figures 5.17c and 5.18c are spoiled by numerical round-off errors, and the same appears to be true for the points with $|m''| \gtrsim 10$ and $V_0 = 1$ or $V_0 = 0.1$ in figure 5.18a. In the opposite case of small V_0 and large $\hbar$ the numerical results nicely agree with the convergence to zero as predicted by the relation (5.93).

Apart from the numerical problems, the figures give a clear indication of the typical behaviour of the matrix elements. The $|W_{m,m+2m''}|$ are peaked around $m'' = 0$ and decay more or less exponentially on both sides. Figure 5.18 also illustrates that in general for any site m the speed of this decay is not the same for $m'' < 0$ and $m'' > 0$, in agreement with the inequality (5.87). By taking suitable limits (with respect to the parameters), the $|W_{m,m+2m''}|$ decay fast enough to allow equation (5.84) to approach a tight binding equation. Some aspects of the case of those parameter combinations which do not give rise to tight binding are addressed in subsection 5.3.4 below.

This leads to another observation which marks a noteworthy difference to the rotor. Above, I have discussed which conditions have to be met for equation (5.84) to become a tight binding equation. Strictly speaking, equation (5.84) with the cosine kick potential cannot become a tight binding equation under any circumstances, because equation (5.88) indicates that *no* site is coupled to its *nearest* neighbour, and the closest interacting sites are characterized by indices m, m' with $|m - m'| = 2$. Even more, the dynamics on the sites with even indices and on the sites with odd indices are completely decoupled. This means that the cosine-kicked oscillator must be described not by a single, but by *a pair of* discrete SCHRÖDINGER equations which are interwoven with but independent of each other:

$$\epsilon_{2m}\overline{u}_{E,2m} + \sum_{m'=0}^{\infty} W_{2m,2m'}\,\overline{u}_{E,2m'} = 0 \tag{5.94a}$$

$$\epsilon_{2m+1}\overline{u}_{E,2m+1} + \sum_{m'=0}^{\infty} W_{2m+1,2m'+1}\,\overline{u}_{E,2m'+1} = 0. \tag{5.94b}$$

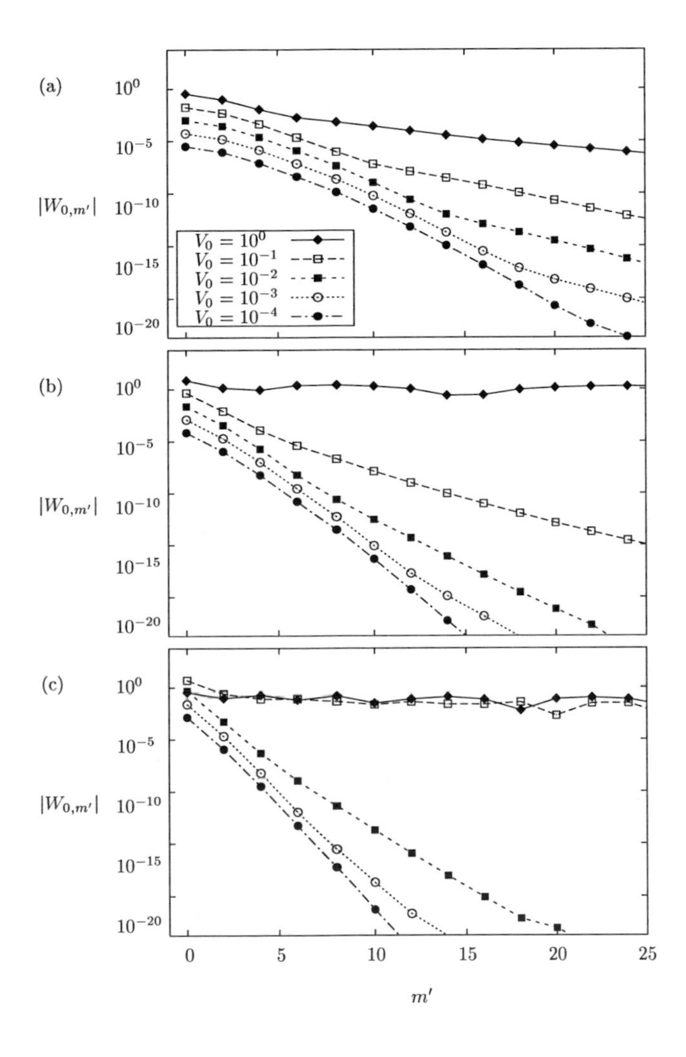

Figure 5.17: Absolute values of the hopping matrix elements $W_{m,\, m'=m+2m''}$ (5.86) of the kicked harmonic oscillator for $m = 0$. (a) $\hbar = 1.0$; (b) $\hbar = 0.1$; (c) $\hbar = 0.01$. Note that $W_{m,m'} = 0$ for $m + m'$ odd.

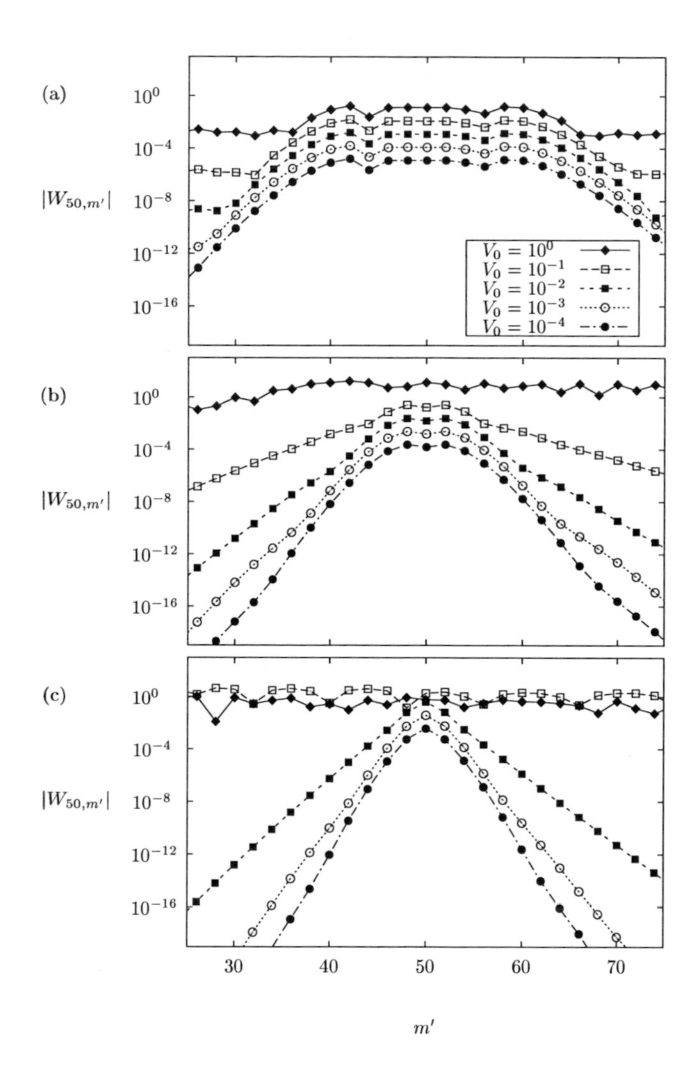

Figure 5.18: The same as figure 5.17, but for $m = 50$.

In each of these two subsystems tight binding is possible within the approximations discussed above. Of course, if one has tight binding in *both* subsystems at the same time, then the entire system is tightly bound and prone to localization.

On the other hand, by choosing a different kick potential as a replacement for the cosine potential (1.18),

$$\tilde{V}(x) \;:=\; -2\hbar \arctan\left(\sqrt{\frac{2}{\hbar}}\,\tilde{\lambda}\,x - \tilde{\eta}\right) \tag{5.95}$$

with arbitrary nonzero real constants $\tilde{\eta}$ and $\tilde{\lambda}$, a single true tight binding model for the kicked harmonic oscillator can be obtained. Again, as in the case of the alternate kick potential (5.64) of the quantum kicked rotor, the diagonal energies (5.81) remain unchanged by the introduction of this alternate potential. The new hopping matrix elements

$$\tilde{W}_{m,m'} \;:=\; -\left\langle m \left| \tan\frac{\tilde{V}(\hat{x})}{2\hbar} \right| m' \right\rangle, \quad m, m' \in \mathbb{N}_0, \tag{5.96}$$

corresponding to $\tilde{V}(x)$ (cf. equation (5.75)) are given by

$$\tilde{W}_{m,m'} \;=\; \frac{1}{\sqrt{\pi}\,2^{m+m'}\,m!\,m'!} \int\limits_{-\infty}^{\infty} \mathrm{d}x\, e^{-x^2} \left(\sqrt{2}\,\tilde{\lambda}\,x - \tilde{\eta}\right) \mathrm{H}_m(x)\,\mathrm{H}_{m'}(x), \tag{5.97}$$

where the integration can be accomplished by employing the recurrence relation (3.42) and the orthonormality relation

$$\int\limits_{-\infty}^{\infty} \mathrm{d}x\, e^{-x^2}\,\mathrm{H}_m(x)\,\mathrm{H}_{m'}(x) \;=\; \sqrt{\pi}\,2^m\,m!\,\delta_{m,m'} \tag{5.98}$$

of the HERMITE polynomials [AS72]. It yields

$$\tilde{W}_{m,m'} \;=\; \begin{cases} \sqrt{m}\,\tilde{\lambda} & m' = m-1 \\ -\tilde{\eta} \quad \text{for} & m' = 0 \\ \sqrt{m+1}\,\tilde{\lambda} & m' = m+1 \\ 0 & \text{else}\,. \end{cases} \tag{5.99}$$

The result is the tight binding equation

$$\epsilon_m \overline{u}_{E,m} + \tilde{\lambda}\left(\sqrt{m}\,\overline{u}_{E,m-1} + \sqrt{m+1}\,\overline{u}_{E,m+1}\right) \;=\; \tilde{\eta}\,\overline{u}_{E,m}, \tag{5.100}$$

coupling each site to its nearest neighbour sites. Note that for the kicked harmonic oscillator no alternate potential could be constructed that yields a tight binding system coupling each site to its respective second-nearest neighbours only, as motivated by equations (5.94).

The alternate kick potential (5.95) is intended here for nothing more than demonstrating how a tight binding system describing the quantum kicked harmonic oscillator can be constructed; in particular I do not discuss the classical dynamics of the harmonic oscillator with this kick potential here. However, motivated by the results described in [Jun95], one might speculate about the existence of classical — and perhaps quantum mechanical? — stochastic webs even for this aperiodic kick potential (in cases of *resonance* with respect to T). In the following I do not discuss the alternate system specified by the potential (5.95) any further.

In this subsection I have shown that the quantum kicked harmonic oscillator can be modelled by a discrete SCHRÖDINGER equation which is similar to the model used in ANDERSON's theory. Provided certain conditions are met, it is also possible to obtain an approximative tight binding model as in the case of the rotor. In the following two subsections I discuss how these findings can be used to prove ANDERSON localization in the oscillator.

5.3.2 Localization Established

As discussed in the previous subsection, both discrete SCHRÖDINGER equations (5.94a, 5.94b) in a natural way give rise to the approximating tight-binding equation

$$(\epsilon_m + W_{m,m})\,\overline{u}_{E,m} + W_{m,m-2}\,\overline{u}_{E,m-2} + W_{m,m+2}\,\overline{u}_{E,m+2} \;=\; 0, \qquad (5.101)$$

provided the $W_{m,m\pm 2m''}$ tend to zero with growing $|m''|$ fast enough. I now proceed to the discussion of an appropriate reformulation of this equation by means of transfer matrices. It is straightforward to apply the results to the alternate system described by equation (5.100). It is also possible to generalize this reformulation to discrete SCHRÖDINGER equations that cannot be described by a tight binding equation (5.101) because of matrix elements $W_{m,m+2m''}$ decaying only slowly with m''; in these cases more terms of the series in equations (5.94) need to be considered. I come back to this important point in subsection 5.3.4 below.

In terms of the transfer matrix formalism of subsection 5.1.2, the definitions

$$\mathcal{T}_m^{(1)} \;:=\; \begin{pmatrix} -\dfrac{\epsilon_m + W_{m,m}}{W_{m,m+2}} & -\dfrac{W_{m,m-2}}{W_{m,m+2}} \\ 1 & 0 \end{pmatrix} \qquad (5.102)$$

and

$$\vec{u}^{(1)}_{E,m} := \begin{pmatrix} \overline{u}_{E,m} \\ \overline{u}_{E,m-2} \end{pmatrix} \tag{5.103}$$

in conjunction with the equation of motion

$$\vec{u}^{(1)}_{E,m+2} = \mathcal{T}^{(1)}_m \, \vec{u}^{(1)}_{E,m} \tag{5.104}$$

seem to suggest themselves for modelling equation (5.101).

However, this formulation is disadvantageous for several reasons. First, it rests on the assumption that $W_{m,m+2} \neq 0$ *for all* m, because $W_{m,m+2}$ makes up the denominators of two of the matrix elements of the transfer matrix (5.102). Unfortunately, this assumption cannot be checked analytically for all combinations of V_0 and $\hbar$, because a closed solution of the integral in equation (5.86) is not available. Already a single $\widetilde{m}$ with $W_{\widetilde{m},\widetilde{m}+2} = 0$ could spoil the whole theory. Even more,

$$W_{m,m\pm 2} \neq 0 \quad \forall\, m \tag{5.105}$$

is a necessary condition[11] for the tight binding equation (5.101) to provide a meaningful model of the kicked harmonic oscillator. For later reference, I call those systems (i.e. those combinations of the parameters V_0 and $\hbar$) not satisfying the condition (5.105) *pathological.* While in the course of the computations leading to figures 5.17 and 5.18 I have not encountered any numerical evidence for such an m contradicting (5.105), this is still a matter of principle and should be addressed as such. Figure 5.19 shows some examples of the typical oscillatory behaviour of $W_{m,m\pm 2}$ found by numerical integration; the numerical data plotted in the figure is accurate enough to be sure that even those matrix elements which come close to zero are nevertheless clearly nonzero.

The workaround for the case of a vanishing (second-) nearest neighbour matrix element sketched in the footnote on page 141 only works in the context of the kicked rotor, because of the translation invariance of the discrete SCHRÖDINGER equation for that model. This is in contrast to the translation *non*invariance of the discrete SCHRÖDINGER equation of the kicked harmonic oscillator, i.e. the explicit m-dependence of the oscillator's hopping matrix element $W_{m,m+2}$ for the interaction with the nearest neighbour at the m-th site. Divisions by $W_{m,m\pm 2}$ occur in *all* transfer matrix models of the kicked harmonic oscillator, including those to be discussed below, and, as a matter of principle, cannot be avoided. In which sense the condition (5.105) can be taken to be satisfied is addressed in some more detail in subsection 5.3.4.

[11] Note that $W_{m,m+2} \neq 0 \; \forall m$ implies $W_{m,m\pm 2} \neq 0 \; \forall m$, because $W_{m,m'} = W_{m',m}$.

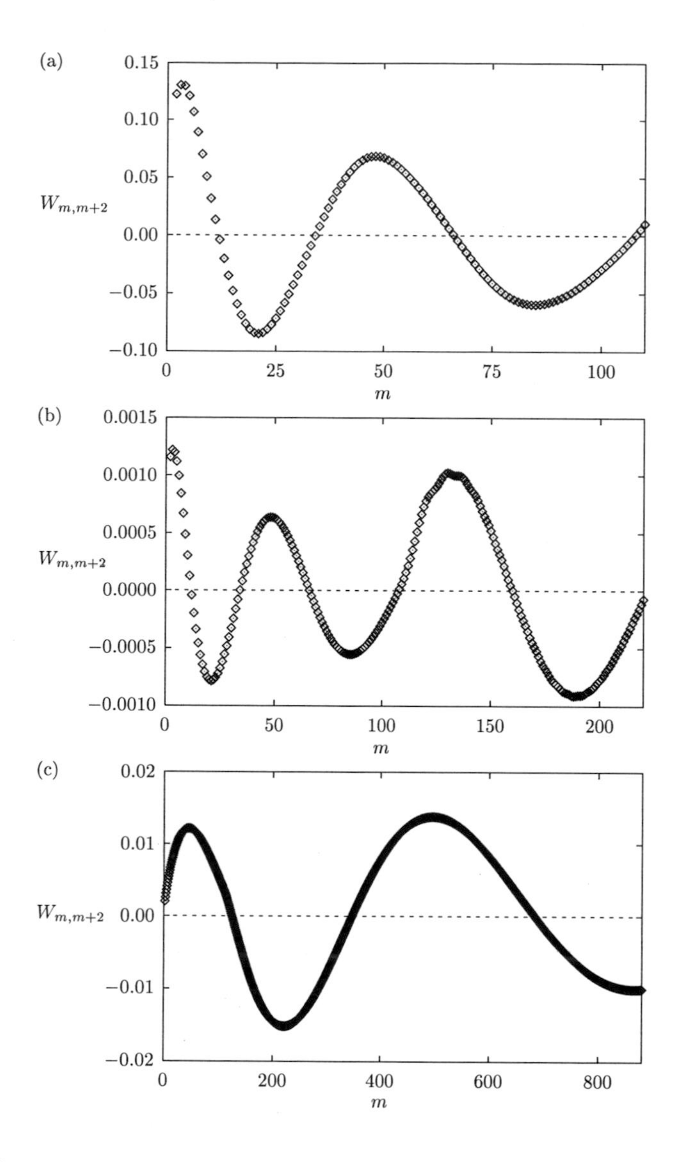

Figure 5.19: Hopping matrix elements for nearest neighbour interaction: $W_{m,m+2} = W_{m+2,m}$. (a) $V_0 = 1.0$, $\hbar = 1.0$; (b) $V_0 = 0.01$, $\hbar = 1.0$; (c) $V_0 = 0.01$, $\hbar = 0.1$. (The lack of smoothness for $100 < m < 150$ in figure (b) appears to be due to convergence problems of the numerical algorithm and could probably be avoided by making a somewhat increased computational effort.)

A second, and more important, disadvantage is that the transfer matrices of equation (5.102) do not satisfy the unimodularity requirement of FURSTENBERG's theorem:

$$\det\left(\mathcal{T}_m^{(1)}\right) = \frac{W_{m,m-2}}{W_{m,m+2}} \neq 1 \tag{5.106}$$

in general, as a consequence of the asymmetry (5.87) of the matrix elements of the kicked harmonic oscillator. As a result, if these particular $\mathcal{T}_m^{(1)}$ were used, FURSTENBERG's theorem would not be applicable and nothing could be said about the asymptotic behaviour of the $|\overline{u}_{E,m}|$.

Finally, backpropagation towards the left cannot simply be accomplished as in the case of the rotor by using the same transfer matrices $\mathcal{T}_m^{(1)}$ as for propagation to the right and setting

$$\vec{v}_{E,m}^{(1)} := \begin{pmatrix} \overline{u}_{E,m-2} \\ \overline{u}_{E,m} \end{pmatrix}, \tag{5.107}$$

because generally this gives

$$\vec{v}_{E,m-2}^{(1)} \neq \mathcal{T}_m^{(1)}\, \vec{v}_{E,m}^{(1)}. \tag{5.108}$$

The last-mentioned disadvantage is shared by all transfer matrix models of the kicked harmonic oscillator. Essentially, this is again a consequence of the asymmetry (5.87).

In the light of these shortcomings of the model specified by the matrix (5.102), the question needs to be addressed if and how the localization results of subsection 5.1.2 carry over to the oscillator.

One way to approach this problem is to define the more general vectors

$$\vec{u}_{E,m}^{(2)} := \begin{pmatrix} \overline{u}_{E,m} \\ \Theta_{m-2}\,\overline{u}_{E,m-2} \end{pmatrix} \tag{5.109}$$

with yet undetermined constants $\Theta_m \in \mathbb{C}$, and to construct new $\mathcal{T}_m^{(2)}$ in such a way that they are unimodular from the beginning. After a little algebra one finds that

$$\mathcal{T}_m^{(2)} := \begin{pmatrix} -\dfrac{\epsilon_m + W_{m,m}}{W_{m,m+2}} & -\dfrac{1}{\Theta_m} \\ \Theta_m & 0 \end{pmatrix} \tag{5.110}$$

are suitable unimodular transfer matrices if the Θ_m satisfy the recurrence relation

$$\Theta_m = \frac{W_{m,m+2}}{W_{m,m-2}} \Theta_{m-2} ; \tag{5.111}$$

the choice of the initial conditions Θ_0 and Θ_1 is free. Using $W_{m,m+2} = W_{m+2,m}$, and choosing $\Theta_0 = W_{20}$ and $\Theta_1 = W_{31}$,

$$\Theta_m = W_{m,m+2} \tag{5.112}$$

is obtained. No explicit closed formula for the Θ_m can be given because the same is true for $W_{m,m+2}$. In this way I obtain

$$\vec{u}^{(2)}_{E,m} = \begin{pmatrix} \overline{u}_{E,m} \\ W_{m-2,m}\, \overline{u}_{E,m-2} \end{pmatrix} \tag{5.113}$$

and

$$\mathcal{T}^{(2)}_m = \begin{pmatrix} -\dfrac{\epsilon_m + W_{m,m}}{W_{m,m+2}} & -\dfrac{1}{W_{m,m+2}} \\ W_{m,m+2} & 0 \end{pmatrix} . \tag{5.114}$$

Equations (5.113, 5.114) and the dynamical equation

$$\vec{u}^{(2)}_{E,m+2} = \mathcal{T}^{(2)}_m\, \vec{u}^{(2)}_{E,m} \tag{5.115}$$

combined provide an equivalent reformulation of the tight binding equation (5.101); they are a meaningful approximation to the discrete SCHRÖDINGER equation (5.94) as long as the hopping matrix elements for nearest neighbour interaction are all nonzero ("nonpathological") and the $W_{m,m\pm 2m''}$ decay fast enough with growing $|m''|$.

For proving localization in the kicked harmonic oscillator it suffices to consider the transfer dynamics in positive m-direction, because the lattice on which the discrete SCHRÖDINGER equation of the oscillator is defined is infinite towards the right only, as opposed to the bi-infinite lattice in the case of the rotor. So localization towards the left is automatically built into the system by the condition $m \geq 0$, once FURSTENBERG's theorem has been shown to apply. Nevertheless it is instructive to establish the analogy with the conventional tight binding models as far-reaching as possible. In fact, using the models discussed below, exponential localization is typically found in both directions — cf. figures 5.11a, 5.11b. What is more, a discrete point spectrum of energy eigenvalues describing the localized dynamics can only be expected in the case of exponential localization in both directions — cf. subsection 5.1.2.

In the rotor case, the same transfer matrix can be used for right- and leftward dynamics on the lattice, provided the two different sets of vectors $\vec{u}_{E,m}$, $\vec{v}_{E,m}$ of equations (5.62) are used. For the oscillator, on the other hand, the nontriviality of the off-diagonal matrix elements of the $\mathcal{T}_m^{(2)}$ necessitates the introduction of another set of transfer matrices for leftward transfer. No vectors $\vec{u}_{E,m}^{(2)}$, $\vec{v}_{E,m}^{(2)}$ could be constructed that allow to model the tight binding equation (5.101) using the same matrix for propagation into both directions. In a sense, the situation in the oscillator case is reversed, as compared to the rotor: the simplest formulation of backpropagation is obtained by using the same vectors $\vec{u}_{E,m}^{(2)}$ as for propagation in positive m-direction, but another set of transfer matrices, namely (quite naturally) the inverses of the $\mathcal{T}_m^{(2)}$:

$$\vec{u}_{E,m-2}^{(2)} = \mathcal{T}_m^{(2,\mathrm{left}_1)} \, \vec{u}_{E,m}^{(2)} \tag{5.116}$$

with

$$\mathcal{T}_m^{(2,\mathrm{left}_1)} := \left(\mathcal{T}_{m-2}^{(2)}\right)^{-1} = \begin{pmatrix} 0 & \dfrac{1}{W_{m-2,m}} \\ -W_{m-2,m} & -\dfrac{\epsilon_{m-2} + W_{m-2,m-2}}{W_{m-2,m}} \end{pmatrix}. \tag{5.117}$$

Alternatively, if for the sake of analogy one wants to use the vectors

$$\vec{v}_{E,m}^{(2)} := \begin{pmatrix} W_{m-2,m}\, \overline{u}_{E,m-2} \\ \overline{u}_{E,m} \end{pmatrix} \tag{5.118}$$

then one has

$$\vec{v}_{E,m-2}^{(2)} = \mathcal{T}_m^{(2,\mathrm{left}_2)} \, \vec{v}_{E,m}^{(2)} \tag{5.119}$$

with

$$\mathcal{T}_m^{(2,\mathrm{left}_2)} := \begin{pmatrix} -\dfrac{\epsilon_{m-2} + W_{m-2,m-2}}{W_{m-2,m}} & -W_{m-2,m} \\ \dfrac{1}{W_{m-2,m}} & 0 \end{pmatrix}. \tag{5.120}$$

Summarizing, the transfer matrices $\mathcal{T}_m^{(2)}$, $\mathcal{T}_m^{(2,\mathrm{left}_1)}$ and $\mathcal{T}_m^{(2,\mathrm{left}_2)}$ can be used to describe the right- and leftward tight-binding dynamics on the lattice in all but the pathological cases. Up to this point no statement has been made concerning the respective localization properties that might follow from this description. In the following I discuss just this aspect of the dynamics.

The above construction of a tight-binding model for the kicked harmonic oscillator is made to be as close to the corresponding model for the kicked rotor as possible. However, comparing equations (5.114, 5.117, 5.120) with their rotor counterpart (5.61) makes it clear that the harmonic oscillator case is considerably more intricate, due to the presence of $W_{m,m}$ and $W_{m,m+2}$ in the matrix elements. This makes the application of FURSTENBERG's theorem more difficult than in the canonical rotor case.

In many cases, the transfer matrices $\mathcal{T}_m$ defined in equations (5.114, 5.117, 5.120) satisfy the assumptions of FURSTENBERG's theorem: the $\mathcal{T}_m$ are all unimodular by construction, the respective groups generated by the sets $\{\mathcal{T}_m,\ m \in \mathbb{N}\}$ are noncompact and irreducible in the sense of appendix B, and all available numerical evidence supports the assumption that the $\mathcal{T}_m$ considered give rise to "sufficiently well-behaved" measures μ as defined by the integrability condition (B.2).

The last remark may be made more explicit by discussing in some more detail the assumptions of FURSTENBERG's theorem for the $\mathcal{T}_m$ considered here. From appendix B, the integrability condition is

$$\int_G \mathrm{d}\mu(M)\, \log(\|M\|_\mathrm{m}) \;<\; \infty, \tag{B.2}$$

where G is the matrix group to which the $\mathcal{T}_m$ belong, μ is the measure according to which the $\mathcal{T}_m$ are distributed in this group, and $\|M\|_\mathrm{m} = \max\limits_{i,j} |M_{ij}|$ is the maximum norm of the matrix M. For more details on FURSTENBERG's theorem see appendix B.

In order to establish the validity of the condition (B.2) with respect to the maximum norm, the distributions of values of the matrix elements of the $\mathcal{T}_m$ need to be checked. The off-diagonal matrix elements of $\mathcal{T}_m^{(2)}$, $\mathcal{T}_m^{(2,\mathrm{left}_1)}$ and $\mathcal{T}_m^{(2,\mathrm{left}_2)}$ are determined by the nearest neighbour interactions $W_{m,m+2}$, for which no closed formula is available, as discussed earlier. Nevertheless, the corresponding distribution of values can be studied numerically. Figures 5.20b, 5.20c and 5.21b, 5.21c present distributions of $\left(\mathcal{T}_m^{(2)}\right)_{12} = -1/W_{m,m+2}$ and $\left(\mathcal{T}_m^{(2)}\right)_{21} = W_{m,m+2}$ for $V_0 = 0.01$ and two different values of $\hbar$. Obviously, the distributions are nicely peaked, and make it thus plausible that, by integration in the sense of (B.2), they contribute to $\mathcal{T}_m^{(2)}$ satisfying the integrability condition. It would be desirable to strengthen this argument by considering other parameter combinations as examples as well, but unfortunately the numerical effort for this task is very high; already the

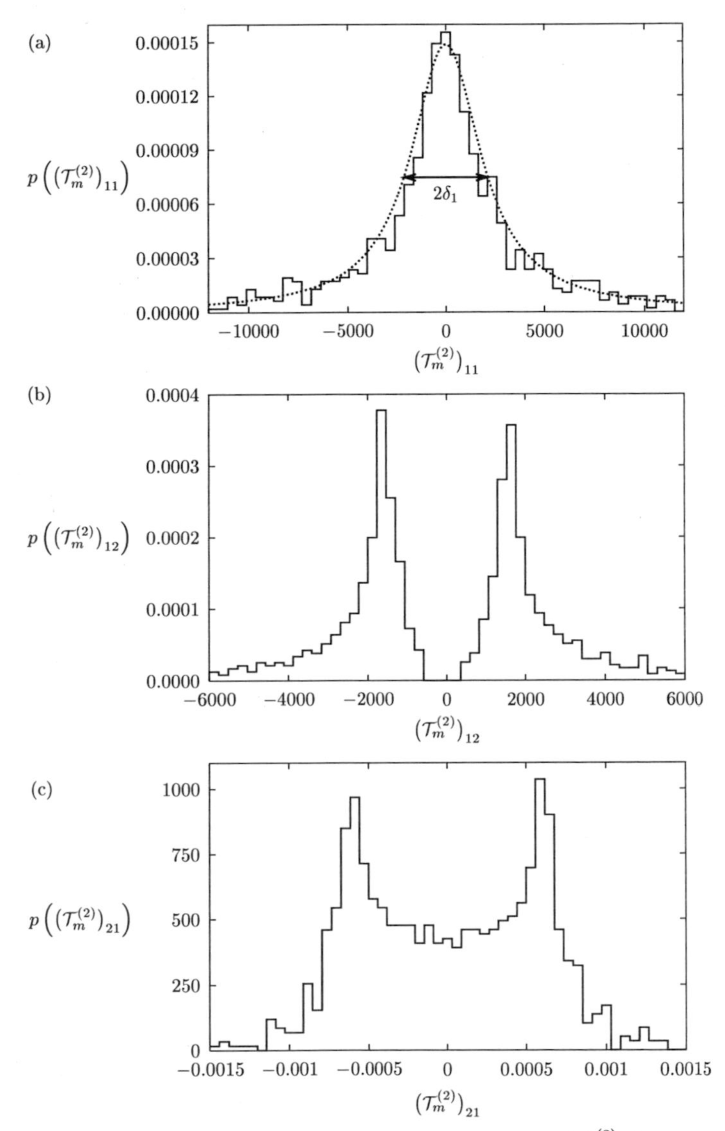

Figure 5.20: Distribution of the nonzero matrix elements of $\mathcal{T}_m^{(2)}$. Histograms for $2 \leq m \leq 1000$, values sampled in 51 intervals. $\hbar = 1.0$, $V_0 = 0.01$, $E = 1.0$, $T = 1.0$. The dotted line in figure (a) marks an approximating Lorentzian (5.41) with a numerically determined value of δ: $\delta_1 = 2141$.

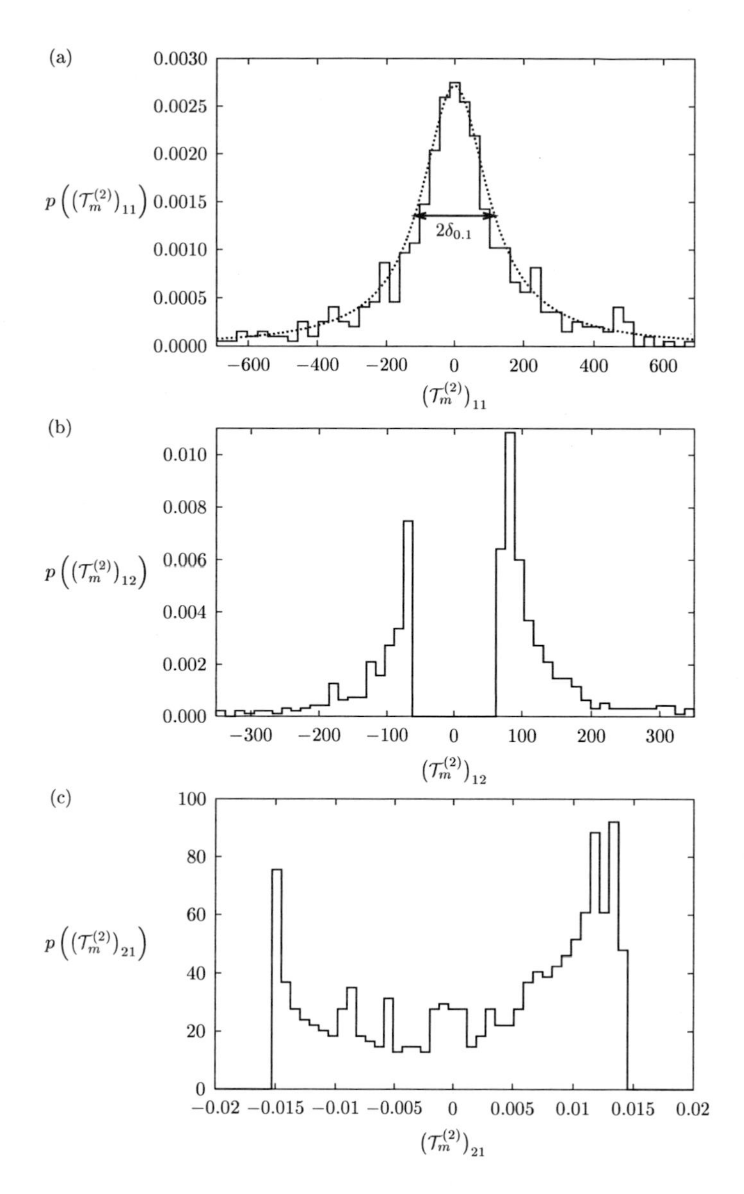

Figure 5.21: The same as figure 5.20, but for $\hbar = 0.1$. Here, $\delta_{0.1} = 117$ in figure (a).

computation of the matrix elements needed for figures 5.20 and 5.21 required a considerable amount of computer time. Still, it is natural to assume that a similarly unproblematic behaviour of the $W_{m,m+2}$ is given for other values of V_0 and $\hbar$ as well. A more stringent reasoning leading to (B.2) being satisfied is presented in subsection 5.3.3 below.

It remains to investigate the nontrivial diagonal matrix elements of the $\mathcal{T}_m$, namely

$$\left(\mathcal{T}_m^{(2)}\right)_{11} = \left(\mathcal{T}_{m+2}^{(2,\mathrm{left}_1)}\right)_{22} = \left(\mathcal{T}_{m+2}^{(2,\mathrm{left}_2)}\right)_{11} = -\frac{\epsilon_m + W_{m,m}}{W_{m,m+2}}. \tag{5.121}$$

Since these are functions not only of the hopping matrix elements $W_{m,m'}$, but also of the diagonal energies ϵ_m, the results depend on the parameters E and T, too, and general analytical conclusions become even more difficult. On the other hand, E and T enter into the matrix elements only via the (quasi-) random number generator (5.81), and as long as T takes on nonresonant values, the actual values of the two parameters should be expected to be more or less irrelevant. One can at least expect that the familiar Lorentzian shape of the distribution of ϵ_m (cf. figure 5.16a) does not get distorted too much by the combination with $W_{m,m}$ and $W_{m,m+2}$ in equation (5.121). For the particular case shown in figure 5.16a (with $E = 1.0$, $T = 1.0$, $\hbar = 1.0$), this expectation is verified in figure 5.20a, where a histogram for the distribution of the $\left(\mathcal{T}_m^{(2)}\right)_{11}$ is shown; figure 5.21a shows the same for $\hbar = 0.1$.[12] The figures make it clear that — at least for these combinations of parameters, but probably in a more general sense — the values of the $\left(\mathcal{T}_m^{(2)}\right)_{11}$ are still nicely peaked around a finite value and likely give rise to a smooth and sufficiently fast decaying μ. Within the accuracy of the histograms, the distributions of the matrix elements involving the diagonal energies even take on quite exactly the form of a Lorentzian, for which compliance with (B.2) is analytically shown in appendix B. But note that the distributions are broad, with half widths of the order of magnitude of 10^3 and 10^2, respectively, obviously a consequence of the small values the $W_{m,m+2}$ take on.

Summarizing, although a rigorous proof is lacking, there is good reason — including some convincing numerical evidence — to assume that typically the condition (B.2) is satisfied for the matrices $\mathcal{T}_m^{(2)}$, $\mathcal{T}_m^{(2,\mathrm{left}_1)}$ and $\mathcal{T}_m^{(2,\mathrm{left}_2)}$ modelling the kicked harmonic oscillator.

As a result, I have shown that at least for some combinations of values of the parameters, FURSTENBERG's theorem can be applied to the transfer matrix formulation of the kicked harmonic oscillator in a way which is largely analogous

[12] Note that due to the intricacy of the numerical evaluation of the $W_{m,m'}$, making it very computer time consuming, only a much smaller number of values of the $\left(\mathcal{T}_m^{(2)}\right)_{11}$ could be taken into account here than in figure 5.16a (10^3 values versus 10^5 values), thus giving rise to the quite coarse-grained histograms in figure 5.20 and 5.21.

to the conventional procedure in the case of the kicked rotor; it may be assumed that similar results probably hold for many other values of the parameters. By the same reasoning as in subsection 5.1.3 it can be concluded that the norms of the vectors $\vec{u}^{(2)}_{E,m}$ and $\vec{v}^{(2)}_{E,m}$ generated by the respective equations of motion decay exponentially, which in turn implies the same for the absolute values of the expansion coefficients $|\overline{u}_{E,m}|$ of the quasienergy states discussed in subsection 5.3.1. This finally establishes the result that the nonresonant kicked harmonic oscillator exhibits ANDERSON localization, provided the parameters are such that the conditions are met which I have discussed above. When this is the case then generic sequences of states $|\psi_n\rangle$ generated by the quantum map (5.72) consist of exponentially localized states. The states are localized with respect to the basis of harmonic oscillator eigenstates, and therefore they are localized in phase space as well. Figure 5.11 stresses the first aspect of localization; figures 5.8 through 5.10 and the examples in section C.4 of the appendix demonstrate localization in phase space.

As discussed in subsection 5.1.2, a lower bound for the speed of the decay of $|\overline{u}_{E,m}|$ is established via the LIAPUNOV exponent of the corresponding transfer matrices. From this point of view it does not come as a surprise that in the case of the oscillator the transfer matrices $\mathcal{T}^{(2)}_m$ and $\mathcal{T}^{(2,\mathrm{left}_1)}_m$ or $\mathcal{T}^{(2,\mathrm{left}_2)}_m$ do not coincide as they do in the case of the rotor. This is again a consequence of the asymmetry expressed by the inequality (5.87) and can lead to different speeds of convergence to zero of $|\overline{u}_{E,m}|$ on the right and on the left. But note that FURSTENBERG's theorem provides a lower bound on the speed of convergence only; the speed may be actually larger, and in special cases it *can* coincide on the right and on the left.

Inevitably, the above discussion of the distribution of values of the matrix elements — especially of the off-diagonal matrix elements — of the transfer matrices is to some degree heuristic, and therefore the same is true, unfortunately, with respect to the distribution μ of the transfer matrices. In particular the arguments in favour of nicely distributed off-diagonal matrix elements leading to FURSTENBERG-integrability are but qualitative. This problem is addressed — and solved — in subsection 5.3.3 by considering a different class of transfer matrices.

5.3.3 Localization in All but the Pathological Cases

In the previous subsection implicitly the "classical" approach to the modelling by means of transfer matrices was followed: only the first component of the vector equations (5.115, 5.116, 5.119) was constructed to be equivalent to the tight binding equation (5.101), while the second component was made to be a *trivial* identity by construction; in addition, only the three amplitudes $\overline{u}_{E,m+2}$, $\overline{u}_{E,m}$ and $\overline{u}_{E,m-2}$ were allowed to contribute for any given m. In the present subsection I use a more general starting point for deriving transfer matrices which are more suitable for proving localization than $\mathcal{T}_m^{(2)}$, $\mathcal{T}_m^{(2,\text{left}_1)}$ and $\mathcal{T}_m^{(2,\text{left}_2)}$.

Generalized or *higher order transfer matrix models* reformulating the tight binding equation (5.101) can be introduced by considering N-dimensional transfer matrices (rather than the only two-dimensional transfer matrices as used up to this point) with some suitable $N \geq 2$:

$$\left(\mathcal{T}_m^{[N]}\right)_{i,j} \in \mathbb{C} \quad \text{with} \quad 1 \leq i, j \leq N. \tag{5.122a}$$

The components of the corresponding vectors $\vec{u}_{E,m}^{[N]}$ may be chosen as linear combinations of as many amplitudes $\overline{u}_{E,m}$ as desired. For example, if the M nearest contributing neighbour sites are to be taken into account, I can use

$$\left(\vec{u}_{E,m}^{[N]}\right)_i = \sum_{\substack{m'=-M \\ m' \text{ even}}}^{M} v_{m,i,m'} \, \overline{u}_{E,m+m'}, \tag{5.122b}$$

with even $M \geq 2$ and suitable coefficients $v_{m,i,m'} \in \mathbb{R}$. Finally, for the equation of motion one has

$$\vec{u}_{E,m+2}^{[N]} = \mathcal{T}_m^{[N]} \, \vec{u}_{E,m}^{[N]}, \tag{5.122c}$$

as usual.

The definition of the transfer matrix model is then completed by specifying the parameters $\left(\mathcal{T}_m^{[N]}\right)_{i,j}$ and $v_{m,i,m'}$. This has to be done in such a way that the following conditions are satisfied:

(i) $$\det\left(\mathcal{T}_m^{[N]}\right) = 1, \tag{5.122d}$$
in order to make FURSTENBERG's theorem applicable.

(ii) For at least one i there must be an $\widetilde{m}'$ such that for all m

$$v_{m,i,m'} = \begin{cases} 1 & \text{if } m' = \widetilde{m}' \\ 0 & \text{else.} \end{cases} \tag{5.122e}$$

This ensures that the asymptotic behaviour of the $\overline{u}_{E,m}$ can be inferred from the asymptotic behaviour of the $\vec{u}_{E,m}^{[N]}$: after application of FURSTENBERG's theorem one has the result $\|\vec{u}_{E,m}^{[N]}\|$ decays exponentially with $m \to \infty$, wherefrom the same can be concluded for $\left(\vec{u}_{E,m}^{[N]}\right)_i = \overline{u}_{E,m+\widetilde{m}'}$ by virtue of equation (5.122e).

(iii) *At least one* of the N subequations of (5.122c) must be equivalent to the tight binding equation (5.101).

(iv) The other subequations of (5.122c) may contain arbitrary identities. Either these identities are trivial, or they are (possibly trivial) linear or nonlinear combinations of the tight binding equation.

The importance of condition (iii) is easily underestimated. It avoids the construction of models that are only *consistent* with the tight binding equation, without being *equivalent* to it. Such a situation occurs, for example, if one specifies the parameters in such a way that the only nontrivial subequation of (5.122c) is a nontrivial linear combination of (5.101); then it is not guaranteed that the solutions $\overline{u}_{E,m}$ of such a model also satisfy the tight binding equation.

This scheme allows to construct a multitude of different transfer matrix models. Once a model has been defined in this way, the remaining conditions of FURSTENBERG's theorem need to be checked, thus possibly leading to a proof of localization, if the model has been constructed properly.

Following this scheme, one can discuss the transfer matrix model with $N = 3$ and $M = 2$, for example. I have tried to choose the 18 parameters of this model in such a way that the resulting expressions are as simple as possible. After some (lengthy) algebra, the transfer matrix

$$\mathcal{T}_m^{(3)} := \begin{pmatrix} -\dfrac{\epsilon_m + W_{m,m}}{W_{m,m+2}} + \dfrac{W_{m-2,m}}{W_{m-2,m-4}} & \dfrac{\epsilon_{m-2} + W_{m-2,m-2}}{W_{m-2,m-4}} - \dfrac{W_{m,m-2}}{W_{m,m+2}} & 1 \\ 1 & 0 & 0 \\ 0 & 1 & 0 \end{pmatrix} \tag{5.123a}$$

is obtained, acting on the vectors

$$\vec{u}_{E,m}^{(3)} := \begin{pmatrix} \overline{u}_{E,m} \\ -\dfrac{W_{m,m+2}}{W_{m,m-2}}\overline{u}_{E,m+2} - \dfrac{\epsilon_m + W_{m,m}}{W_{m,m-2}}\overline{u}_{E,m} \\ -\dfrac{W_{m-2,m}}{W_{m-2,m-4}}\overline{u}_{E,m} - \dfrac{\epsilon_{m-2} + W_{m-2,m-2}}{W_{m-2,m-4}}\overline{u}_{E,m-2} \end{pmatrix} \tag{5.123b}$$

within the equation of motion

$$\vec{u}^{(3)}_{E,m+2} = \mathcal{T}^{(3)}_m \, \vec{u}^{(3)}_{E,m}. \tag{5.123c}$$

The system (5.123) is suited for propagation in positive m-direction. Backpropagation can be achieved by using

$$\mathcal{T}^{(3,\text{left})}_m = \left(\mathcal{T}^{(3)}_{m-2}\right)^{-1}$$

$$= \begin{pmatrix} 0 & 1 & 0 \\ 0 & 0 & 1 \\ 1 & \frac{\epsilon_{m-2} + W_{m-2,m-2}}{W_{m-2,m}} - \frac{W_{m-4,m-2}}{W_{m-4,m-6}} & -\frac{\epsilon_{m-4} + W_{m-4,m-4}}{W_{m-4,m-6}} + \frac{W_{m-2,m-4}}{W_{m-2,m}} \end{pmatrix} \tag{5.124a}$$

and

$$\vec{u}^{(3)}_{E,m-2} = \mathcal{T}^{(3,\text{left})}_m \, \vec{u}^{(3)}_{E,m}. \tag{5.124b}$$

It is easy to see — by explicitly writing down all three subequations — that the first subequation of the equation of motion (5.123c) contains the tight binding equation *three* times (in the sense of (iv)), the second subequation is equivalent to the tight-binding equation (satisfying condition (iii)), and the third subequation is just a trivial identity.

The $\mathcal{T}^{(3)}_m$ do not share the worst disadvantage of the $\mathcal{T}^{(2)}_m$ discussed in the previous subsection. Most matrix elements take on the trivial values 0 or 1, which are uncritical with respect to the integrability condition (B.2); the only nontrivial matrix elements, $\left(\mathcal{T}^{(3)}_m\right)_{11} = -\left(\mathcal{T}^{(3,\text{left})}_{m+2}\right)_{32}$ and $\left(\mathcal{T}^{(3)}_m\right)_{12} = -\left(\mathcal{T}^{(3,\text{left})}_{m+2}\right)_{33}$, are characterized by Lorentzian distributions of values, as figure 5.22 shows. In fact, for both values of $\hbar$ considered in this figure, the approximating Lorentzians are nearly identical to those in figures 5.20a and 5.21a, respectively, where they approximate the distributions $p\big(\big(\mathcal{T}^{(2)}_m\big)_{11}\big)$. This indicates that the quotient of two consecutive nearest neighbour matrix elements — which marks the only difference between $\left(\mathcal{T}^{(2)}_m\right)_{11}$ and $\left(\mathcal{T}^{(3)}_m\right)_{1j}$ — does not take on values large enough to make $p\big(\big(\mathcal{T}^{(2)}_m\big)_{11}\big)$ and $p\big(\big(\mathcal{T}^{(3)}_m\big)_{1j}\big)$ differ significantly, an observation that is supported by figure 5.23, which displays typical values of $W_{m,m+2}/W_{m,m-2}$ for several values of V_0 and $\hbar$. Apparently the distribution of all nontrivial matrix elements of the $\mathcal{T}^{(3)}_m$ is governed essentially by the Lorentzian distribution generated by the (quasi-) random number generator (5.81).

Summarizing, for the parameters considered explicitly, and most probably for *all* parameter combinations belonging to the nonpathological systems with $W_{m,m+2} \neq 0$ for all m, $\mathcal{T}_m^{(3)}$ and $\mathcal{T}_m^{(3,\mathrm{left})}$ satisfy the integrability condition (B.2) and therefore give rise to quantum localization of the T-nonresonant kicked harmonic oscillator; the localization mechanism is identified as that of classical ANDERSON localization.

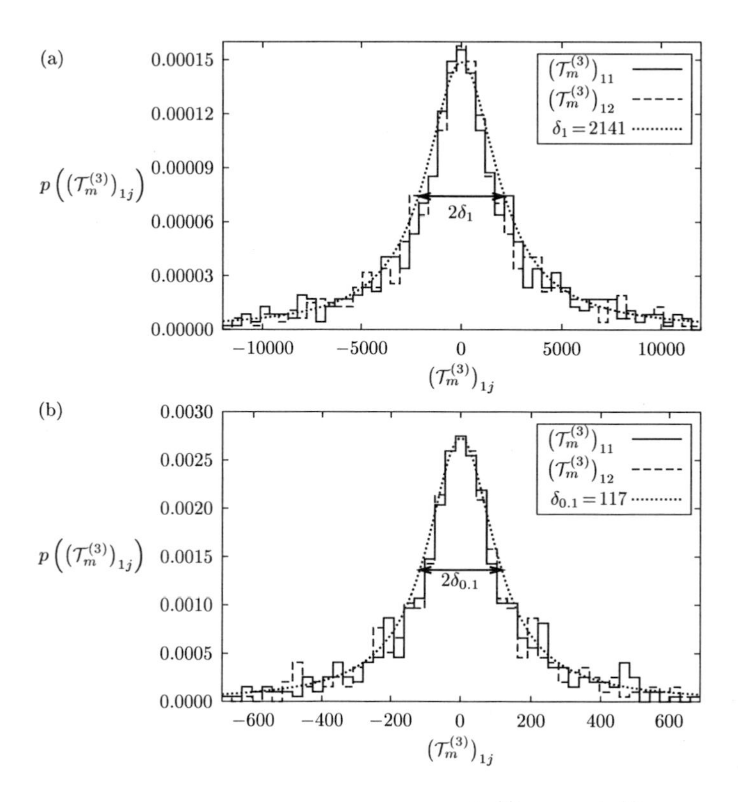

Figure 5.22: Distribution of the matrix elements $\left(\mathcal{T}_m^{(3)}\right)_{11}$ and $\left(\mathcal{T}_m^{(3)}\right)_{12}$. Histograms for $2 \leq m \leq 1000$, values sampled in 51 intervals. The dotted lines mark approximating Lorentzians (5.41) with numerically determined values of δ. $V_0 = 1.0$, $E = 1.0$, $T = 1.0$. (a) $\hbar = 1.0$; $\delta_1 = 2141$. (b) $\hbar = 0.1$; $\delta_{0.1} = 117$.

5.3.4 Discussion

In this subsection I discuss some of the problems of the theory developed in the previous subsections, and I outline how these problems can be either avoided or solved.

One of the issues that need to be addressed is to what degree quantum localization can be expected when the matrix elements $W_{m,m+m''}$ do not decay rapidly enough with m''. This problem is not unique to the quantum kicked harmonic oscillator; the same can happen with respect to the quantum kicked rotor: figure 5.6 indicates that for larger values of V_0 the quality of the tight binding approximation deteriorates there, too. In the literature, this is obviously not regarded as an obstacle to ANDERSON localization in the kicked rotor, based on the tight binding equation; at least, this problem is not addressed in the standard literature. But see [Kle97] for a discussion of this aspect, with a positive result on localization in a particular ANDERSON model with "long range hopping".

Judging from figures 5.17 and 5.18, such a case of slow decay should be expected in the quantum kicked harmonic oscillator for larger values of V_0 and $\hbar$. Even in situations like this, localization is generically given. This can be seen by replacing the tight binding equation (5.101) with a similar approximation to the

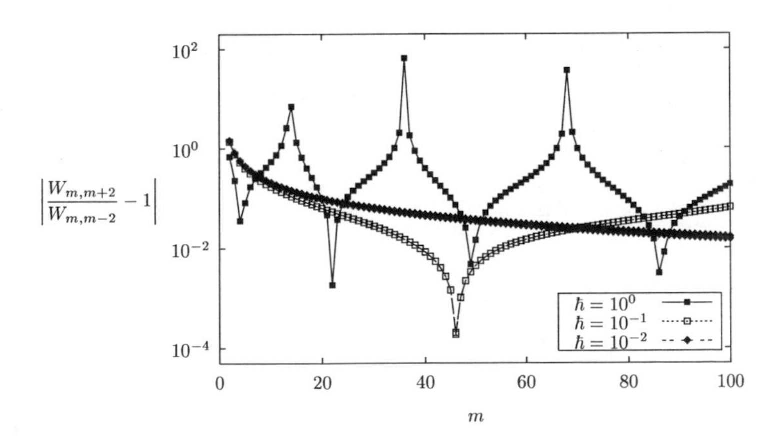

Figure 5.23: $|W_{m,m+2}/W_{m,m-2} - 1|$ with $\hbar = 1.0/0.1/0.01$. For each value of $\hbar$, data for three values of V_0 are plotted, coinciding within the accuracy of this plot: $V_0 = 10^{-2}/10^{-3}/10^{-4}$.

discrete SCHRÖDINGER equation (5.84), the only difference being that now the interaction with neighbouring sites is assumed to be characterized by a suitably longer range $m''_{\max}$ (with even $m''_{\max} \geq 2$):

$$\epsilon_m \overline{u}_{E,m} + \sum_{\substack{m''=-\min(m,m''_{\max}) \\ m'' \text{ even}}}^{m''_{\max}} W_{m,m+m''}\, \overline{u}_{E,m+m''} = 0. \tag{5.125}$$

Replacing the tight binding equation, this "not-so-tight-binding equation" can be used in the scheme for constructing transfer matrix models as described in subsection 5.3.3.

The same technique helps to avoid the troubles that come with the pathological cases where $W_{m,m+2} = 0$ or $W_{m,m-2} = 0$ for some m. The problem of unavoidably having to divide by such a zero, as encountered in subsections 5.3.2 and 5.3.3, simply vanishes for suitably large $m''_{\max}$ if there exists nontrivial interaction between some sites at all.

The other, probably more satisfying, answer to the question of the pathologically vanishing nearest neighbour interactions is the following. The matrix elements for nearest neighbour interaction can be regarded as the values of a continuous function $f(z; V_0, \hbar)$, sampled at equidistant intervals on the z-axis. Figure 5.19 gives the impression that f is well-behaved in the sense that all its intersections with the z-axis are transversal, and that the zeroes of f have no accumulation point on the z-axis. Then it follows that *generically* pathological cases do not occur [Str01a]. In all those degenerate cases with $W_{m,m+2} = 0$ or $W_{m,m-2} = 0$ for some m, an arbitrarily small variation of either V_0 or $\hbar$ should suffice to eliminate the pathological situation. The disadvantage of this argument, though, is its heuristic nature, due to the lack of an explicit, integrated formula for the $W_{m,m'}$.

ANDERSON localization of the kicked harmonic oscillator appears to be a quite general, and perhaps even generic phenomenon: in the preceding subsections, I have motivated the fact that the most important ingredients of all transfer matrices are the diagonal energies (5.81) which are independent of the kick potential. Only the *period* T of the kicks enters equation (5.81) in a truly significant way. The matrix elements (5.83), on the other hand, have been seen to be rather irrelevant for the localization properties — as long as they give rise to tight or at least narrow binding — and the matrix elements mark the only point where properties of the kick potential enter the theory. What is more, in the matrix elements the kick potential effectively gets randomized by being used as the argument of the tangent only — cf. equation (5.75). In this sense, quantum localization of the nonresonant kicked harmonic oscillator should be expected for *generic* kick functions, not only for the cosine kick potential (1.18). With respect to the value of T, one may also speak of genericity of localization in an additional sense: under variation of T, the resonant values (5.85) not leading to ANDERSON localization are obtained with probability zero.

It is interesting to note the fundamental difference between the conditions of nonresonance that lead to quantum localization in the two model systems considered in this chapter: The resonance condition (5.67) for the quantum kicked rotor is obviously a consequence of the *quantum* nature of the system, as the expression (5.67) involves the PLANCK constant $\hbar$ and has no classical counterpart. On the other hand, nonresonance in the context even of the *quantum* version of the kicked harmonic oscillator involves just the classical resonance condition (1.22) on the kick period T. It is one of the advantages of the particular scaling (1.15, 1.16, 2.4) of the dynamical variables and parameters employed here, that it clearly exposes this property of the dynamics, in contrast to the scaling used elsewhere [BRZ91]. So, using the abovementioned scaling, quantum localization in the kicked harmonic oscillator comes about as a genuine quantum phenomenon — without a classical analogue — that is entirely controlled by a classical parameter: the question of nonresonance of the kick period T alone decides on the localization of quantum states in the kicked harmonic oscillator. In particular, the existence of localization is independent of the strength of the kicks. Then, the other parameters V_0 and $\hbar$, via the matrix elements (5.86), control the strength of the interaction between the "sites" and determine if there is — approximately — tight binding or rather some medium range interaction; in the same way they control the localization length of the system.

A final remark on the numerical results in section 5.2 and section C.4 of the appendix: for some of the presented examples, the quantum map has been iterated for a very large number of times. Up to and beyond 10^5 kicks have been considered, while the number of harmonic oscillator eigenfunctions taken into account for the basis was only between 3000 and 6000. This was possible just because of the quantum localization of the systems considered in that and the present section; otherwise, for example for the resonant systems discussed in chapter 4, such basis sizes would have allowed kick numbers only typically not exceeding a few thousand.

Conclusion

This study is intended to make a contribution to the still developing theory of quantum chaos, by investigating an important model system that in classical mechanics is characterized by weak chaos: the kicked harmonic oscillator.

Well-known classical properties of this model system that also manifest themselves in quantum mechanics — above all the stochastic webs generated by this system and the diffusive dynamics within the channels of these webs — make the investigation of its quantum dynamics a nontrivial task.

For the numerical iteration of the quantum map I have chosen to use the matrix representation of the FLOQUET operator with respect to the eigenbasis of the unkicked, i.e. free, harmonic oscillator. While from a numerical point of view it is difficult to evaluate the expressions for the matrix elements of the FLOQUET operator for the kick potential used here, once this has been accomplished one has a very powerful and numerically stable method for iterating the quantum map even for very long times. In practically all cases considered, the numerical error could safely be attributed to the insufficient size $m_{\max}$ of the basis used; other sources of numerical error — which cannot be avoided for finite differences methods, for example — did not play a major role.

In quantum phase space, the eigenstates $|m\rangle$ of the harmonic oscillator are nicely localized around the corresponding classical trajectories of the harmonic oscillator. This also means that the whole basis $\{|m\rangle \mid 0 \leq m \leq m_{\max}\}$ is localized in a circular region centered about the origin of phase space. Therefore, by systematically increasing $m_{\max}$, quantum states in an increasingly large region of phase space can be completely described using this basis. Furthermore, although the $|m\rangle$ are tailor-made for the *free* harmonic oscillator, they also turn out to be well suited to describe the quantum states that are subject to the nontrivial dynamics of the *kicked* harmonic oscillator.

The robust numerical method, using this basis with values of $m_{\max}$ up to 6000, was fundamental to exploring the quantum dynamics of the model system both for the *resonance* and *nonresonance* cases. It allowed to establish the result that, in the first case, the quantum map generates quantum stochastic webs which are very similar to their classical counterparts in several ways. Both types of web are characterized by the same topology and symmetry, and in the channels of both types of web the dynamics is unbounded. While in the classical case the dynamics in the channels is diffusive, the quantum dynamics there is either diffusive or even ballistic. This result is especially interesting as it is in disagreement with the general hypothesis on quantum chaos that classically chaotic systems with diffusive dynamics are characterized by quantum suppression of diffusion — at least for initial states $|\psi_0\rangle$ that are located in those regions of phase space (of a weakly chaotic system) being characterized by classically chaotic dynamics.

It has to be mentioned that for obtaining this result the numerical algorithm had to be pushed to its limit, with the consequence of declining accuracy. It would have been desirable to further increase the basis size $m_{\max}$ in order to improve on the accuracy and to be able to follow the unbounded dynamics for longer times, but this would have required a considerable additional numerical effort. On the other hand, the result of quantum non-suppression of chaos in the resonance case is supported by a convincing analytical argument, such that its validity can hardly be questioned. However, clarification of this issue might be an appropriate object of future work on the kicked harmonic oscillator.

The numerical results in the case of *non*resonance are in agreement with the conventional quantum suppression of diffusion hypothesis, as this is exactly what is found for the model system considered here. This numerical result of quantum localization of the nonresonant kicked harmonic oscillator is given an analytical explanation in terms of the theory of ANDERSON localization. While the general idea of proving localization by mapping the problem onto the well-understood ANDERSON model is motivated by the theory of quantum localization in the kicked rotor, it should be noted that establishing this mapping for the kicked harmonic oscillator was much less straight-forward and required a number of additional considerations. For example, the kicked harmonic oscillator's hopping matrix elements for interaction between the sites of an ANDERSON-like lattice are much more intricate than those of the kicked rotor. Nevertheless, localization could be explained analytically in this way.

Resonances are of particular importance throughout this study. Classically, resonances with respect to T are important because they give rise to classical stochastic webs. The fact that quantum theory predicts the existence of quantum stochastic webs, too, and that these quantum webs are obtained in exactly the same cases of resonance with respect to T, is a nice example of quantum-classical correspondence, which has been confirmed numerically in this study. Finally, two theories have been discussed in this study that are entirely independent of each other: the theoretical foundation of quantum stochastic webs and the theory of ANDERSON localization. From a T-resonance point of view it is a nice and reassuring feature of the latter that it is consistent with the former in that it predicts localization for all values of T with the exception of the resonant T that give rise to stochastic webs. So both theories combined give the complete quantum picture in a complementary way, and there is no contradiction between the two.

By correspondence, even in the quantum case the resonances with respect to T might be understood to be an essentially classical phenomenon, because they are closely associated with the existence of classical stochastic webs. However, these classical resonances turn out to be true *quantum resonances* as well, since they are obtained through even two independent truly quantum mechanical theories, without explicitly assuming resonance in the classical case. — In addition to the resonances with respect to T there exists another class of resonances, with respect to $\hbar$, which is of importance in the analytical discussion of quantum web formation. These $\hbar$-resonances have no obvious classical counterpart, such that they manifest another class of true quantum resonances in the quantum kicked harmonic oscillator.

Finally, I want to point to the fact that the numerical work needed to come to the above conclusions was indeed extensive. First, for investigating the quantum webs above all computers with large memory were needed. Second, in order to study the localization phenomena veritable long-time simulations were performed for which high speed of computation was the prime requirement. Therefore, only the availability of a large number of fast workstations made it possible to study the model system for many different parameter values and initial conditions.

The results obtained here are but a small contribution to the theory of quantum chaos, concentrating on a particular — though important — model system. In order to get a better understanding of quantum chaos, more classically chaotic model systems need to be studied with respect to their quantum behaviour. It might be of particular interest to consider other web-generating systems, some of which have been discussed in chapter 1.

For now, a *general* and *comprehensive* theory of quantum chaos, applicable to all systems, regardless of the degree of their classical chaoticity, and describing all their essential features, is still lacking.

Wie's dich auch aufzuhorchen treibt,
Das Dunkel, das Rätsel, die Frage bleibt.

Die Frage bleibt
THEODOR FONTANE

Appendix A

Quantum Phase Space Distribution Functions

Many ways for obtaining a two-variable function from a one-variable one have been proposed ...

G. Torres-Vega
J. Chem. Phys. **98** (1993) 7040.

How can a quantum mechanical dynamical system be compared with its counterpart in classical mechanics? This is one of the principal questions in the theory of *quantum chaos*. Typically one studies quantum systems corresponding to classical dynamical systems which are known to be — using the language of nonlinear dynamics — weakly or hard *chaotic*. The object of these studies is to compare the outcomes of the two dynamical theories (classical and quantum mechanics) as effectively as possible, for example in order to understand which quantum properties correspond to certain classical phenomena.

In classical mechanics the concept of phase space has been used very successfully for a long time, especially when analyzing the *dynamics* of a system. Using this concept, the state of the system is represented by a point — or a distribution of points — in a space that is spanned by *both* the position *and* the momentum coordinates. In quantum mechanics, on the other hand, the state of the system can for example be formulated in terms of *either* the position *or* the momentum representation, where the wave functions (which are fields, as opposed to the points in classical phase space) exclusively depend on the position or momentum coordinates, but not on both sets of variables at the same time. Among other consequences this means that in the framework of a typical quantum mechanical representation one needs only half the number of dynamical variables as compared with the corresponding classical phase space. This raises the question of how these two apparently so different concepts of representing a state can be compared with each other.

The distribution function introduced by WIGNER in 1932 is an important tool for the investigation of this correspondence problem [Wig32]. Using the WIGNER distribution function, it is possible to describe quantum mechanical phenomena in a language which is *as classical as possible*, by employing a suitable quantum analogue of classical phase space. In the classical limit $\hbar = 0$ the quantum mechanical phase space turns into its classical counterpart, and the distribution function defined in quantum phase space becomes the classical LIOUVILLE probability density in the same limit. This correspondence of the classical and quantum concepts paves the way for a straightforward comparison of the results of the two dynamical theories. In BERRY's words: "WIGNER's picture is peculiarly well suited to the present problem, because it is in phase-space that the distinction between classical regular and irregular motion manifests itself most clearly, and so one can hope that the analogous quantal distinctions will reveal themselves with corresponding clarity in the form of [quantum phase space distribution functions] ..." [Ber83].

The above reference to the LIOUVILLE probability density also indicates what interpretation one is aiming at when introducing quantum distribution functions: They are supposed to serve as quantum replacements of the classical phase space probability density. A number of different terms are used — more or less synonymously — for quantum distribution functions in order to emphasize this interpretation: (quasi-) probability distribution, (quasi-) probability density, distribution function, phase space distribution. The significance of the prefix *quasi* is discussed in section A.5.

Following the WIGNER distribution function many more (and mostly quite different) quantum distribution functions have been introduced since 1932; the most important ones of these are due to KIRKWOOD [Kir33], GLAUBER [Gla63b, Gla65] and HUSIMI [Hus40]. COHEN has developed a systematic classification of most of these distribution functions [Coh66];[1] comprehensive accounts of this theory can be found in [BJ84, HOSW84, Lee95]. For the following discussion I mainly use the terminology of [Lee95].

For simplicity of notation I consider one-dimensional systems and *pure* states only, i.e. states which can be described by the specific density operator $\hat{\rho}(t) = |\psi(t)\rangle \langle\psi(t)|$. Most of the following statements can also be formulated for mixed

[1] An entirely different approach to quantum distribution functions is described in [TV93a, TV93b, TVF93, TVZ+96]. In the present study I do not consider this particular variant of the theory.

states described by arbitrary density operators, and many of the results can be generalized to higher dimensions as well.

A.1 Definition of Quantum Phase Space Distribution Functions

One possible motivation for introducing quantum distribution functions is their utility for the comparison of classical and quantum mechanics, as mentioned above. In addition to this there is another, equally important motivation for studying these functions: they can be used to compute expectation values in a comparatively simple way, where the computational simplification is mainly due to the fact that the corresponding formulae depend on scalars only, as opposed to conventional expressions of quantum expectation values which typically involve operators.

Consider a classical observable $A(x,p)$ that depends on the position and momentum variables x and p. The expectation value of A can be computed as

$$\langle A\rangle_t = \int \mathrm{d}x \int \mathrm{d}p\, A(x,p)\, F(x,p,t), \tag{A.1}$$

where the classical phase space probability density $F(x,p,t)$ describes the state of the system at time t, and can be obtained, for example, by solving the LIOUVILLE equation. (All integrals in this appendix are from $-\infty$ to ∞, and the index $_t$ explicitly indicates the time dependence of the expectation value.) The objective of the following considerations is to find a quantum mechanical expression which is analogous to equation (A.1).

Rather than discussing a general quantum observable $\hat{A}(\hat{x},\hat{p})$, with the position and the momentum operators $\hat{x}$ and $\hat{p}$, I begin by considering a particular operator instead, namely $e^{i(\xi\hat{x}+\eta\hat{p})}$, with constants $\xi,\eta\in\mathbb{R}$. Exponentials of this type are used below in the FOURIER expansion (A.7) to construct any other operator. In order to obtain an expression analogous to equation (A.1) the operator $e^{i(\xi\hat{x}+\eta\hat{p})}$ somehow has to be substituted by a corresponding scalar expression. For example one could set

$$\left\langle e^{i(\xi\hat{x}+\eta\hat{p})}\right\rangle_t = \int \mathrm{d}x \int \mathrm{d}p\, e^{i(\xi x+\eta p)}\, F_1(x,p,t) \tag{A.2a}$$

with a distribution function $F_1(x,p,t)$. Similarly, one could have

$$\left\langle e^{i\xi\hat{x}}e^{i\eta\hat{p}}\right\rangle_t = \int \mathrm{d}x \int \mathrm{d}p\, e^{i(\xi x+\eta p)}\, F_2(x,p,t) \tag{A.2b}$$

as well, with another distribution function $F_2(x,p,t)$. But due to the fact that $\hat{x}$ and $\hat{p}$ do not commute one has

$$e^{i(\xi\hat{x}+\eta\hat{p})} = e^{i\xi\hat{x}}e^{i\eta\hat{p}}e^{\frac{1}{2}i\hbar\xi\eta} \neq e^{i\xi\hat{x}}e^{i\eta\hat{p}}, \tag{A.3}$$

as is easily confirmed using the BAKER-CAMPBELL-HAUSDORFF formula.[2] Therefore, in general the distribution functions F_1 and F_2 are not identical, the reason being that in the integrands of equations (A.2a) and (A.2b) the same scalar function $e^{i(\xi x+\eta p)}$ has been associated with the two different operators $e^{i(\xi\hat{x}+\eta\hat{p})}$ and $e^{i\xi\hat{x}}e^{i\eta\hat{p}}$, respectively.

In order to avoid this ambiguity one first chooses a complex-valued *kernel function* $f(\xi,\eta)$; then the scalar $e^{i(\xi x+\eta p)}$ is defined to be associated with the operator $f(\xi,\eta)e^{i(\xi\hat{x}+\eta\hat{p})}$ *exclusively*,

$$f(\xi,\eta)e^{i(\xi\hat{x}+\eta\hat{p})} \mathrel{\widehat{=}} e^{i(\xi x+\eta p)}, \tag{A.5}$$

thereby establishing a one-to-one correspondence between scalars and operators. In this way the above-mentioned ambiguity is shifted towards the definition of the kernel $f(\xi,\eta)$. Depending on the rule of association specified by this function, one can define different distribution functions $F^f(x,p,t)$ via

$$\left\langle f(\xi,\eta)e^{i(\xi\hat{x}+\eta\hat{p})}\right\rangle_t =: \int \mathrm{d}x \int \mathrm{d}p \; e^{i(\xi x+\eta p)} F^f(x,p,t). \tag{A.6}$$

The quantization rule (A.5) not only defines how to associate exponential operators with scalars, but is much more general, as it can be applied to each term of the FOURIER expansion of any operator $\hat{A}$,

$$\hat{A}(\hat{x},\hat{p}) = \frac{1}{2\pi}\int \mathrm{d}\xi \int \mathrm{d}\eta \; \tilde{A}(\xi,\eta)e^{i(\xi\hat{x}+\eta\hat{p})}. \tag{A.7}$$

Therefore the scalar function associated with $\hat{A}$ obviously is

$$A^f(x,p) := \frac{1}{2\pi}\int \mathrm{d}\xi \int \mathrm{d}\eta \; \frac{\tilde{A}(\xi,\eta)}{f(\xi,\eta)} e^{i(\xi x+\eta p)}, \tag{A.8}$$

[2] A simplified BAKER-CAMPBELL-HAUSDORFF (BCH) formula

$$e^{\hat{A}}e^{\hat{B}} = e^{\hat{A}+\hat{B}}e^{\frac{1}{2}[\hat{A},\hat{B}]} \tag{A.4}$$

holds in the special case when the operators $\hat{A}$ and $\hat{B}$ both commute with their commutator $[\hat{A},\hat{B}]$ (which is true in the present case, because $[\hat{x},\hat{p}] = i\hbar$ is a c-number). See [Per93] for a proof, and [Wil67, Ote91] for more on BCH formulae.

which is defined unambiguously.[3] Using the representation (A.8) of the classical observable associated with the operator $\hat{A}$, it is then straightforward to compute its quantum mechanical expectation value:

$$\langle \hat{A} \rangle_t = \frac{1}{2\pi} \int d\xi \int d\eta \int dx \int dp \, \frac{\tilde{A}(\xi,\eta)}{f(\xi,\eta)} e^{i(\xi x + \eta p)} F^f(x,p,t) \tag{A.9a}$$

$$= \int dx \int dp \; A^f(x,p) \, F^f(x,p,t), \tag{A.9b}$$

where equation (A.9b) is the desired expression analogous to equation (A.1).

An *explicit* expression for the distribution function F^f can be obtained from the *implicit* definition (A.6) by FOURIER transformation. For a system in the state $|\psi(t)\rangle$ at time t, the expectation values are given by $\langle \cdot \rangle_t = \langle \psi(t) \,|\cdot|\, \psi(t) \rangle$, and one gets

$$\int d\xi \int d\eta \; \langle \psi(t) \left| e^{i(\xi\hat{x}+\eta\hat{p})} \right| \psi(t) \rangle \, f(\xi,\eta) \, e^{-i(\xi x+\eta p)} = 4\pi^2 F^f(x,p,t),$$

which by insertion of the identity operator $\mathbb{1} = \int dx'' \, |x''\rangle \langle x''|$ can also be written as

$$F^f(x,p,t) = \frac{1}{4\pi^2} \int d\xi \int d\eta \int dx'' \; \langle x'' | \psi(t) \rangle \langle \psi(t) \left| e^{i(\xi\hat{x}+\eta\hat{p})} \right| x'' \rangle \, f(\xi,\eta) \, e^{-i(\xi x+\eta p)}. \tag{A.10}$$

The exponential $e^{i\eta\hat{p}}$ acts as a translation operator in position space (cf. equation (5.23));

$$e^{i\eta\hat{p}} \, |x''\rangle = \hat{T}(-\hbar\eta) \, |x''\rangle = |x'' - \hbar\eta\rangle \,, \tag{A.11}$$

such that

$$e^{i(\xi\hat{x}+\eta\hat{p})} \, |x''\rangle = e^{\frac{1}{2}i\hbar\xi\eta} e^{i\xi\hat{x}} e^{i\eta\hat{p}} \, |x''\rangle = e^{-\frac{1}{2}i\hbar\xi\eta} e^{i\xi x''} \, |x'' - \hbar\eta\rangle \,. \tag{A.12}$$

[3] The question needs to be addressed if the theory, and in particular the evaluation of integrals like that in equation (A.8), could be spoiled by kernel functions $f(\xi,\eta)$ that have zeroes or are singular for some values of ξ, η. I do not discuss this issue here, but refer the reader to [SPM99] where it is shown that the formalism can be applied smoothly even in such notorious cases.

Note that the kernel function can also be chosen, more generally, as a functional of the quantum state $|\psi\rangle$ of the system itself: $f(\xi,\eta,\psi)$. COHEN gives an example for such a kernel function f^{C} that, despite being quite complicated, leads to the very simple COHEN *distribution function* $F^{C}(x,p,t) = |\langle q\,|\psi(t)\rangle|^2 \, |\langle p\,|\psi(t)\rangle|^2$ that nicely combines the position and momentum representations of $|\psi\rangle$ in an intuitive way [Coh66]. However, since choosing a ψ-dependent f has a number of unfavourable consequences — for example, equation (A.8) indicates that in this case the function A^{C} associated with the operator $\hat{A}$ becomes ψ-dependent, too — I do not further discuss kernels that are functionals of $|\psi\rangle$.

Inserting this into equation (A.10) and substituting $x' := x'' - \hbar\eta/2$, I finally obtain a convenient explicit formula for the distribution function $F^f(x,p,t)$:

$$
\boxed{\begin{aligned} F^f(x,p,t) \;=\; & \frac{1}{4\pi^2}\int\mathrm{d}\xi\int\mathrm{d}\eta\int\mathrm{d}x'\;\left\langle x'+\tfrac{1}{2}\hbar\eta\right|\psi(t)\rangle\times \\ & \times\left\langle\psi(t)\left|x'-\tfrac{1}{2}\hbar\eta\right.\right\rangle f(\xi,\eta)\,e^{i\xi(x'-x)}e^{-i\eta p}. \end{aligned}} \tag{A.13}
$$

A.2 Special Distribution Functions

In the same way in which the kernel $f(\xi,\eta)$ determines the association between operators and scalars, it also defines an *operator ordering* [Meh77, AM77]. In this context an expansion (A.7) of an operator is called *ordered* if it is written as a superposition of terms that are of the form $f(\xi,\eta)e^{i(\xi\hat{x}+\eta\hat{p})}$. The essential point of this definition is that the multiplication of the exponential $e^{i(\xi\hat{x}+\eta\hat{p})}$ with the kernel function yields a characteristic composition of exponential operators, the explicit form of which depends on the particular choice of f.

The implications of this definition become clearer when some specific examples are considered:

- For the kernel function

$$
f^{\mathrm{W}}(\xi,\eta) \;:=\; 1 \tag{A.14}
$$

one obtains from equation (A.13), integrating over ξ and x' and substituting $x'' := \hbar\eta/2$, the WIGNER *distribution function* [Wig32]:

$$
F^{\mathrm{W}}(x,p,t) \;=\; \frac{1}{\pi\hbar}\int\mathrm{d}x''\;\langle\psi(t)\,|x-x''\rangle\,\langle x+x''\,|\psi(t)\rangle\;e^{-\frac{2ipx''}{\hbar}}. \tag{A.15}
$$

It corresponds to the WEYL ordering of operators [Wey31, DGS72],

$$
f^{\mathrm{W}}(\xi,\eta)e^{i(\xi\hat{x}+\eta\hat{p})} \;=\; e^{i(\xi\hat{x}+\eta\hat{p})} \;\widehat{=}\; e^{i(\xi x+\eta p)}, \tag{A.16}
$$

and by equations (A.7, A.8) gives rise to the WEYL *transform* of the operator $\hat{A}$:

$$
A^{\mathrm{W}}(x,p) \;=\; \hbar\int\mathrm{d}\eta\;\left\langle x+\tfrac{1}{2}\hbar\eta\right|\hat{A}\left|x-\tfrac{1}{2}\hbar\eta\right\rangle e^{-i\eta p}. \tag{A.17}
$$

The last two terms of expression (A.16) indicate the association of operators and scalars that is defined by f^{W}, as discussed in section A.1.

Using the WIGNER distribution function, the overlap of two states $|\psi_1\rangle$ and $|\psi_2\rangle$ described by the WIGNER functions F_1^{W} and F_2^{W} can be computed as the overlap of the corresponding WIGNER functions in phase space:

$$|\langle\psi_1|\psi_2\rangle|^2 = 2\pi\hbar\int\mathrm{d}x\int\mathrm{d}p\, F_1^{\mathrm{W}}(x,p,t)\, F_2^{\mathrm{W}}(x,p,t). \tag{A.18}$$

Besides the HUSIMI distribution function (which is introduced in section A.3 below), the WIGNER function is the most commonly used quantum phase space distribution function.

- The kernel function

$$f^{\mathrm{S}}(\xi,\eta) := e^{-\frac{1}{2}i\hbar\xi\eta} \tag{A.19}$$

defines the *standard ordering* of operators; it is characterized by all $\hat{x}$-dependent terms preceding the $\hat{p}$-dependent terms,

$$f^{\mathrm{S}}(\xi,\eta)e^{i(\xi\hat{x}+\eta\hat{p})} = e^{i\xi\hat{x}}e^{i\eta\hat{p}} \mathrel{\widehat{=}} e^{i(\xi x+\eta p)}, \tag{A.20}$$

which leads to the *standard-ordered distribution function*[4]:

$$F^{\mathrm{S}}(x,p,t) = \frac{1}{\sqrt{2\pi\hbar}}\langle\psi(t)|x\rangle\,\langle p|\psi(t)\rangle\, e^{\frac{ipx}{\hbar}}. \tag{A.21}$$

Essentially, $F^{\mathrm{S}}(x,p,t)$ is the product of the position and momentum representations of the state $|\psi(t)\rangle$.

- Setting

$$f^{\mathrm{AS}}(\xi,\eta) := \frac{1}{f^{\mathrm{S}}(\xi,\eta)} \tag{A.22}$$

and thus having all $\hat{p}$-dependent terms precede the $\hat{x}$-dependent terms, one gets the *antistandard-ordered* distribution function (also known as the KIRKWOOD *distribution function* [Kir33] or RIHACZEK *distribution function* [Rih68]):

$$F^{\mathrm{AS}}(x,p,t) = \frac{1}{\sqrt{2\pi\hbar}}\langle x|\psi(t)\rangle\,\langle\psi(t)|p\rangle\, e^{-\frac{ipx}{\hbar}}. \tag{A.23}$$

It is obtained from its standard-ordered counterpart $F^{\mathrm{S}}(x,p,t)$ by complex conjugation.

[4]This slightly inaccurate naming convention is common practice. In order to avoid misunderstandings I want to stress here that in the narrow sense it is only operators that are (or are not) standard-ordered. For distribution functions (standard) ordering is not defined at all. Therefore, ***the standard-ordered distribution is not standard-ordered.*** Similar statements hold for all the other orderings of operators and their associated distribution functions.

- For systems that can be described as a harmonic oscillator with mass m_0 and frequency ω_0 — in this appendix no scaling as in section 2.1 is employed, such that the parameters m_0 and ω_0 are retained in the formulae — or as an ensemble of harmonic oscillators, the *normal-ordered* and the *antinormal-ordered distribution functions* $F^{\mathrm{N}}(x,p,t)$ and $F^{\mathrm{AN}}(x,p,t)$ are useful.[5] They are defined by requiring operators to be (anti-) standard-ordered not with respect to $\hat{x}$, $\hat{p}$, but with respect to the ladder operators, i.e. with respect to the creation operator $\hat{a}^\dagger$ and the annihilation operator $\hat{a}$, where $\hat{a}$ is given by

$$\hat{a} := \frac{1}{\sqrt{2\hbar m_0\omega_0}}\left(m_0\omega_0\hat{x} + i\hat{p}\right), \tag{A.24}$$

as usual. (Equation (2.30) is obtained from this definition in the case of the scaling (1.15, 2.4), i.e. by formally setting $m_0 = \omega_0 = 1$.)

This (anti-) standard ordering of operators is achieved by using the kernels

$$f^{\mathrm{N}}\ (\xi,\eta) := \exp\left(\frac{\hbar\xi^2}{4m_0\omega_0} + \frac{\hbar m_0\omega_0\eta^2}{4}\right) \tag{A.25a}$$

$$f^{\mathrm{AN}}(\xi,\eta) := \frac{1}{f^{\mathrm{N}}(\xi,\eta)}, \tag{A.25b}$$

as can easily be confirmed by direct computation:[6]

$$f^{\mathrm{N}}\ (\xi,\eta)\, e^{i(\xi\hat{x}+\eta\hat{p})} = e^{v\hat{a}^\dagger}e^{-v^*\hat{a}} \tag{A.26a}$$

$$f^{\mathrm{AN}}(\xi,\eta)\, e^{i(\xi\hat{x}+\eta\hat{p})} = e^{-v^*\hat{a}}e^{v\hat{a}^\dagger}, \tag{A.26b}$$

where the complex parameter v is defined as

$$v(\xi,\eta) := \sqrt{\frac{\hbar}{2m_0\omega_0}}\left(-m_0\omega_0\eta + i\xi\right). \tag{A.27}$$

The corresponding distribution functions are

$$F^{\mathrm{N}}\ (x,p,t) = \frac{1}{4\pi^2}\int\!\mathrm{d}\xi\int\!\mathrm{d}\eta\int\!\mathrm{d}x'\ \langle x'|\psi(t)\rangle\ \langle\psi(t)|e^{v\hat{a}^\dagger}e^{-v^*\hat{a}}|x'\rangle\ e^{-i(\xi x+\eta p)} \tag{A.28a}$$

[5]Typical applications may be found in quantum optics; see [Vou94, KH95b] for some examples. Another field where F^{N} and F^{AN} are utilized frequently is the modelling of a heat bath by an ensemble of harmonic oscillators [Coh94, HB95, HB96]. Cf. also the monographs by Louisell [Lou64, Lou73] and Dineykhan et al. [DEGN95].

[6]For more information on (anti-) normal ordering of operators see [DEGN95].

and

$$F^{\text{AN}}(x,p,t) = \frac{1}{4\pi^2}\int \mathrm{d}\xi \int \mathrm{d}\eta \int \mathrm{d}x' \,\langle x'|\psi(t)\rangle\,\langle \psi(t)|e^{-v^*\hat{a}}e^{v\hat{a}^\dagger}|x'\rangle\; e^{-i(\xi x+\eta p)}. \tag{A.28b}$$

The normal-ordered distribution F^{N} is also called GLAUBER-SUDARSHAN *distribution function* or *P-function* [Gla63b, Gla65, Sud63], while the antinormal-ordered distribution F^{AN} is sometimes referred to as the *Q-function* [Gla65].

Defining $|x',p'\rangle$ as a Gaussian wave packet centered at $(x',p')^t$ in phase space,[7]

$$\langle x\,|x',p'\rangle \;=\; \sqrt[4]{\frac{m_0\omega_0}{\pi\hbar}}\,e^{-\frac{m_0\omega_0}{2\hbar}}(x-x')^2\,e^{\frac{ip'x}{\hbar}} \tag{A.29}$$

with $x',p'\in\mathbb{R}$, the antinormal-ordered distribution function can be written in a particularly concise way:

$$F^{\text{AN}}(x,p,t) \;=\; \frac{1}{2\pi\hbar}\,|\,\langle x,p\,|\psi(t)\rangle\,|^2, \tag{A.30}$$

as is easily verified by substituting formula (A.29) into equation (A.30) and comparing the result with the definition (A.13) for $f(\xi,\eta)=f^{\text{AN}}(\xi,\eta)$.

Therefore, F^{AN} is essentially obtained by computing the convolution of the state $|\psi(t)\rangle$ with a Gaussian in position space. The physical meaning of this convolution becomes clearer in section A.5. From equation (A.30) it is also clear that F^{AN} is non-negative; this is of importance for the interpretation of F^{AN} as a (quasi-) probability distribution function in section A.5. F^{AN} in the form of equation (A.30) finds its main application in the discussion of the HUSIMI distribution function in section A.3.

Since there are infinitely many different functions $f(\xi,\eta)$ that can be chosen as kernel functions, there exist equally many different distribution functions $F^f(x,p,t)$. The important point is that all these different F^f are equivalent: each of them contains the same information about the state $|\psi(t)\rangle$, and each F^f can be used to compute the expectation values (A.9a) of any operator $\hat{A}(\hat{x},\hat{p})$ that is expanded as in equation (A.7).

The connection between the WIGNER distribution function and the antinormal-ordered distribution function is an important example:

$$F^{\text{AN}}(x,p,t) \;=\; \frac{1}{\pi\hbar}\int \mathrm{d}x' \int \mathrm{d}p'\; e^{-\frac{m_0\omega_0}{\hbar}(x-x')^2-\frac{1}{\hbar m_0\omega_0}(p-p')^2}\,F^{\text{W}}(x',p',t). \tag{A.31}$$

[7] A Gaussian wave packet like $|q',p'\rangle$ is often referred to as a *coherent state*. Such a state is characterized by $\langle\hat{x}\rangle = x'$, $\langle\hat{p}\rangle = p'$, $\Delta\hat{x} = \sqrt{\hbar/(2m_0\omega_0)}$ and $\Delta\hat{p} = \sqrt{\hbar m_0\omega_0/2}$, such that $|q',p'\rangle$ is a state with minimum uncertainty product: $(\Delta\hat{x})(\Delta\hat{p}) = \hbar/2$. In section A.3 I discuss coherent states from a more general point of view and an important generalization of this concept.

This unveils the antinormal-ordered distribution function as the convolution of the WIGNER function with a Gaussian in phase space; I discuss this fact in some more detail in section A.5.

Formula (A.31) may be proved in the following way: using similar arguments as those leading to equation (A.13), with the definition (A.6) I first rewrite $F^{\rm AN}(x,p,t)$ as

$$F^{\rm AN}(x,p,t) = \frac{1}{4\pi^2}\int {\rm d}\xi \int {\rm d}\eta \, \left\langle f^{\rm AN}(\xi,\eta)\, e^{i(\xi\hat{x}+\eta\hat{p})}\right\rangle_t \, e^{-i(\xi x+\eta p)} \tag{A.32}$$

and obtain for the expectation value in the integrand:

$$\begin{aligned}\left\langle f^{\rm AN}(\xi,\eta)\, e^{i(\xi\hat{x}+\eta\hat{p})}\right\rangle_t &= f^{\rm AN}(\xi,\eta)\left\langle f^{\rm W}(\xi,\eta)\, e^{i(\xi\hat{x}+\eta\hat{p})}\right\rangle_t \\ &= f^{\rm AN}(\xi,\eta)\int {\rm d}x' \int {\rm d}p' \, e^{i(\xi x'+\eta p')} F^{\rm W}(x',p',t);\end{aligned} \tag{A.33}$$

equation (A.31) then follows by integration over ξ and η.

Other conversion formulae between arbitrary different distribution functions $F^{f_1}(x,p,t)$ and $F^{f_2}(x,p,t)$ with $f_1 \neq f_2$ can be obtained by similar computations. Some formulae of this type are listed in [Lee95].

A.3 Minimum Uncertainty States and the Husimi Distribution Function

In this section I return to the issue of coherent states which have been mentioned merely in passing in section A.2. After a brief account of GLAUBER's theory of coherent states [Gla63a, Gla63b, Gla63c, Gla65, Gla66] I discuss an important generalization and its application in the theory of quantum phase space distribution functions.

A.3.1 Coherent States

A *coherent state* $|\alpha\rangle$ is an eigenstate of the annihilation operator $\hat{a}$ with respect to the eigenvalue $\alpha \in \mathbb{C}$:

$$\hat{a}\,|\alpha\rangle = \alpha\,|\alpha\rangle\,. \tag{A.34}$$

These states $|\alpha\rangle$ are especially useful when dealing with operators that are expanded in terms of the ladder operators $\hat{a}^\dagger$, $\hat{a}$: an operator in this form can easily be applied to a coherent state by treating $\hat{a}$ as the c-number α, i.e. by evaluating expressions of the form $f(\hat{a})\,|\alpha\rangle = f(\alpha)\,|\alpha\rangle$ and $\langle\alpha|\,g(\hat{a}^\dagger) = \langle\alpha|\,g(\alpha^*)$. Situations like this are frequently encountered in quantum optics. An example for the application of coherent states in this field is the modelling of stationary vibrational states of a laser-driven ion in an ion trap [DMFV96a, DMFV96b].

An explicit formula for coherent states can be found by expanding $|\alpha\rangle$ in terms of the eigenstates of the harmonic oscillator, or *number states* $|n\rangle$, $n \in \mathbb{N}_0$, substituting into the definition (A.34) and equating the coefficients of the $|n\rangle$. The result is

$$|\alpha\rangle \;=\; e^{-\frac{1}{2}|\alpha|^2} \sum_{n=0}^{\infty} \frac{1}{\sqrt{n!}} \alpha^n \,|n\rangle \tag{A.35}$$

for all $\alpha \in \mathbb{C}$. Here, the factor $e^{-\frac{1}{2}|\alpha|^2}$ is chosen in such a way that $|\alpha\rangle$ is normalized, since for all $\alpha, \beta \in \mathbb{C}$ one has:

$$\langle\alpha\,|\beta\rangle \;=\; e^{\alpha^*\beta-\frac{1}{2}(|\alpha|^2+|\beta|^2)}, \tag{A.36}$$

and therefore $\langle\alpha\,|\alpha\rangle = 1$. Equations (A.34) and (A.35) show that every $\alpha \in \mathbb{C}$ is an eigenvalue of $\hat{a}$ and thus defines a coherent state $|\alpha\rangle$. Note that the $|\alpha\rangle$ are not pairwise orthogonal, as the scalar product $\langle\alpha\,|\beta\rangle$ is not zero for $\alpha \neq \beta$.

In order to come to a more intuitive understanding of coherent states it is useful to consider WEYL's unitary *displacement operator* [MM71]

$$\hat{D}(\alpha) \;:=\; e^{\alpha\hat{a}^\dagger-\alpha^*\hat{a}} \;=\; e^{\frac{i}{\hbar}(p'\hat{x}-x'\hat{p})} \tag{A.37}$$

with the real parameters x' and p', which are essentially the real and imaginary parts of α:

$$\alpha \;=\; \alpha(x',p') \;=:\; \frac{1}{\sqrt{2\hbar m_0\omega_0}}\,(m_0\omega_0 x' + ip')\,. \tag{A.38}$$

Formally, α is obtained from the definition of the annihilation operator (A.24) by exchanging the operators $\hat{x}$, $\hat{p}$ for the scalars x', p'.

As indicated by its name, the displacement operator acts on a state by shifting it in phase space. Using the BAKER-CAMPBELL-HAUSDORFF formula (A.4) one can derive the product representation

$$\hat{D}(\alpha) \;=\; e^{-\frac{i}{2\hbar}x'p'}\,e^{\frac{i}{\hbar}p'\hat{x}}\,e^{-\frac{i}{\hbar}x'\hat{p}} \tag{A.39}$$

of $\hat{D}(\alpha)$, which turns out to be the composition of the two translation operators $e^{-\frac{i}{\hbar}x'\hat{p}}$ and $e^{\frac{i}{\hbar}p'\hat{x}}$ — in addition to these there is also the phase factor $e^{-i\frac{x'p'}{2\hbar}}$ which is irrelevant for the interpretation of $\hat{D}(\alpha)$. The translation operators each move a wave packet in position and momentum space by x' and p', respectively,

$$\left\langle x \middle| e^{-\frac{i}{\hbar}x'\hat{p}} \middle| \psi \right\rangle = \langle x - x' | \psi \rangle \tag{A.40a}$$

$$\left\langle p \middle| e^{\frac{i}{\hbar}p'\hat{x}} \middle| \psi \right\rangle = \langle p - p' | \psi \rangle , \tag{A.40b}$$

such that in phase space

$$\hat{D}(x', p') := \hat{D}\big(\alpha(x', p')\big) \tag{A.41}$$

moves the wave packet by the vector $(x', p')^t$. The translation operator $\hat{T}(\cdot)$ of equation (A.11) is a special case of the more general $\hat{D}(\cdot, \cdot)$: $\hat{T}(x') = \hat{D}(x', 0)$.

Using the displacement operator one can now write a coherent state as

$$|\alpha\rangle = e^{-\frac{1}{2}|\alpha|^2} e^{\alpha \hat{a}^\dagger} |0\rangle = \hat{D}(\alpha) |0\rangle . \tag{A.42}$$

This implies that all coherent states can be generated by $\hat{D}(\alpha)$ acting on the coherent state defined by $\alpha = 0$. This particular coherent state $|\alpha = 0\rangle$ is identical with the ground state $|n = 0\rangle$ of the harmonic oscillator, as can be concluded from equation (A.35).

With this result it is now clear how the set of all coherent states $\{|\alpha\rangle, \alpha \in \mathbb{C}\}$ is to be interpreted: it is obtained by moving the ground state of the harmonic oscillator to all points of the phase plane. Therefore the well-known properties of $|n = 0\rangle$,

$$\langle x | 0 \rangle = \sqrt[4]{\frac{1}{\pi \sigma^2}}\, e^{-\frac{1}{2\sigma^2}x^2} \quad \text{with} \quad \sigma^2 = \frac{\hbar}{m_0 \omega_0} \tag{A.43}$$

(see equation (2.39)), carry over to each of the coherent states. With the parameters x', p' of equation (A.38) one then has for all states $|\alpha\rangle$:

$$\langle \hat{x} \rangle = x' \tag{A.44a}$$

$$\langle \hat{p} \rangle = p' \tag{A.44b}$$

$$\Delta \hat{x} = \frac{1}{\sqrt{2}} \sigma \tag{A.44c}$$

$$\Delta \hat{p} = \frac{1}{\sqrt{2}} \frac{\hbar}{\sigma} . \tag{A.44d}$$

In particular, the coherent states are *minimum uncertainty states*, i.e. they are characterized by the smallest possible product of standard deviations of the position and momentum operators as given by the HEISENBERG uncertainty relation:

$$(\Delta\hat{x})(\Delta\hat{p}) = \frac{\hbar}{2}. \tag{A.45}$$

In addition, setting $m_0\omega_0 = 1$ (for example by a suitable scaling as in section 2.1) one obtains coherent states for which the standard deviations of position and momentum are the same:

$$\Delta\hat{x} = \Delta\hat{p} = \sqrt{\frac{\hbar}{2}}. \tag{A.46}$$

In the form (A.35) coherent states were first constructed by SCHRÖDINGER, who intended to use them to demonstrate the "continuous transition from micro- to macromechanics" [Sch26]. He showed that a wave packet (A.35), when specified as the initial state submitted to the potential of the harmonic oscillator, does not broaden during the dynamics.[8] With the according Hamiltonian

$$\hat{H}_{\text{ho}} = \frac{1}{2m_0}\hat{p}^2 + \frac{1}{2}m_0\omega_0^2\hat{x}^2 = \hbar\omega_0\left(\hat{a}^\dagger\hat{a} + \frac{1}{2}\right) \tag{A.47}$$

($\hat{H}_{\text{ho}}$ is identical with $\hat{H}_{\text{free}}$ of equations (2.29a) and (2.31) before scaling) one obtains for the time evolution of $|\alpha\rangle$ after time t:

$$e^{-\frac{i}{\hbar}\hat{H}_{\text{ho}}t}|\alpha\rangle = e^{-\frac{i}{2}\omega_0 t}\left|e^{-i\omega_0 t}\alpha\right\rangle, \tag{A.48}$$

that is, again a (different) coherent state, except for a phase factor. Therefore the uncertainties $\Delta\hat{x}$, $\Delta\hat{p}$ are conserved, which is just the expected behaviour for a classical particle.[9] Further information about the time evolution of coherent states may be found in [Ger92].

[8] But note that SCHRÖDINGER's generalizing interpretation in [Sch26] of this feature of the harmonic oscillator is erroneous. See [Str01b] for some background material on this issue.

[9] Moreover, for the present case of the harmonic oscillator EHRENFEST's theorem allows to conclude that the dynamics of the mean values of the position and momentum operators coincide with the classical dynamics of the observables x and p, namely a harmonic oscillation bounded by turning points which are determined by the energy $E = \hbar\omega_0|\alpha|^2$:

$$\langle\hat{x}\rangle = \sqrt{\frac{2E}{m_0\omega_0^2}}\cos(\omega_0 t - \varphi_0) \tag{A.49a}$$

$$\langle\hat{p}\rangle = -\sqrt{2m_0E}\sin(\omega_0 t - \varphi_0), \tag{A.49b}$$

with the phase shift $\varphi_0 = \arg(\alpha)$.

The equations (A.39–A.43) also show that the state $|x', p'\rangle$ of equation (A.29) in fact is a coherent state up to a phase factor,

$$\langle x\,|\alpha\rangle \;=\; \sqrt[4]{\frac{m_0\omega_0}{\pi\hbar}}\,e^{-\frac{m_0\omega_0}{2\hbar}(x-x')^2}\,e^{i\frac{xp'}{\hbar}}\,e^{-i\frac{x'p'}{2\hbar}} \;=\; \langle x\,|x', p'\rangle\; e^{-i\frac{x'p'}{2\hbar}}, \tag{A.50}$$

and with equation (A.30) one has for the antinormal-ordered distribution function:

$$F^{\mathrm{AN}}(x,p,t) \;=\; \frac{1}{2\pi\hbar}\,\big|\,\langle\alpha(x,p)\,|\psi(t)\rangle\,\big|^2. \tag{A.51}$$

For this reason the antinormal-ordered distribution function is often called the *coherent state representation* of the state $|\psi(t)\rangle$ (see e.g. [ABB96]).

It is interesting to note that the set of coherent states $\{|\alpha\rangle\,, \alpha\in\mathbb{C}\}$ is *overcomplete*:

$$\frac{1}{\pi}\iint \mathrm{d}^2\alpha\;|\alpha\rangle\,\langle\alpha| \;=\; \mathbb{1} \quad\text{with}\quad \mathrm{d}^2\alpha := \mathrm{d}(\mathrm{Re}\alpha)\,\mathrm{d}(\mathrm{Im}\alpha) \tag{A.52}$$

— in contrast, for instance, to the set $\{|n\rangle\,, n\in\mathbb{N}_0\}$ of eigenstates of the harmonic oscillator, which is "only" complete. *Over*completeness of $\{|\alpha\rangle\}$ is indicated by the factor $1/\pi$ in equation (A.52). Over*completeness* of $\{|\alpha\rangle\}$ guarantees that every state $|\psi\rangle$ can be expanded into a superposition of coherent states; but due to the nonorthogonality of these states as expressed by equation (A.36) this expansion is not unique, in general.[10]

Using the overcompleteness property (A.52) of the coherent states and the expression (A.51) for F^{AN} it is easy to show that the antinormal-ordered distribution is normalized in the sense of

$$\int\limits_{-\infty}^{\infty}\mathrm{d}x\int\limits_{-\infty}^{\infty}\mathrm{d}p\,F^{\mathrm{AN}}(x,p,t) \;=\; 1. \tag{A.53}$$

For more information on coherent states I refer the reader to the specialist literature: the monograph [KS85] remains *the* standard work in this field; some more recent studies are for example [WK93, KWZ94, ZK94], where among other questions the problem of coherent states in finite-dimensional HILBERT spaces is addressed. Finally, in [Nie97a] NIETO presents an interesting overview of the historic development of the theory of coherent states as well as of the squeezed states which I discuss in the following subsection.

[10] By imposing certain additional conditions on the expansion coefficients it is still possible to achieve a unique expansion in terms of coherent states. More on this GLAUBER *expansion* can be found in [Per93].

A.3.2 Squeezed States

Besides the coherent states there exist other, more general types of states with minimum uncertainty product (A.45). Among these, the coherent states are distinguished by having the same uncertainties with respect to the position and momentum operators for $m_0\omega_0 = 1$, as expressed by equation (A.46). Therefore it seems reasonable to consider a broader class of states that satisfy equation (A.45) but not equation (A.46). These *squeezed states* [Ken27, Yue76, HG88] facilitate a very compact definition of the HUSIMI distribution function in subsection A.3.3 below. The following exposition of the subject does not begin with a discussion of standard deviations, but is organized on the analogy of subsection A.3.1. The above observations concerning $\Delta\hat{x}$ and $\Delta\hat{p}$ can then be concluded from the definitions.

In subsection A.3.1 the coherent states have been introduced as the eigenstates of the annihilation operator $\hat{a}$. The essential steps of that definition can be followed just as well with respect to the *generalized annihilation operator*

$$\hat{b} := \mu\hat{a} + \nu\hat{a}^{\dagger}, \quad \mu, \nu \in \mathbb{C} \tag{A.54}$$

as a replacement for $\hat{a}$ — the results of subsection A.3.1 may then be obtained by restricting to the special case $\mu = 1$, $\nu = 0$. As in that case I discuss the eigenstates $|\beta\rangle_{\mathrm{s}}$ of $\hat{b}$ with respect to the eigenvalue $\beta \in \mathbb{C}$:

$$\hat{b}\,|\beta\rangle_{\mathrm{s}} = \beta\,|\beta\rangle_{\mathrm{s}}\,. \tag{A.55}$$

The motivation for referring to these states as *squeezed states*[11] is outlined below on page 207.

Again, an explicit formula for $|\beta\rangle_{\mathrm{s}}$ may be found by expanding it with respect to the eigenstates $|n\rangle$ of the harmonic oscillator. From the ansatz

$$|\beta\rangle_{\mathrm{s}} = \sum_{n=0}^{\infty} c_n\,|n\rangle\,, \quad c_n \in \mathbb{C} \tag{A.56}$$

one gets for arbitrary c_0 the following recurrence relation for the coefficients c_n:

$$c_1 = \frac{\beta}{\mu}c_0 \tag{A.57a}$$

$$c_n = \frac{1}{\sqrt{n}}\frac{\beta}{\mu}c_{n-1} - \sqrt{\frac{n-1}{n}}\frac{\nu}{\mu}c_{n-2} \quad \text{for} \quad n \geq 2, \tag{A.57b}$$

[11] In a further generalizing step *higher order squeezed states* have been introduced as the eigenstates of powers of the generalized ladder operators $\hat{b}^n$ and $(\hat{b}^{\dagger})^m$ with $n, m \geq 2$. See [Mar97, Nie97b] and references therein, where some applications are discussed as well.

where the case $\mu = 0$ has to be excluded; I do not discuss this particular case here any further, since for $\mu = 0$ normalizable eigenstates of $\hat{b}$ do not exist anyway.

If μ, ν satisfy $|\nu| < |\mu|$ then $|\beta\rangle_{\mathrm{s}}$ can be normalized: For any $k > 2$ one can choose constants $C \in \mathbb{R}^+$ and $n_0 \in \mathbb{N}$ such that for $n > n_0$ it follows from $|c_n|^2 \leq Cn^{-k}$ that $|c_{n+1}|^2 \leq C(n+1)^{-k}|\nu|/|\mu|$. For $|\nu| < |\mu|$ one can then show by induction and application of the majorant criterion that the series $\sum_{n=0}^{\infty} |c_n|^2$ converges, thus giving the finite norm of $|\beta\rangle_{\mathrm{s}}$. In the following, c_0 is always chosen in such a way that $|\beta\rangle_{\mathrm{s}}$ is normalized. As an interim result, analogous to the findings concerning α in the previous subsection, one notes that every $\beta \in \mathbb{C}$ is an eigenvalue of $\hat{b}$, as long as $|\nu| < |\mu|$ holds.

For the above considerations any values μ and ν satisfying $|\nu| < |\mu|$ can be chosen. Moreover, the recurrence relation (A.57) shows that $|\beta\rangle_{\mathrm{s}}$ depends on the quotients β/μ and ν/μ only, such that without loss of generality either μ or ν can be chosen without any further restriction. In order to achieve as close an analogy between the operators $\hat{a}$ and $\hat{b}$ as possible, I require μ and ν to satisfy

$$|\mu|^2 - |\nu|^2 = 1, \tag{A.58}$$

whereby the condition for normalizability is automatically met, and the special case of coherent states, $\mu = 1$ and $\nu = 0$, is included as well. The choice (A.58) leads to

$$[\hat{b}, \hat{b}^\dagger] = [\hat{a}, \hat{a}^\dagger] = 1; \tag{A.59}$$

this is one example for properties that carry over from the ordinary ladder operators $\hat{a}, \hat{a}^\dagger$ to the generalized ladder operators $\hat{b}, \hat{b}^\dagger$. The analogy can be carried further by defining *generalized number states* $|n\rangle_{\mathrm{g}}$,

$$|0\rangle_{\mathrm{g}} := |0\rangle \tag{A.60a}$$

$$|n\rangle_{\mathrm{g}} := \frac{1}{\sqrt{n}} \hat{b}^\dagger |n-1\rangle_{\mathrm{g}} \quad \text{for} \quad n \geq 1, \tag{A.60b}$$

which form an orthonormal set and behave in the same way with respect to $\hat{b}$ as the ordinary $|n\rangle$ do with respect to $\hat{a}$:

$$\hat{b}^\dagger |n\rangle_{\mathrm{g}} = \sqrt{n+1}\, |n+1\rangle_{\mathrm{g}} \tag{A.61a}$$

$$\hat{b}\, |n\rangle_{\mathrm{g}} = \sqrt{n} \quad |n-1\rangle_{\mathrm{g}}. \tag{A.61b}$$

The generalized number states can be utilized to give an explicit expression for the squeezed states that is analogous to formula (A.35) for the coherent states:

$$|\beta\rangle_{\mathrm{s}} = e^{-\frac{1}{2}|\beta|^2} \sum_{n=0}^{\infty} \frac{1}{\sqrt{n!}} \beta^n |n\rangle_{\mathrm{g}}. \tag{A.62}$$

Here, the expansion coefficients are explicitly known, as opposed to the merely implicit relation (A.57). I do not make any further use of the generalized number states in the present discussion.

Using the eigenvalue equation (A.55) and the $\hat{b}$-representation of $\hat{x}$ and $\hat{p}$ that can be derived from equations (A.24) and (A.54),

$$\hat{x} = \sqrt{\frac{\hbar}{2m_0\omega_0}}\left((\mu-\nu)\hat{b}^\dagger + (\mu^*-\nu^*)\hat{b}\right) \tag{A.63a}$$

$$\hat{p} = i\sqrt{\frac{\hbar m_0\omega_0}{2}}\left((\mu+\nu)\hat{b}^\dagger - (\mu^*+\nu^*)\hat{b}\right), \tag{A.63b}$$

the uncertainties of the position and momentum operators for the squeezed state $|\beta\rangle_{\mathrm{s}}$ are obtained as

$$\Delta\hat{x} = \sqrt{\frac{\hbar}{2m_0\omega_0}}\,|\mu-\nu| \tag{A.64a}$$

$$\Delta\hat{p} = \sqrt{\frac{\hbar m_0\omega_0}{2}}\,|\mu+\nu|, \tag{A.64b}$$

thus giving the uncertainty product

$$\frac{\hbar}{2} \leq (\Delta\hat{x})(\Delta\hat{p}) = \frac{\hbar}{2}\left|\mu^2-\nu^2\right|. \tag{A.65}$$

Therefore, for states characterized by $|\mu^2-\nu^2| = 1$ — one example again being the coherent states with $\mu = 1$, $\nu = 0$ — the uncertainty product takes on its minimum value. It is these "squeezed states in the narrow sense" that I referred to in the introduction of the present subsection.

The use of the term *squeezed states* for the eigenstates $|\beta\rangle_{\mathrm{s}}$ of $\hat{b}$, including those with uncertainty product larger than $\hbar/2$, is motivated by the comparison of equation (A.64) with the uncertainties (A.44c, A.44d) of the coherent states $|\alpha\rangle$: depending on μ and ν, the $\Delta\hat{x}$, $\Delta\hat{p}$ for $|\beta\rangle_{\mathrm{s}}$ can be made smaller than those for $|\alpha\rangle$; in other words, the former can be "squeezed" [KS95]. In addition, in general $\Delta\hat{x}$ and $\Delta\hat{p}$ are not equal even if $m_0\omega_0 = 1$; this is also contrasted by the coherent states for which equation (A.46) holds, expressing just this equality. More on squeezing — with respect to the Husimi distribution — can be found in section A.6.

For the time evolution of the squeezed states with respect to the harmonic oscillator Hamiltonian (A.47) one has — in close analogy to equation (A.48) —

$$e^{-\frac{i}{\hbar}\hat{H}_{\mathrm{ho}}t}\,|\beta\rangle_{\mathrm{s}}^{\mu,\nu} = e^{-\frac{i}{2}\omega_0 t}\left|e^{-i\omega_0 t}\beta\right\rangle_{\mathrm{s}}^{\mu,\exp(-2i\omega_0 t)\nu}, \tag{A.66}$$

where the notation $|\beta\rangle_s^{\mu,\nu}$ explicitly refers to those μ, ν with respect to which $\hat{b}$, and thereby the squeezed state as well, is defined (see equation (A.54)). Combined with the formulae (A.64) this result shows that in general the uncertainties of position and momentum of squeezed states are not constant with time, but oscillate with frequency $2\omega_0$. Thus constancy of $\Delta\hat{x}$ and $\Delta\hat{p}$ for coherent states turns out to be a remarkable special case.

In the literature (e.g. in [HG88, Lee95]), the most frequently studied special case of squeezed states is that one that finally leads to the definition of the HUSIMI distribution function. The starting point for this discussion is the attempt to rewrite the generalized annihilation operator

$$\hat{b} \;=\; \mu\hat{a} + \nu\hat{a}^\dagger \;=\; \frac{1}{\sqrt{2\hbar m_0\omega_0}}\Big(m_0\omega_0(\mu+\nu)\hat{x} + i(\mu-\nu)\hat{p}\Big) \tag{A.67}$$

in a form analogous to equation (A.24), the only difference being that the oscillator frequency ω_0 is now replaced with some $\kappa \in \mathbb{C}$:

$$\hat{b} \;=\; \frac{1}{\sqrt{2\hbar m_0\kappa}}\left(m_0\kappa\hat{x} + i\hat{p}\right). \tag{A.68}$$

From this it follows that $\mu^2 - \nu^2 = 1$ is a necessary condition for a formula (A.68) to hold. With equation (A.65) one can conclude that all squeezed states corresponding to such a $\hat{b}$ are minimum uncertainty states. And since $|\mu|^2 - |\nu|^2 = 1$ according to equation (A.58), μ and ν must be real numbers. This in turn means that κ is real as well:

$$\kappa \;=\; (\mu+\nu)^2\omega_0 \;=\; \left(1 + 2\nu^2 + 2\nu\sqrt{\nu^2+1}\right)\omega_0 \quad \text{with} \quad \nu \in \mathbb{R}. \tag{A.69}$$

The reverse is also true: squeezed states that correspond to real values of μ and ν have the minimal uncertainty product, and their associated generalized annihilation operator $\hat{b}$ can always be written as in equation (A.68). This special case of squeezed states is essential for the considerations in the next subsection.

A.3.3 The Husimi Distribution Function

With the results of the previous subsections it is now but a small step to the definition of the HUSIMI distribution function. Replacing in equation (A.27) the frequency ω_0 of the harmonic oscillator with an arbitrary $\kappa \in \mathbb{R}$ (the interrelation of which with squeezed states is illustrated by equation (A.69)), a complex number w is defined:

$$w(\xi,\eta) \;:=\; \sqrt{\frac{\hbar}{2m_0\kappa}}\left(-m_0\kappa\eta + i\xi\right). \tag{A.70}$$

At this point the discussion has moved quite a distance away from the original harmonic oscillator: direct contact with it was lost when the generalized ladder operators $\hat{b}$, $\hat{b}^\dagger$ were introduced in subsection A.3.2 in order to replace the original $\hat{a}$, $\hat{a}^\dagger$ in an *ad hoc* fashion; the oscillator frequency ω_0 has been substituted by the abstract quantity κ, and the only remaining original parameter of the oscillator is its mass m_0. Therefore it is reasonable to leave the original oscillator terminology behind altogether and define the new real parameter

$$\zeta := m_0 \kappa \tag{A.71}$$

for the sake of simplicity, as m_0 and κ enter the equations (A.68, A.70) for $\hat{b}$ and w via this product only. For reasons which are explained in section A.6 and motivated by equations (A.64), ζ is called the *squeezing parameter*. One obtains:

$$\hat{b}\,(\zeta) \quad = \quad \frac{1}{\sqrt{2\hbar\zeta}}\,(\zeta\hat{x} + i\hat{p}) \tag{A.72a}$$

$$w(\xi, \eta; \zeta) \quad = \quad \sqrt{\frac{\hbar}{2\zeta}}\,(-\zeta\eta + i\xi)\,. \tag{A.72b}$$

With these $\hat{b}$ and w replacing $\hat{a}$ and v in equation (A.28b), one finally arrives at the "celebrated" Husimi *distribution function* [Hus40]:

$$F^{\mathrm{H}}(x, p, t; \zeta) = \frac{1}{4\pi^2} \int \mathrm{d}\xi \int \mathrm{d}\eta \int \mathrm{d}x' \, \langle x' | \psi(t) \rangle \, \langle \psi(t) | e^{-w^* \hat{b}} e^{w \hat{b}^\dagger} | x' \rangle \, e^{-i(\xi x + \eta p)}. \tag{A.73}$$

$F^{\mathrm{H}}(x, p, t; \zeta)$ depends on the single parameter ζ only, and it is formally — after the above exchange of operators and parameters — identical with the antinormal-ordered distribution function (A.28b) which has *two* parameters, namely m_0 and ω_0. As in the case of F^{AN}, usually not the "complicated" formula (A.73) is used but its more concise version

$$F^{\mathrm{H}}(x, p, t; \zeta) = \frac{1}{2\pi\hbar} \left| {}_{\mathrm{s}}\langle \beta \, | \psi(t) \rangle \right|^2 \tag{A.74}$$

with

$$\beta = \beta(x, p; \zeta) =: \frac{1}{\sqrt{2\hbar\zeta}}\,(\zeta x + ip)\,, \tag{A.75}$$

which is completely analogous to equations (A.51, A.38). For the proof of equation (A.74) one uses the representation (A.62) of the squeezed states $|\beta\rangle_{\mathrm{s}}$ and follows the same steps as in the proof of equation (A.51).

For this calculation, which is an application of the formalism outlined in section A.1, the kernel function $f^{\mathrm{H}}(\xi,\eta;\zeta)$ defining $F^{\mathrm{H}}(x,p,t;\zeta)$ is needed. It is easy to see that this kernel function is

$$f^{\mathrm{H}}(\xi,\eta;\zeta) := \exp\left\{-\left(\frac{\hbar\xi^2}{4\zeta}+\frac{\hbar\zeta\eta^2}{4}\right)\right\}, \tag{A.76}$$

since with this f^{H} the general definition (A.13) of distribution functions in fact yields equation (A.73).

For the practical evaluation of the formula (A.74) for the HUSIMI function normally the position representation of the squeezed state $|\beta\rangle_{\mathrm{s}}$,

$$\langle x'\,|\beta\rangle_{\mathrm{s}} = \sqrt[4]{\frac{\zeta}{\pi\hbar}}\,e^{-\frac{\zeta}{2\hbar}(x'-x)^2}\,e^{i\frac{x'p}{\hbar}}\,e^{-i\frac{xp}{2\hbar}}\,, \tag{A.77}$$

is used which is analogous to equation (A.50).

In line with the formalism of section A.1, as a byproduct the definition (A.76) in a natural way leads to the introduction of yet another distribution function which is the counterpart of the HUSIMI function in the same way in which the antinormal-ordered distribution function is the counterpart of the normal-ordered distribution: with the definition

$$f^{\mathrm{AH}}(\xi,\eta;\zeta) := \frac{1}{f^{\mathrm{H}}(\xi,\eta;\zeta)} \tag{A.78}$$

an "anti-HUSIMI distribution function" $F^{\mathrm{AH}}(x,p,t;\zeta)$ is obtained that — to the best of my knowledge — has not yet been mentioned in the literature (although LEE comes close to F^{AH} when he defines the "anti-HUSIMI transform" of an operator [Lee95]).

Finally, it remains to specify the operator orderings themselves that are associated with these distribution functions. Not surprisingly it turns out that the anti-HUSIMI and the HUSIMI distribution functions correspond to standard ordering and anti-standard ordering with respect to the generalized ladder operators $\hat{b}^\dagger$, $\hat{b}$, respectively:

$$f^{\mathrm{AH}}(\xi,\eta;\zeta)\,e^{i(\xi\hat{x}+\eta\hat{p})} = e^{w\hat{b}^\dagger}e^{-w^*\hat{b}} \tag{A.79a}$$

$$f^{\mathrm{H}}\;(\xi,\eta;\zeta)\,e^{i(\xi\hat{x}+\eta\hat{p})} = e^{-w^*\hat{b}}e^{w\hat{b}^\dagger}. \tag{A.79b}$$

In section A.5 I discuss in some more detail the significance of the parameter ζ and how this parameter can be used in the course of the analysis of a physical state $|\psi\rangle$.

A.4 Dynamics

Since $F^f(x,p,t)$, for each f, contains exactly the same information as the HILBERT space vector $|\psi(t)\rangle$ it is possible to give a complete formulation of quantum mechanics by means of phase space distribution functions only, that is without referring to quantum state vectors or wave functions. In such a framework $F^f(x,p,t)$ describes the state of the system at time t, whereas the time evolution of $F^f(x,p,t)$ is determined by the equation of motion

$$\begin{aligned}\frac{\partial F^f(x,p,t)}{\partial t} &= \frac{2}{\hbar}\,\frac{f\left(-i\frac{\partial}{\partial x_2}-i\frac{\partial}{\partial x_1},-i\frac{\partial}{\partial p_2}-i\frac{\partial}{\partial p_1}\right)}{f\left(-i\frac{\partial}{\partial x_2},-i\frac{\partial}{\partial p_2}\right)f\left(-i\frac{\partial}{\partial x_1},-i\frac{\partial}{\partial p_1}\right)}\times\\ &\quad\times\sin\left\{\frac{\hbar}{2}\left(\frac{\partial}{\partial x_1}\frac{\partial}{\partial p_2}-\frac{\partial}{\partial x_2}\frac{\partial}{\partial p_1}\right)\right\}\times\\ &\quad\times\tilde{H}^f(x_1,p_1)\,F^f(x_2,p_2,t)\Big|_{x_1=x_2=x,\,p_1=p_2=p}\,,\end{aligned}\qquad\text{(A.80)}$$

which takes the role normally occupied by the SCHRÖDINGER equation in the conventional formulation of quantum mechanics. Here, $\tilde{H}^f(x,p)$ is the f *transform* of the Hamiltonian; it is obtained by first ordering the Hamiltonian $\hat{H}(\hat{x},\hat{p})$ according to the kernel function $1/f(-\xi,-\eta)$, as described in section A.1, and then replacing the operators $\hat{x}$, $\hat{p}$ with the scalars x, p. COHEN was the first to formulate the equation of motion (A.80) [Coh66]; a derivation starting from the VON NEUMANN equation can be found in [Lee95].

As an example, for the special case of the WIGNER distribution function equation (A.80) becomes

$$\begin{aligned}\frac{\partial F^{\mathrm{W}}(x,p,t)}{\partial t} &= \frac{2}{\hbar}\sin\left\{\frac{\hbar}{2}\left(\frac{\partial}{\partial x_1}\frac{\partial}{\partial p_2}-\frac{\partial}{\partial x_2}\frac{\partial}{\partial p_1}\right)\right\}\times\\ &\quad\times\tilde{H}^{\mathrm{W}}(x_1,p_1)\,F^{\mathrm{W}}(x_2,p_2,t)\Big|_{x_1=x_2=x,\,p_1=p_2=p}\,.\end{aligned}\qquad\text{(A.81)}$$

Obviously, the equations of motion (A.80) and (A.81) are quite awkward to work with. This is the most important reason why in practical applications quantum mechanics hardly ever is discussed with distribution functions completely replacing quantum states and wave functions. Rather, even when the final task is to obtain distribution functions, typically the well-established version of quantum mechanics is used, i.e. one starts by solving — numerically, if necessary — the

SCHRÖDINGER equation and then inserts the solution $|\psi(t)\rangle$ into equation (A.13) or (A.74), for instance, in order to compute the desired $F^f(x,p,t)$.

But in addition to the computation of distribution functions, the equation of motion for WIGNER's function is of importance for the investigation of the semiclassical approximation with $\hbar \approx 0$. For Hamiltonians of the type $H(x,p) = p^2/2m_0 + V(x)$ equation (A.81) can be ordered with respect to powers of $\hbar$:

$$\frac{\partial F^{\mathrm{W}}}{\partial t} = -\frac{p}{m_0}\frac{\partial F^{\mathrm{W}}}{\partial x} + \sum_{n=0}^{\infty} \frac{1}{(2n+1)!}\left(\frac{\hbar}{2i}\right)^{2n} \frac{\partial^{2n+1} V}{\partial x^{2n+1}} \frac{\partial^{2n+1} F^{\mathrm{W}}}{\partial p^{2n+1}}. \tag{A.82}$$

This equation was first derived by WIGNER in 1932 [Wig32]. Using the POISSON bracket $\{\cdot,\cdot\}$, it can be rewritten in the form

$$\frac{\partial F^{\mathrm{W}}}{\partial t} = \left\{H, F^{\mathrm{W}}\right\} + \mathcal{O}(\hbar^2), \tag{A.83}$$

with $\mathcal{O}(\hbar^2)$ as usual denoting terms that are at least quadratic in $\hbar$ and that give rise to the quantum mechanical corrections to the classical phase space distribution function $F(x,p,t)$ as obtained from the classical LIOUVILLE equation

$$\frac{\partial F}{\partial t} = \{H, F\}. \tag{A.84}$$

Therefore in the classical limit $\hbar = 0$ the equation of motion (A.83) for F^{W} formally becomes the LIOUVILLE equation (A.84); this makes WIGNER's function a useful tool for the investigation of the correspondence issue. Furthermore, for potentials $V(x)$ that do not contain powers of x exceeding 2, the dynamics of F^{W} is completely classical, since in this case the equations (A.83) and (A.84) coincide. But note that F^{W} itself depends on $\hbar$, such that the similarity of these quantum mechanical and classical equations of motion is formal in the first place, and the details of obtaining equation (A.84) from (A.83) in the classical limit are nontrivial in general.

I return to equations (A.82) and (A.83) in section A.5 when I discuss the interpretation of $F^f(x,p,t)$ as a (quasi-) probability distribution function in quantum phase space.

A.5 On the Interpretation as Probability Densities

The motivation of the theory outlined in this appendix was to construct distribution functions which can play the same role in quantum mechanics as the conventional phase space probability densities do in classical mechanics (cf. pages 191ff). Therefore it now needs to be checked to what extent the functions $F^f(x,p,t)$ actually have the typical properties of probability densities.

A "genuine" phase space probability density $F(x,p,t)$ is characterized by the following properties:[12]

(i) Reality: $F(x,p,t) \in \mathbb{R}$ for all x, p and t.

(ii) Positivity: $F(x,p,t) \geq 0$ for all x, p and t.

(iii) By integration over p and x, $F(x,p,t)$ must give the correct quantum mechanical *marginal probability densities* in position and momentum space, respectively:[13]

$$\int \mathrm{d}p \; F(x,p,t) = |\langle x \,|\psi(t)\rangle\,|^2 \tag{A.85a}$$

$$\int \mathrm{d}x \; F(x,p,t) = |\langle p \,|\psi(t)\rangle\,|^2 . \tag{A.85b}$$

The first two conditions are natural, since all probabilities that are to be computed using F must be real-valued and non-negative as well. The third condition is necessary in order to obtain agreement with the usual Copenhagen interpretation of quantum mechanics, where the wave functions $\langle x \,|\psi(t)\rangle$ and $\langle p \,|\psi(t)\rangle$ take the role of probability amplitudes.

The properties (i)–(iii) are not automatically satisfied for all distribution functions F^f, but must be checked for each individual f. They are by no means necessary consequences of the definitions in sections A.1 and A.2. On the contrary, it can be shown that for nonzero $\hbar$ *none* of the known F^f satisfies all three conditions. This may be interpreted as a consequence of HEISENBERG's uncertainty relation and was already observed by WIGNER in 1932 [Wig32]. This shortcoming of all distribution functions F^f is the reason for using the prefix *quasi* in terms like quasiprobability density or quasiprobability distribution function.

WIGNER's distribution function, for example, is real-valued and gives the correct marginal probability distributions for position and momentum, but it can take on negative values. In fact, it typically oscillates very rapidly and with large

---

[12] There are further, more technical conditions that are to be satisfied by a probability density [Lee95]; these additional properties are not needed for the present discussion.

[13] Equations (A.85a, A.85b) are just two specific examples for the marginalization of a distribution function. Generally speaking, a *marginalization* of a distribution function F is a line integral over F in phase space. In equations (A.85a, A.85b), the line integrals are taken along the p and x axes of phase space, respectively. More on marginalizations of distribution functions may be found in [MMT97].

amplitude between positive and negative values under variation of x or p. This behaviour of the WIGNER function is further discussed and demonstrated for some examples in section A.6.

For the antinormal-ordered and the HUSIMI distribution functions one has, using equations (A.51, A.74),

$$0 \leq F^{\mathrm{AN}}(x,p,t) \leq \frac{1}{2\pi\hbar} \tag{A.86a}$$

$$0 \leq F^{\mathrm{H}}(x,p,t;\zeta) \leq \frac{1}{2\pi\hbar}, \tag{A.86b}$$

respectively. Both are real-valued and non-negative. They are even bounded and do not oscillate as rapidly as F^{W}. But, on the other hand, they do not yield the correct probability densities for x and p by marginalization. Again I refer the reader to section A.6 for more details. Table A.1 gives an overview of the essential properties of the distribution functions defined in sections A.2 and A.3.

It is important to keep in mind that despite the different properties of the distribution functions, they are all equivalent in the sense that all of them contain exactly the same information about the state of the system. This can be exploited by choosing that type of distribution function F^f (respectively by choosing that kernel function f) that is most suited to study the particular problem in question.

For the purpose of comparing the classical and quantum mechanical *equations of motion* of a given system, one usually uses WIGNER's function (cf. equation (A.82); see [Bun95] for details). F^{W} is also especially well suited for the

distribution function	real-valued	non-negative	correct marginal distributions
$F^{\mathrm{W}}(x,p,t)$	yes	no	yes
$F^{\mathrm{S}}(x,p,t)$	no	no	yes
$F^{\mathrm{AS}}(x,p,t)$	no	no	yes
$F^{\mathrm{N}}(x,p,t)$	yes	no	no
$F^{\mathrm{AN}}(x,p,t)$	yes	yes	no
$F^{\mathrm{H}}(x,p,t;\zeta)$	yes	yes	no
$F^{\mathrm{AH}}(x,p,t;\zeta)$	yes	no	no

Table A.1: Properties of the WIGNER, standard-ordered, antistandard-ordered, normal-ordered, antinormal-ordered, HUSIMI and anti-HUSIMI distribution functions.

discussion of scattering systems, because in such systems, for large energies, the classical limiting case often is a good approximation to the exact quantum system already; equation (A.82) can then be used to determine the quantum correction terms in a systematic way. On the other hand, if the quantum and classical *phase space (quasi-) probability densities* themselves are in the focus of attention, then most frequently the HUSIMI distribution function F^{H} or the coherent state representation F^{AN} are employed. In section A.6 I discuss some of the advantages of this choice by considering two familiar quantum states as examples.

A.6 Typical Applications

The first point to note when it comes to the practical application of phase space distribution functions is that their numerical computation for a given quantum state $|\psi(t)\rangle$ does not cost as much effort as it might seem at first sight of equations like (A.13). For example, the WIGNER function can be written as

$$F^{\mathrm{W}}(x,p,t) = \frac{1}{\sqrt{2\pi}} \int \mathrm{d}\eta \; \Phi(\eta;x,t)\, e^{-i\eta p} \tag{A.87a}$$

with

$$\Phi(\eta;x,t) = \frac{1}{\sqrt{2\pi}} \left\langle \psi(t) \left| x - \tfrac{1}{2}\hbar\eta \right\rangle \left\langle x + \tfrac{1}{2}\hbar\eta \right| \psi(t) \right\rangle , \tag{A.87b}$$

and for the HUSIMI distribution one has

$$F^{\mathrm{H}}(x,p,t;\zeta) = \sqrt{\frac{\hbar\zeta}{\pi}} \left| \frac{1}{\sqrt{2\pi}} \int \mathrm{d}\eta \; \Phi(\eta;x,t,\zeta)\, e^{-i\eta p} \right|^2 \tag{A.88a}$$

with

$$\Phi(\eta;x,t,\zeta) = e^{-\frac{\zeta}{2\hbar}(x-\hbar\eta)^2} \left\langle \hbar\eta \left| \psi(t) \right\rangle \right. . \tag{A.88b}$$

These expressions show that both F^{W} and F^{H} can be computed efficiently from $\langle x | \psi(t) \rangle$ by FOURIER transformation, and the same is true for other F^f as well. Typically the method of *fast* FOURIER *transformation* (FFT) [CT65, EMR93] is employed here which drastically reduces the numerical effort, as compared with conventional methods of computation.[14]

[14]Using the FFT method for an $n \times n$ grid in phase space, the number of necessary complex operations is reduced from $\mathcal{O}(n^3)$ to $\mathcal{O}(n^2 \log_2 n)$ [Sto99, PTVF94].

In the literature, initially mainly the WIGNER function has been used. But in the last years the situation has changed in favour of the HUSIMI function which is in widespread use by now. In the following I want to motivate this transition with two arguments, the first being analytical while the second is a numerical example of a typical application.

In the context of HUSIMI functions, equation (A.31) can be written as

$$F^{\mathrm{H}}(x,p,t;\zeta) = \frac{1}{\pi\hbar}\int \mathrm{d}x' \int \mathrm{d}p' \, G(x-x',p-p';\zeta) F^{\mathrm{W}}(x',p',t), \tag{A.89}$$

where $G(x,p;\zeta)$ is a Gaussian in phase space:

$$G(x,p;\zeta) = e^{-\frac{\zeta}{\hbar}x^2 - \frac{1}{\hbar\zeta}p^2}. \tag{A.90}$$

The convolution (A.89) of the WIGNER function with $G(x,p;\zeta)$ effectively is an averaging of $F^{\mathrm{W}}(x,p,t)$ in the neighbourhood of the phase space point $(x,p)^t$, whereby the spatially rapidly oscillating WIGNER function gets smoothed. One result in particular of this smoothing process is the fact that the HUSIMI function is non-negative, as opposed to the WIGNER function. Similarly, the smoothing process also transforms the (in general) unbounded WIGNER function into a bounded function (see equation (A.86b)).

The numerical example below shows that the rapid oscillations of the WIGNER function have no obvious counterpart in the classical phase portrait, whereas the phase space patterns of the HUSIMI function follow the classical structures much more closely. In this sense the smoother HUSIMI function can be interpreted as a (quasi-) probability density in a much more intuitive way than the WIGNER function. This property makes F^{H} the first choice distribution function for the purpose of visualization of quantum nonlinear dynamical systems, especially when the classical dynamics is chaotic.

A parallel argument explains why the normal-ordered distribution function is utilized rarely and why the anti-HUSIMI distribution function is not used at all for creating an intuitive picture of a quantum state. Similar to equation (A.89) one has

$$F^{\mathrm{W}}(x,p,t) = \frac{1}{\pi\hbar}\int \mathrm{d}x' \int \mathrm{d}p' \, G(x-x',p-p';\zeta)\, F^{\mathrm{AH}}(x',p',t;\zeta) \tag{A.91}$$

(where one again obtains the normal-ordered case $F^{\mathrm{N}}(x,p,t)$ from $F^{\mathrm{AH}}(x,p,t;\zeta)$ by setting $\zeta = m_0\omega_0$), which implies that the WIGNER function itself can be viewed as a smoothed version of the anti-HUSIMI distribution:

$$F^{\mathrm{AH}} \xrightarrow[\text{via equ. (A.91)}]{\text{averaging}} F^{\mathrm{W}} \xrightarrow[\text{via equ. (A.89)}]{\text{averaging}} F^{\mathrm{H}}. \tag{A.92}$$

The anti-Husimi distribution is thus even rougher than the Wigner function and accordingly counterintuitive.

Finally there exists a strong argument in favour of the Husimi distribution from the experimenters' point of view: Takahashi and Saitô argue that F^{H} is much better suited to any experimental setup than F^{W} since the averaging in equation (A.89) acts in a similar way as the coarse graining effect that is inherent to all experimental measurement processes [TS85]. In a particular experimental setup, on the other hand, the Wigner function still has the striking advantage that *it can be measured directly* at each point of phase space, without the need for any further complex computation: in [MCK93] it is shown that, after some algebra, F^{W} can be written as

$$F^{\mathrm{W}}(x,p,t) \;=\; \frac{1}{\pi\hbar}\sum_{n=0}^{\infty}(-1)^n \left|\langle\psi(t)\big|\hat{D}\big(\alpha(x,p)\big)\big|n\rangle\right|^2 . \tag{A.93}$$

If $|\psi\rangle$ describes the radiation field, then the term $\left|\langle\psi\big|\hat{D}\big(\alpha(x,p)\big)\big|n\rangle\right|^2$ is just the probability density of detecting a photon with the energy $\hbar\omega_0(n+1/2)$ at the phase space point $(x,p)^t$. The parameter $\alpha=\alpha(x,p)$ can be controlled within the experimental setup, such that the photon count statistics for different values of α then gives the Wigner function as determined by equation (A.93). Some reports on measurements following this or related patterns of *quantum-state tomography* [Leo95, Leo96] can be found in [BW96, BW97, Jor97, BRWK99], for example.

At this point one is now in the position to further interpret the additional "squeezing" parameter ζ of the (anti-) Husimi distribution. Equations (A.89, A.91) show that ζ controls the way in which the averaging discussed above is performed, namely by specifying the width of the Gaussian (A.90) in x- and p-direction. The effective area over which is averaged is an ellipse with widths $\Delta x=\sqrt{\hbar/(2\zeta)}$ and $\Delta p=\sqrt{\hbar\zeta/2}$ in x- and p-direction, respectively — compare these values with the uncertainties (A.64) of general squeezed states. Therefore, larger values of ζ yield an averaging area that is squeezed in the direction of x and broadened in the direction of p. The particular choice $\zeta=1.0$, which is used in many applications and corresponds to the coherent state representation (A.51), accordingly specifies a circular averaging area.

In order to exemplify the above statements, in figures A.1–A.3 I compare the most important ways to graphically represent quantum states. The states used for this demonstration are the familiar ground state $|n=0\rangle$ of the harmonic oscillator and its tenth eigenstate $|n=10\rangle$.

Figure A.1a shows a classical phase space trajectory of the harmonic oscillator at the energy $E_0=\hbar\omega_0/2$. The corresponding quantum mechanical wave function in the position representation (figure A.1b) bears no obvious resemblance to the

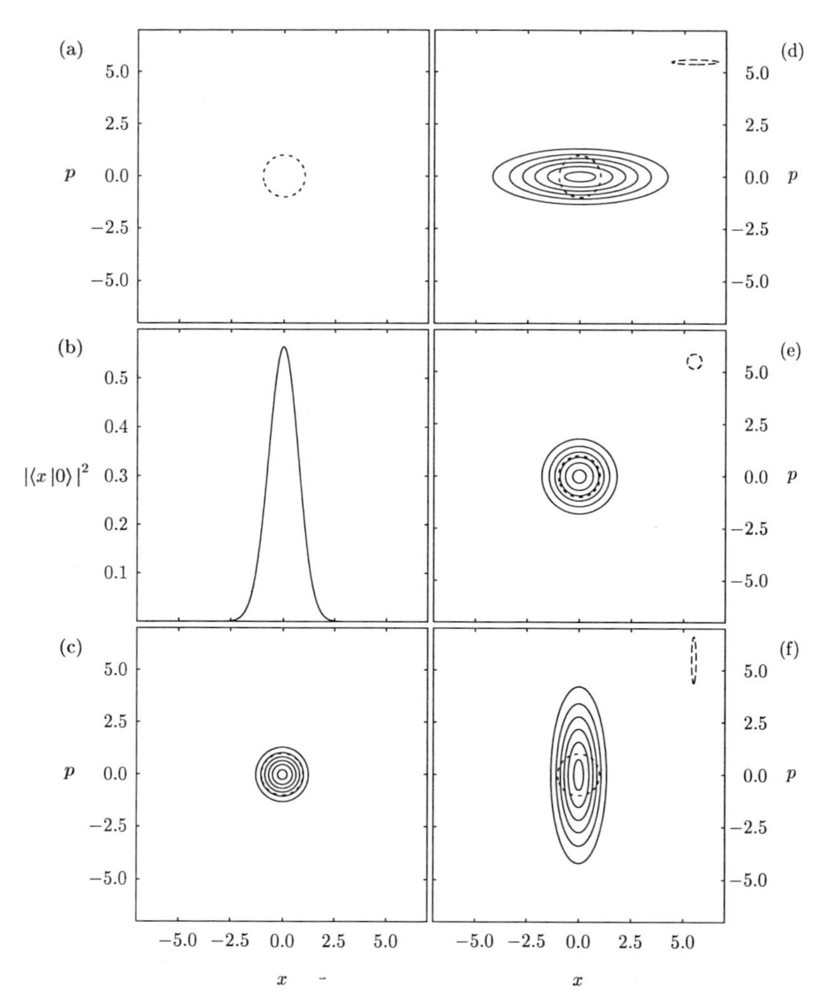

Figure A.1: The ground state $|0\rangle$ of the harmonic oscillator in different representations ($\hbar = 1.0$, $m_0 = 1.0$, $\omega_0 = 1.0$).
(a) Trajectory of the corresponding classical system in phase space. (b) Probability density $|\langle x\,|0\rangle\,|^2$ in the position representation. (c) Contour plot of the WIGNER distribution function. (d)–(f) Contour plots of the HUSIMI distribution function for $\zeta = 0.1/1.0/10.0$. In the upper right corners of the HUSIMI plots the respective averaging areas as defined by the squeezing parameter ζ are shown. In figures (c)–(f) the classical trajectory is indicated by a dashed line as in figure (a), and the contour lines are drawn at $20\%, 35\%, \ldots, 95\%$ of the respective maximum values of the distribution function.

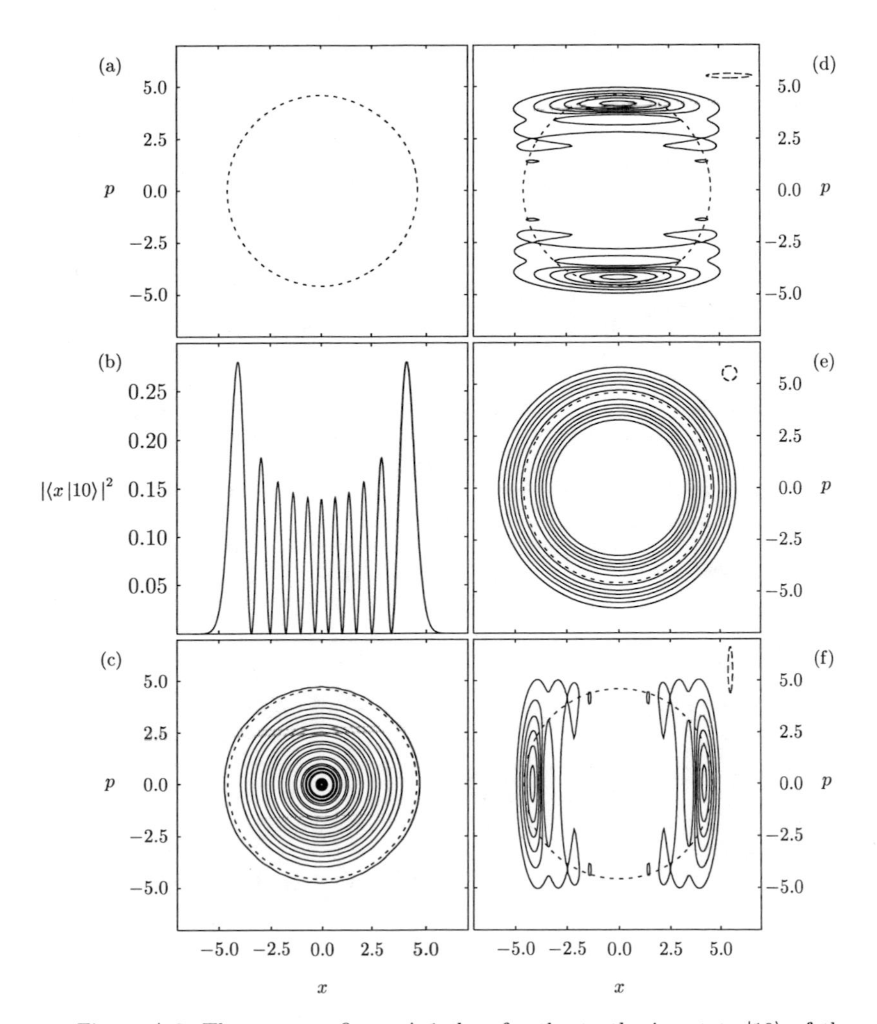

Figure A.2: The same as figure A.1, but for the tenth eigenstate $|10\rangle$ of the harmonic oscillator.

classical trajectory and is therefore of little use for the purpose of comparing the classical and quantum states. For this purpose, figure A.1c is much better suited: the contour plot of the WIGNER function of $|0\rangle$ that is shown here is essentially a bell-shaped function that is peaked at the origin of phase space — near the classical trajectory. The contour plots of the HUSIMI distributions for different values of ζ in figures A.1d–A.1f are not much more informative; all of them show bell-shaped curves centered about the origin. Note that the HUSIMI function for $\zeta = 1.0$ is the one that is closest to the classical trajectory, since for $\zeta \neq 1.0$ one obtains distributions that do not reproduce the classical rotational symmetry of the trajectory.

While the advantages of the HUSIMI distribution are not obvious with respect to the simple ground state $|0\rangle$ they become clearly visible when more complex states are considered. Figure A.2 shows the same kind of plots as figure A.1 but now for the state $|10\rangle$. The highly oscillatory character of the WIGNER function is illustrated in the contour plot A.2c and in its three-dimensional counterpart in figure A.3a. These plots do not show much similarity to the corresponding classical path in figure A.2a. F^{W} exhibits rapid oscillations in radial direction which do not correspond to any classical feature in an obvious way, and F^{W} takes on negative values for some $(x,p)^t$. These are typical properties of the WIGNER function that are observed not only with respect to $|10\rangle$ but with respect to most quantum states. In addition to these shortcomings the WIGNER function of $|10\rangle$ erroneously gives the impression of a state that is more localized than $|10\rangle$ actually is, since the larger portion of the oscillations of F^{W} show up well inside the classical path, as can be seen in figure A.2c.

Only when turning to the HUSIMI distribution in figures A.2d–A.2f — and in figures A.3b, A.3c with the 3D versions — (some aspects of) the classical trajectory can be identified in quantum phase space. This is best achieved for $\zeta = 1.0$: $F^{\mathrm{H}}(x,p;1.0)$ essentially is a circular ridge which takes on its maximum value at the points of the classical trajectory. For $\zeta = 0.1$ and $\zeta = 10.0$ rotational invariance is lost and two characteristic peaks emerge with no obvious counterpart in the classical picture. But even in these cases the location of the classical path can clearly be identified.

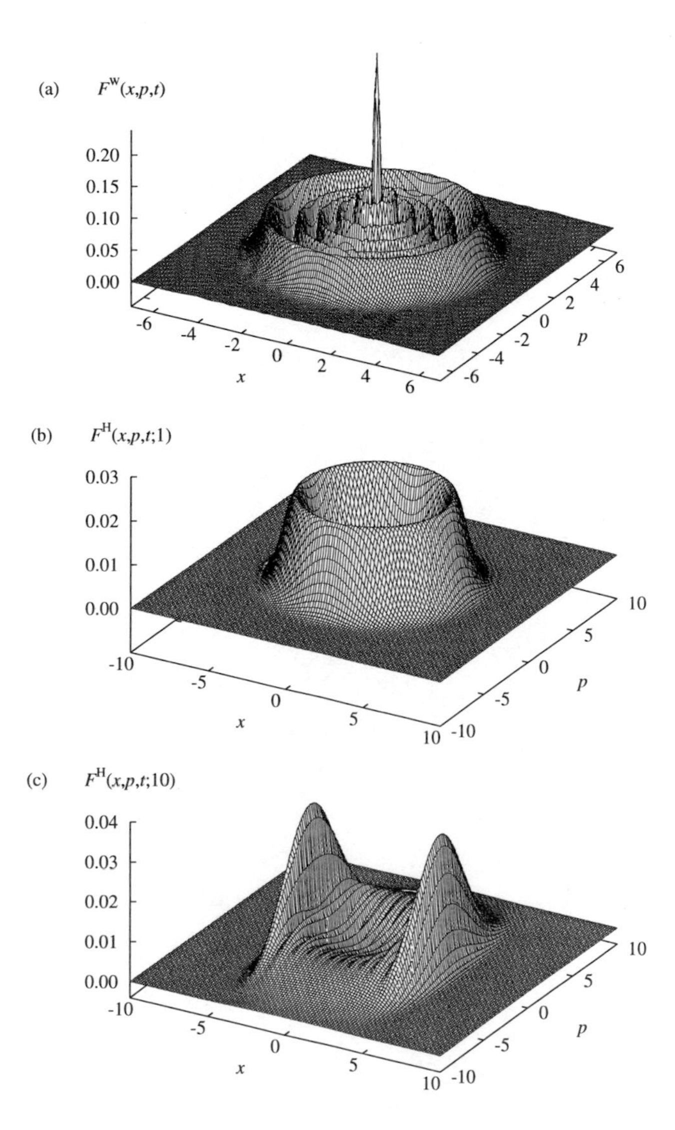

Figure A.3: Three-dimensional plots of some of the distribution functions of the tenth eigenstate $|10\rangle$ of the harmonic oscillator. Figures (a), (b) and (c) are the 3D versions of figures A.2c, A.2e and A.2f, respectively.

Appendix B

Random Products of Unimodular Matrices

It is the nature of all greatness not to be exact.

EDMUND BURKE

Using measure theoretical methods, FURSTENBERG has derived a theorem that is useful for proving ANDERSON localization on one-dimensional lattices as, for example, in chapter 5 of this study. The theorem deals with a specific class of random matrices X_m, $m \in \mathbb{N}$, and states that generically the norms of the vectors $X_m X_{m-1} \cdots X_1 \vec{u}_0$ tend to infinity at an exponential rate with $m \to \infty$, provided that certain — not very restrictive — conditions are met. The theorem is applied in chapter 5, where the role of the X_m is taken by the transfer matrices $\mathcal{T}_m$.

Consider the classical group $\mathrm{SL}(N, \mathbb{R})$ of unimodular real (N, N) matrices with the usual matrix product. A subgroup G of $\mathrm{SL}(N, \mathbb{R})$ is called irreducible if the only subspaces of $\mathbb{R}^N$ left fixed by the matrices of G are $\mathbb{R}^N$ and $\{(0, \ldots, 0)^t\}$; otherwise G is called reducible. Let μ be a measure on $\mathrm{SL}(N, \mathbb{R})$, induced by the distribution of the elements of a given set of unimodular matrices $\{X_m,\ m \in \mathbb{N}\} \subseteq \mathrm{SL}(N, \mathbb{R})$, and let G be the smallest closed subgroup of $\mathrm{SL}(N, \mathbb{R})$ that contains the support of μ.

FURSTENBERG's theorem discloses some details of the asymptotic behaviour of the norms $\|\vec{u}_m\|$ of the vectors

$$\vec{u}_m := X_m X_{m-1} \cdots X_1 \vec{u}_0, \quad m \in \mathbb{N} \tag{B.1}$$

for $\vec{u}_0 \in \mathbb{R}^N$. The vector norm $\|\cdot\|$ used here is the conventional Euclidean 2-norm and not to be confused with the matrix norm $\|\cdot\|_{\mathrm{m}}$ used below.

Theorem. If G is irreducible and μ satisfies

$$\int_G \mathrm{d}\mu(M)\, \log(\|M\|_{\mathrm{m}}) \;<\; \infty, \tag{B.2}$$

then for random matrices $\{X_m\}$ which are independently distributed according to μ, with probability one the limit

$$\lim_{m\to\infty} \frac{\log\|\vec{u}_m\|}{m} \;=:\; \gamma_\mu \tag{B.3}$$

exists for all nonzero vectors $\vec{u}_0 \in \mathbb{R}^N$.
Moreover, if G is noncompact, then γ_μ is strictly positive.

The theorem holds with probability one only, due to the random nature of the X_m. Almost all such sequences of matrices — and thus almost all distributions of matrices obtained in this way — give the desired result, but there are some that do not. The probability of coming by such an exceptional case is zero, though, such that for all practical applications the theorem can be considered to be true. The (rather technical) proof of the theorem is given in FURSTENBERG's paper [Fur63].

In a typical application one is given a set $\{X_m\}$ of unimodular matrices with a distribution μ on $\mathrm{SL}(N,\mathbb{R})$. First one has to identify the corresponding matrix group G; then irreducibility and noncompactness of G need to be confirmed. This, by FURSTENBERG's theorem, establishes the result that for almost all such sets $\{X_m\}$ with a sufficiently well-behaved measure μ, the norms of the vectors $\vec{u}_m$ grow exponentially for large m,

$$\|\vec{u}_m\| \;\sim\; e^{\gamma_\mu m} \quad \text{with} \quad \gamma_\mu > 0, \tag{B.4}$$

provided that the initial vector $\vec{u}_0$ is nonzero. The rate of growth is given by the "LIAPUNOV exponent γ_μ of the set $\{X_m\}$" and does not depend on $\vec{u}_0$. For checking the requirement (B.2) on μ, any matrix norm $\|\cdot\|_{\mathrm{m}}$ can be used; often the maximum norm $\|M\|_{\mathrm{m}} = \max\limits_{i,j} |M_{ij}|$ is most convenient to work with.

For $N = 2$, the most important example is $\mathrm{SL}(2,\mathbb{R})$ itself:

$$G_1 \;:=\; \{M=(M_{ij}),\;\; 1 \le i,j \le 2,\;\; M_{ij} \in \mathbb{R},\;\; \det(M)=1\} \tag{B.5}$$

defines the appropriate group containing all the transfer matrices (5.61) that model the tight binding equation of the kicked rotor. G_1 also plays the same role with respect to the kicked harmonic oscillator and its transfer matrices (5.114, 5.117, 5.120). Clearly, G_1 is irreducible and noncompact and thus for sufficiently well-behaved μ generically gives rise to exponentially growing $\left\|X_m X_{m-1} \cdots X_1 \vec{u}_0\right\|$ for $X_m \in G_1$. This observation is used, for example, in subsection 5.1.3. There, for the ANDERSON-LLOYD model, a measure μ is used that is generated by the Lorentzian $p(\epsilon_m)$ (5.70) with respect to the transfer matrices $\{\mathcal{T}_m\}$ of equation (5.61), such that

$$\begin{aligned}
\int_{G_1} \mathrm{d}\mu(M)\, \log(\|M\|_{\mathrm{m}}) &= \int_{-\infty}^{\infty} \mathrm{d}\epsilon_m\, p(\epsilon_m)\, \log\left(\max\left(\left|\frac{\epsilon_m}{W_1}\right|, 1\right)\right) \\
&= \frac{2}{\pi} \int_{|W_1|}^{\infty} \mathrm{d}\epsilon_m\, \frac{\log \dfrac{\epsilon_m}{|W_1|}}{1+\epsilon_m^2} \qquad \text{(B.6)} \\
&= -\log|W_1| + \frac{i}{\pi}\Big(\mathrm{Li}_2(-i|W_1|) - \mathrm{Li}_2(i|W_1|)\Big)
\end{aligned}$$

is obtained, with the dilogarithm function Li_2 [AS72]. The expression (B.6) is finite for any given nonzero W_1, such that the integrability condition (B.2) is satisfied. Analogous calculations for the measures corresponding to the two-dimensional transfer matrix models defined by equations (5.114, 5.117, 5.120) with respect to the kicked harmonic oscillator show that — in the cases considered in subsection 5.3.2 — the theorem applies to these models as well, because either the matrix elements used there are Lorentzian distributed, or at least their distribution is sufficiently localized for allowing condition (B.2) to hold.

The group given by

$$G_2 := \left\{ \begin{pmatrix} 1 & M_{12} \\ 0 & 1 \end{pmatrix},\ M_{12} \in \mathbb{R} \right\}, \qquad \text{(B.7)}$$

on the other hand, may serve as a counterexample for $N = 2$. It is easy to verify that G_2 indeed defines a subgroup of $\mathrm{SL}(2,\mathbb{R})$. But G_2 is reducible, since $\mathrm{span}\,\{\vec{e}_1\} \subset \mathbb{R}^2$ is mapped onto itself by the matrices of G_2. Therefore, the norms $\left\|X_m X_{m-1} \cdots X_1 \vec{u}_0\right\|$, $X_m \in G_2$ should not be expected to grow exponentially.

For $N = 3$, the theorem is used in subsection 5.3.3 with respect to $\mathrm{SL}(3,\mathbb{R})$ itself: the transfer matrices (5.123a) and (5.124a) belong to the group defined by

$$G_3 := \left\{ M = (M_{ij}),\ 1 \le i,j \le 3,\ M_{ij} \in \mathbb{R},\ \det(M) = 1 \right\} \qquad \text{(B.8)}$$

Again, irreducibility and noncompactness of G_3 are obvious, and FURSTENBERG's theorem applies, once the integrability condition (B.2) has been verified — as in subsection 5.3.3.

Appendix C

A Picture Book of Quantum Stochastic Webs and Localization

Our interpretation of the experimental material rests essentially on the classical concepts.

NIELS BOHR

In this appendix I present a number of typical examples of quantum stochastic webs and localized quantum dynamics that are generated by the quantum map (2.37). See section 3.3 for a more detailed description of how the quantum states $|\psi_n\rangle$ shown in these pictures are generated. In chapters 4 and 5 some important aspects of the quantum dynamics leading to these $|\psi_n\rangle$ are described and explained.

Having studied the dynamics for several different initial states $|\psi_0\rangle$ it has turned out that in the resonance cases (1.33) essentially there are just two important types of $|\psi_0\rangle$: those located in quantum phase space in one of the *meshes* of the classical stochastic web, and those centered in a stochastic region of the classical web, i.e. in a *channel* of the web. All other initial states are combinations of these two types, and regardless of the actual position of the "initial mesh" or "initial channel" chosen for $|\psi_0\rangle$, the dynamics yield comparable results. Therefore, in the following sections two types of initial states are considered: states centered around the origin $(0,0)^t$ of phase space, i.e. in a mesh, and states centered around one of those intersections of stochastic channels that are closest to the origin, at $(0,p_0)^t$ with suitable p_0.

For better comparison, depending on the value of $\hbar$, harmonic oscillator eigenstates

$$|\psi_0\rangle \;=\; |m(\hbar)\rangle \tag{C.1}$$

are chosen as initial states in such a way that the corresponding energies all take on the same value, approximately:

$$E_0 \;=\; \hbar\left(m(\hbar)+\frac{1}{2}\right) \;\approx\; \frac{1}{2} \tag{C.2}$$

(cf. equation (4.4) and the remark following that equation). Similarly, for initial states centered around $(0,p_0)^t$ rather than $(0,0)^t$

$$|\psi_0\rangle \;=\; \hat{D}(0,p_0)\,|m(\hbar)\rangle\,, \tag{C.3}$$

with the translation operator $\hat{D}(\cdot,\cdot)$ defined in equation (4.1b) and $m(\hbar)$ according to expression (C.2), is used.

In chapter 3 I have discussed the way in which the size $m_{\max}$ of the basis $\{|m\rangle \mid 0 \le m \le m_{\max}\}$ used for expanding the quantum states affects the accuracy of the algorithm. Only the phase space region

$$\left\{(x,p)^t \;\middle|\; \sqrt{x^2+p^2} \;\lesssim\; r_{\max}(\hbar,m_{\max}) = \sqrt{\hbar(2m_{\max}+1)}\right\} \tag{C.4}$$

can be expected to be well described by this basis. Table C.1 gives $r_{\max}(\hbar,m_{\max})$ for several values of $\hbar$ and $m_{\max}$. The necessity for using as large a value of $m_{\max}$ as possible in order to be able do describe a large portion of phase space is obvious. For all but a few calculations described in this study $m_{\max} = 6000$ has

$\hbar$	$r_{\max}(\hbar,3000)$	$r_{\max}(\hbar,6000)$
1.0	77.5	109.5
0.1	24.5	34.6
0.01	7.7	11.0
0.001	2.5	3.5

Table C.1: The radius $r_{\max}(\hbar,m_{\max})$ of the region of phase space that can be described using the basis $\{|m\rangle \mid 0 \le m \le m_{\max}\}$, for $m_{\max} = 3000$ and $m_{\max} = 6000$.

been used. Much larger values of m_{max} are not practical with currently available workstations, as both the execution time per kick and the computer memory needed for storing the FLOQUET matrix elements grow quadratically with m_{max}. Note that m_{max} being finite implies that the numerical algorithm cannot yield exactly periodic phase space structures, but at best can approximate them.

For the series of figures shown below, the parameters $\hbar$ and V_0 are varied more or less systematically in order to yield states which are as prototypical as possible. The states $|\psi_n\rangle$ obtained in this way are then converted into their corresponding HUSIMI distributions (cf. appendix A). The lines in the following contour plots of HUSIMI distributions are drawn at 10%, 20%, ..., 90%, 99% of the respective maximum values of $F^{\text{H}}(x,p,nT-0;1)$ for each state $|\psi_n\rangle$.

C.1 Rectangular Quantum Stochastic Webs

Classically, for $T=\pi/2$ ($q=4$) rectangular stochastic webs are obtained: cf. figures 1.7, 1.8, 1.10b and 1.12. The skeletons of these webs extend in the neighbourhood of the square grid given by equation (1.45). In the figures of the present section, this grid is displayed via thin lines, in addition to the contour lines of $F^{\text{H}}(x,p,nT-0;1)$.

C.1.1 $T=\pi/2,\ p_0=0.0$

Figures 4.2 and C.1–C.8. (Figures from chapters 4 and 5 are not repeated in this appendix.)

Figures 4.2 and C.1 nicely show how the quantum web develops with increasing number n of kicks. Obviously, for $|\psi_0\rangle$ centered in a regular region of phase space, $|\psi_n\rangle$ also tends to concentrate in the meshes of the web, rather than in the channels, where the phase space density gets transported away along the classical separatrices rapidly.

In figure C.2, $\hbar$ and V_0 are large enough to avoid the formation of a web-like structure, and the algorithm has to be stopped after only around 500 kicks, because the norm of the computed state has decayed considerably already.

In the remaining figures of this subsection, the initial states are localized enough — due to the smaller values of $\hbar$ — to yield localized states even for a very large number of kicks. An exception to this rule is displayed in figure C.5, where the large value of $V_0=30.0$ leads to the state flowing apart quickly; note that for $n \gtrsim 10000$ the HUSIMI distribution has developed a periodic structure that is quite different from the quantum web of figures 4.2 and C.1.

It is important to keep in mind that although a localized initial state, together with small enough $\hbar$ and V_0, leads to *localized* dynamics this does not rule out the existence of a *global* stochastic web: clearly, the complete web can be obtained by using an initial state with nonzero contributions to $F^{\mathrm{H}}(x, p, -0; 1)$ in each of the meshes of the classical web.

C.1.2 $T = \pi/2$, $p_0 = \pi$

Figures C.9–C.17, 4.3 and 4.7.

In the first three figures no web-like structure can be seen developing, probably due to the large value of $\hbar = 1.0$, which makes the phase space structures quite coarse-grained. This, together with the fact that the initial position $(0, \pi)^t$ lies in the heart of the classically stochastic region, may account for the lack of structure in these pictures.

The rest of the figures in this subsection and figure 4.3 convincingly show how the central portions of the stochastic web get filled by the phase space density evolving with time. This process is accelerated for larger values of V_0, as can be seen in figure C.16, for example. Here, this process takes place very fast, such that after only approximately 500 kicks the algorithm needs to be stopped, due to declining accuracy. The last figure is also interesting in that it quite exactly reproduces the form and orientation of the meshes of the classical stochastic webs as displayed in figure 1.8. In particular, the waviness of the classical webs, i.e. the sinusoidal deviations of the classical skeleton from the rectilinear grid lines (1.45) — see figure 1.12 for an example — are clearly visible quantum mechanically. The comparison of the classical figure 1.8 and the figures of the present section also shows that apparently the web-like structures in phase space survive for larger values of V_0 in quantum mechanics.

Note that not only the central four meshes of the stochastic webs are outlined by the evolving quantum states, although some of the figures seem to give this impression. Rather, the stochastic channels further away are explored, as well: see figures 4.7, C.13 (for $n = 30000$) and C.14 ($n = 47500$), for example. But for finite n, $F^{\mathrm{H}}(x, p, nT - 0; 1)$ takes on rather small values for larger $|x|$ or $|p|$, because in the channels the phase space density rapidly gets transported away to even more distant parts of the web.

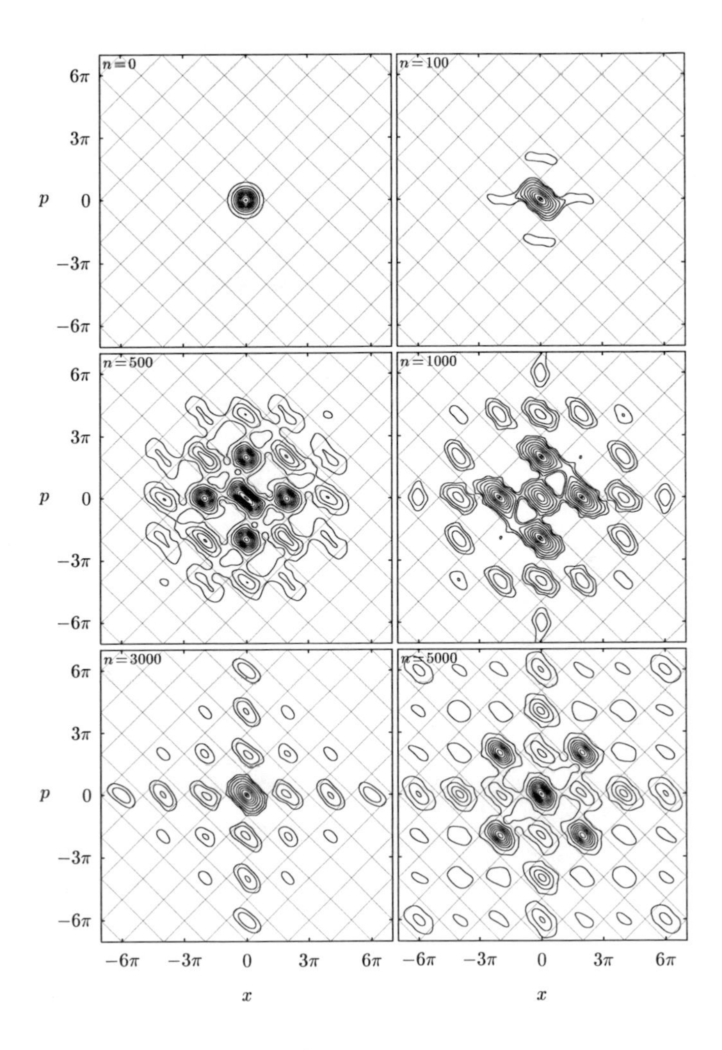

Figure C.1: Contour plots of HUSIMI distributions of $|\psi_n\rangle$.
Parameters: $T = \pi/2$, $\hbar = 1.0$, $V_0 = 1.5$. Initial state: $|\psi_0\rangle = |0\rangle$.

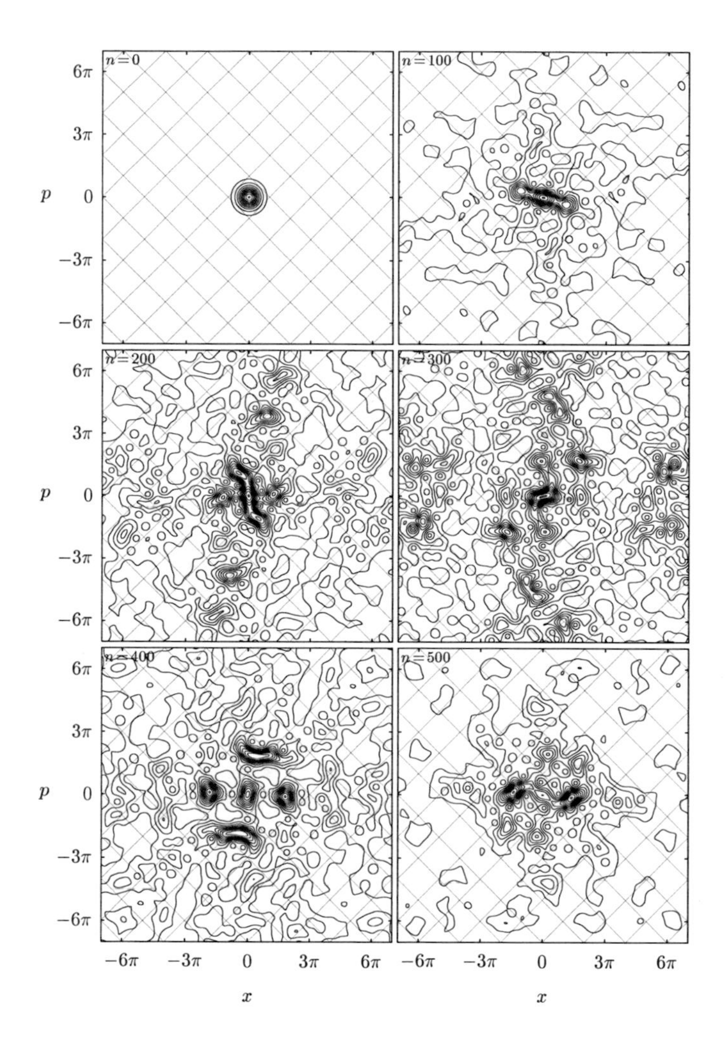

Figure C.2: Contour plots of HUSIMI distributions of $|\psi_n\rangle$.
Parameters: $T = \pi/2$, $\hbar = 1.0$, $V_0 = 3.0$. Initial state: $|\psi_0\rangle = |0\rangle$.

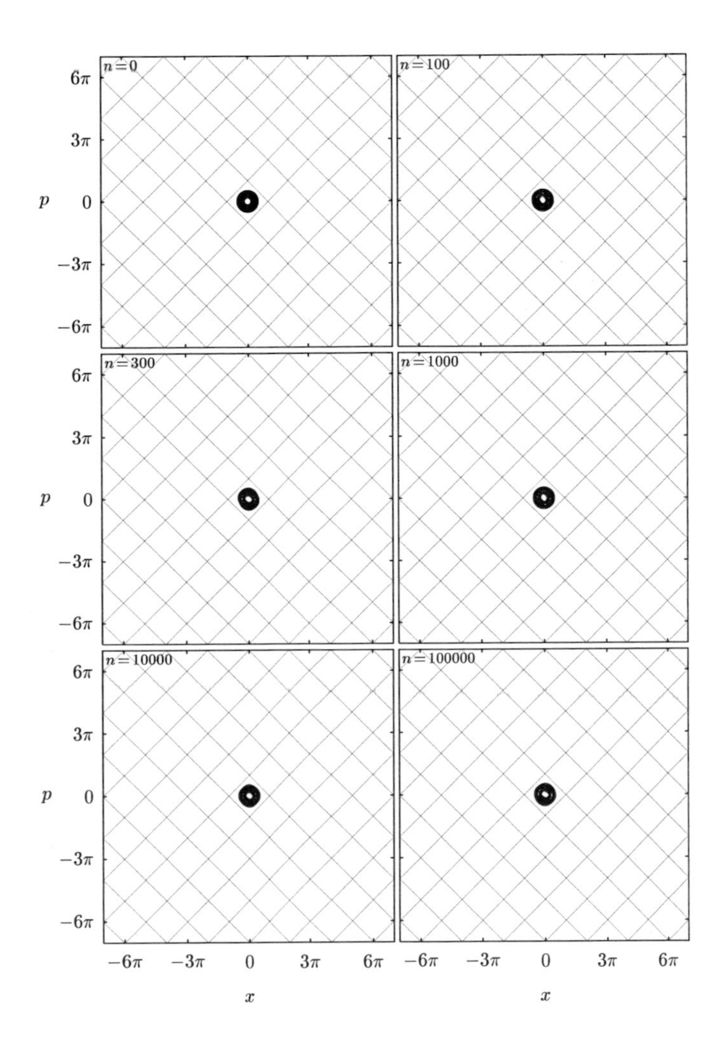

Figure C.3: Contour plots of HUSIMI distributions of $|\psi_n\rangle$.
Parameters: $T = \pi/2$, $\hbar = 0.1$, $V_0 = 1.0$. Initial state: $|\psi_0\rangle = |5\rangle$.

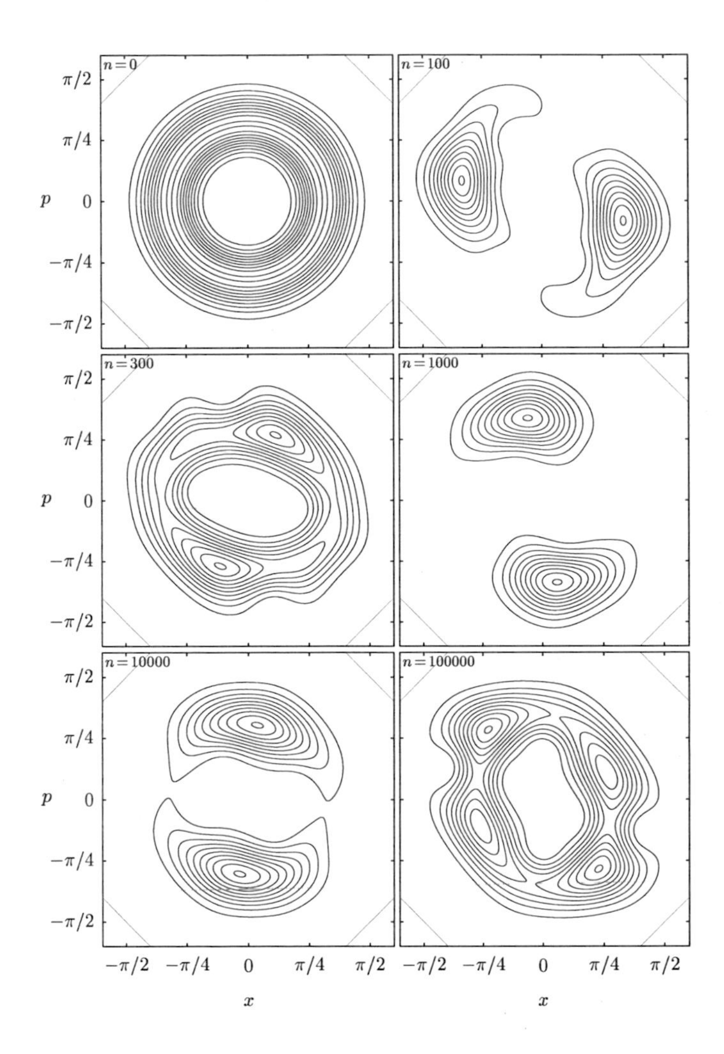

Figure C.4: Contour plots of HUSIMI distributions of $|\psi_n\rangle$.
Parameters: $T = \pi/2$, $\hbar = 0.1$, $V_0 = 5.0$. Initial state: $|\psi_0\rangle = |5\rangle$.

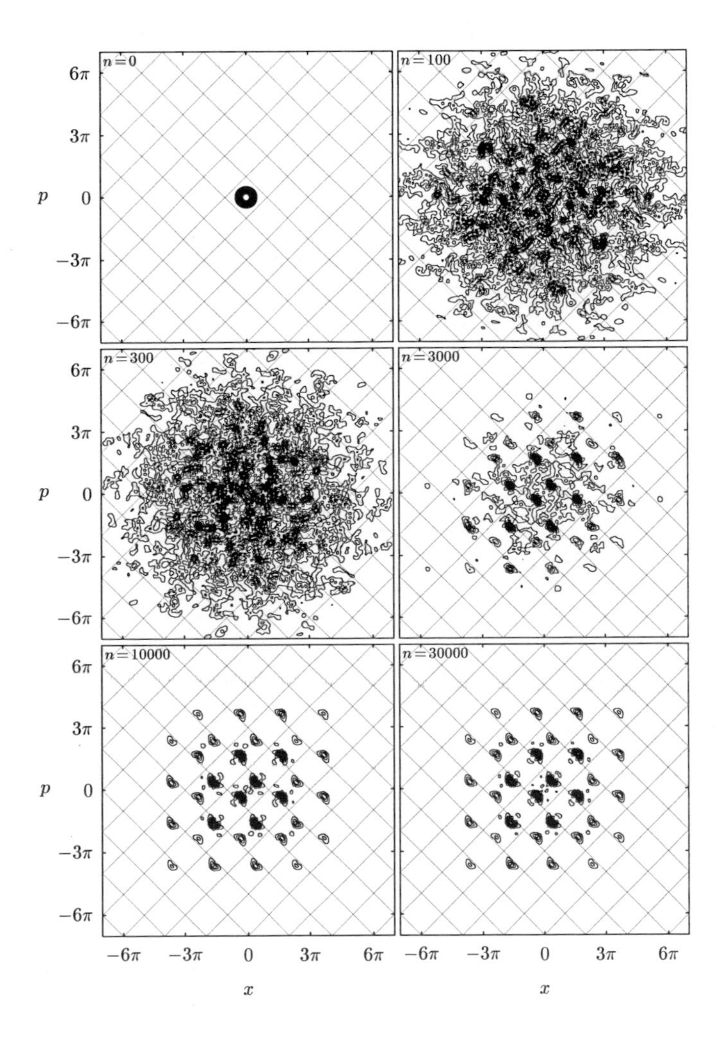

Figure C.5: Contour plots of HUSIMI distributions of $|\psi_n\rangle$.
Parameters: $T = \pi/2$, $\hbar = 0.1$, $V_0 = 30.0$. Initial state: $|\psi_0\rangle = |5\rangle$.

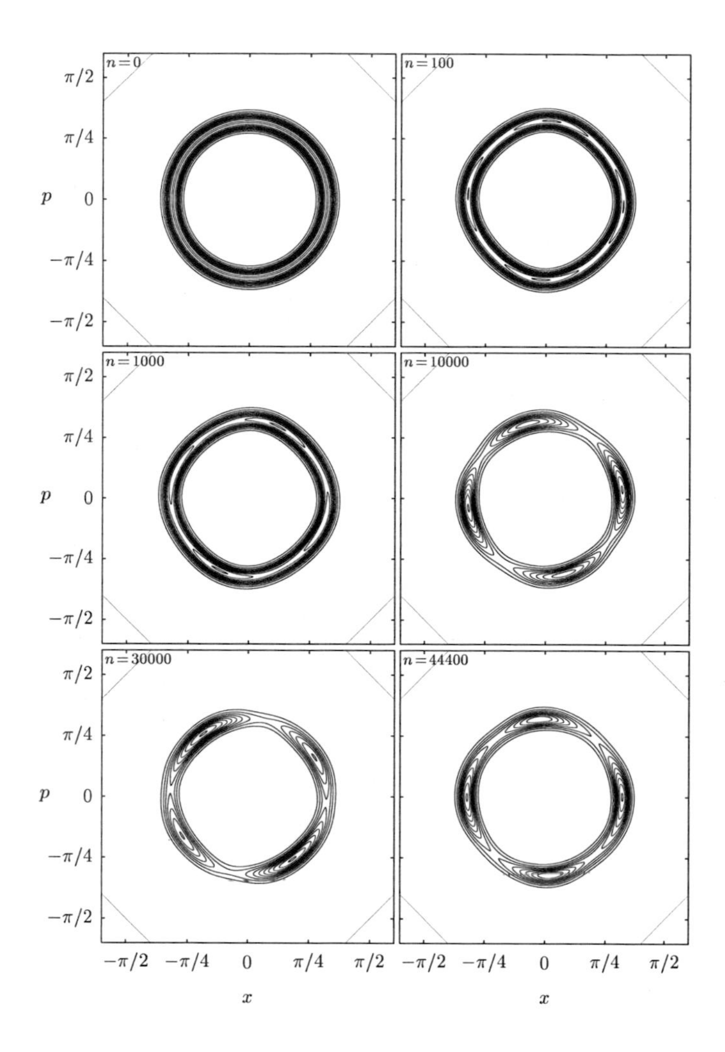

Figure C.6: Contour plots of Husimi distributions of $|\psi_n\rangle$.
Parameters: $T = \pi/2$, $\hbar = 0.01$, $V_0 = 1.0$. Initial state: $|\psi_0\rangle = |50\rangle$.

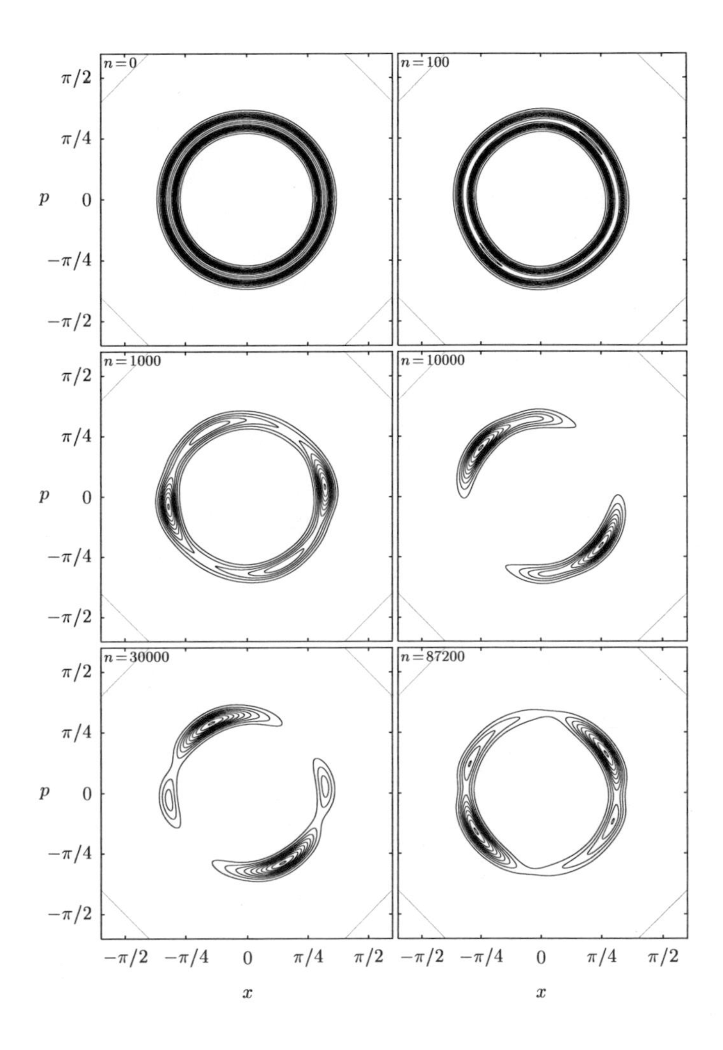

Figure C.7: Contour plots of HUSIMI distributions of $|\psi_n\rangle$.
Parameters: $T = \pi/2$, $\hbar = 0.01$, $V_0 = 5.0$. Initial state: $|\psi_0\rangle = |50\rangle$.

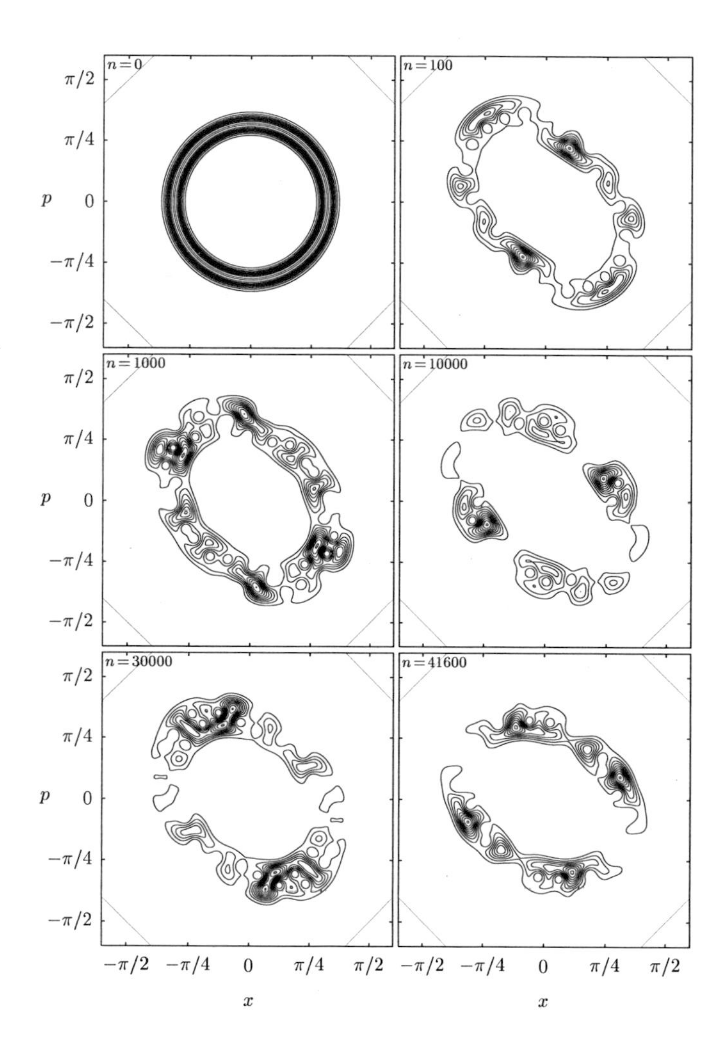

Figure C.8: Contour plots of HUSIMI distributions of $|\psi_n\rangle$.
Parameters: $T = \pi/2$, $\hbar = 0.01$, $V_0 = 50.0$. Initial state: $|\psi_0\rangle = |50\rangle$.

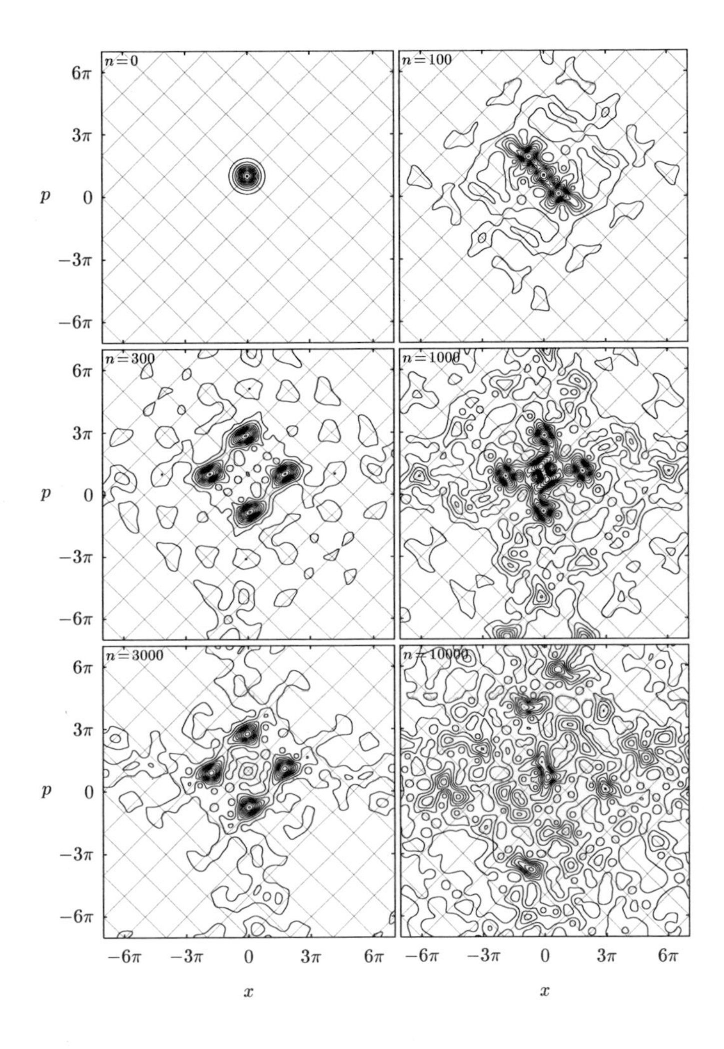

Figure C.9: Contour plots of HUSIMI distributions of $|\psi_n\rangle$.
Parameters: $T = \pi/2$, $\hbar = 1.0$, $V_0 = 1.0$. Initial state: $|\psi_0\rangle = \hat{D}(0, \pi)\,|0\rangle$.

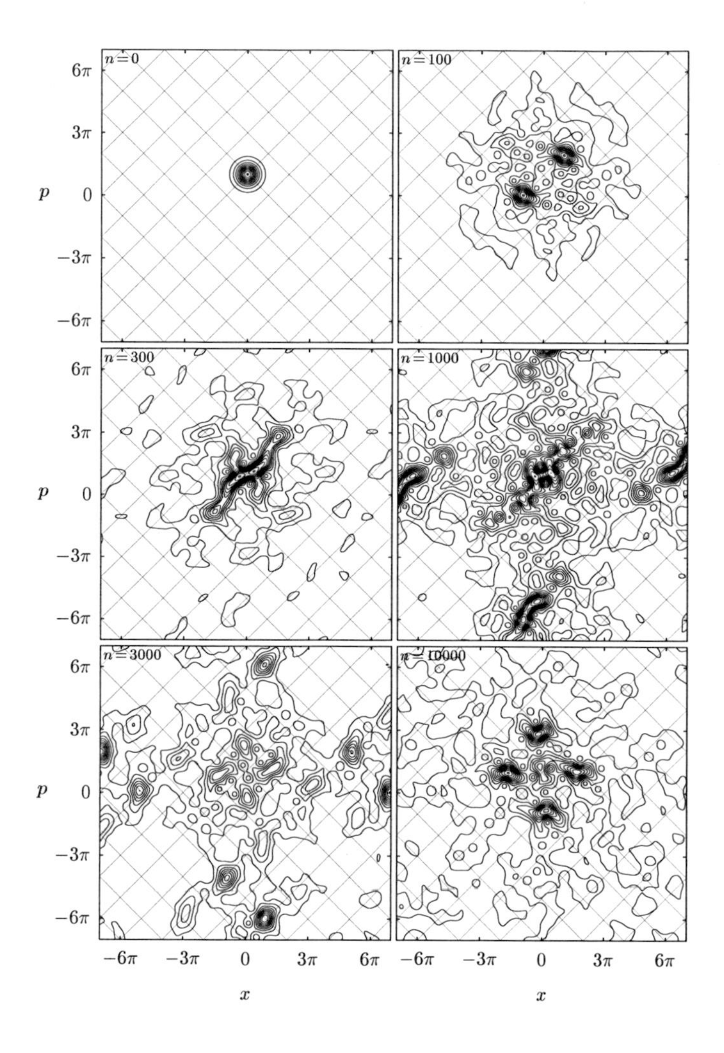

Figure C.10: Contour plots of HUSIMI distributions of $|\psi_n\rangle$.
Parameters: $T = \pi/2$, $\hbar = 1.0$, $V_0 = 1.5$. Initial state: $|\psi_0\rangle = \hat{D}(0, \pi)\,|0\rangle$.

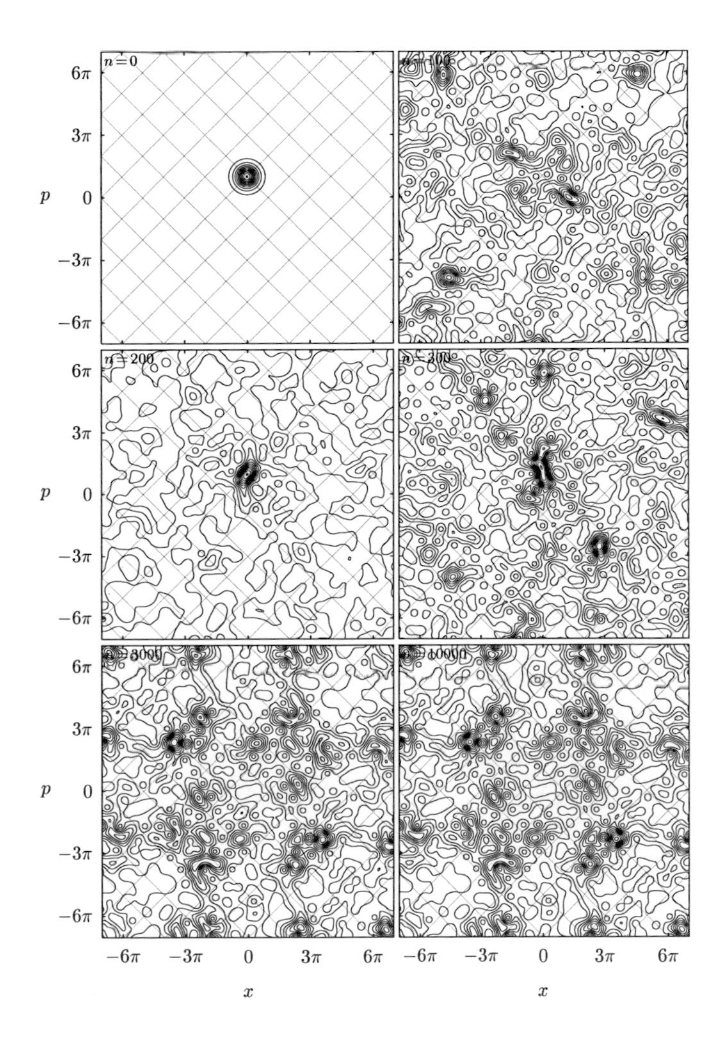

Figure C.11: Contour plots of HUSIMI distributions of $|\psi_n\rangle$.
Parameters: $T = \pi/2$, $\hbar = 1.0$, $V_0 = 20.0$. Initial state: $|\psi_0\rangle = \hat{D}(0, \pi)\,|0\rangle$.

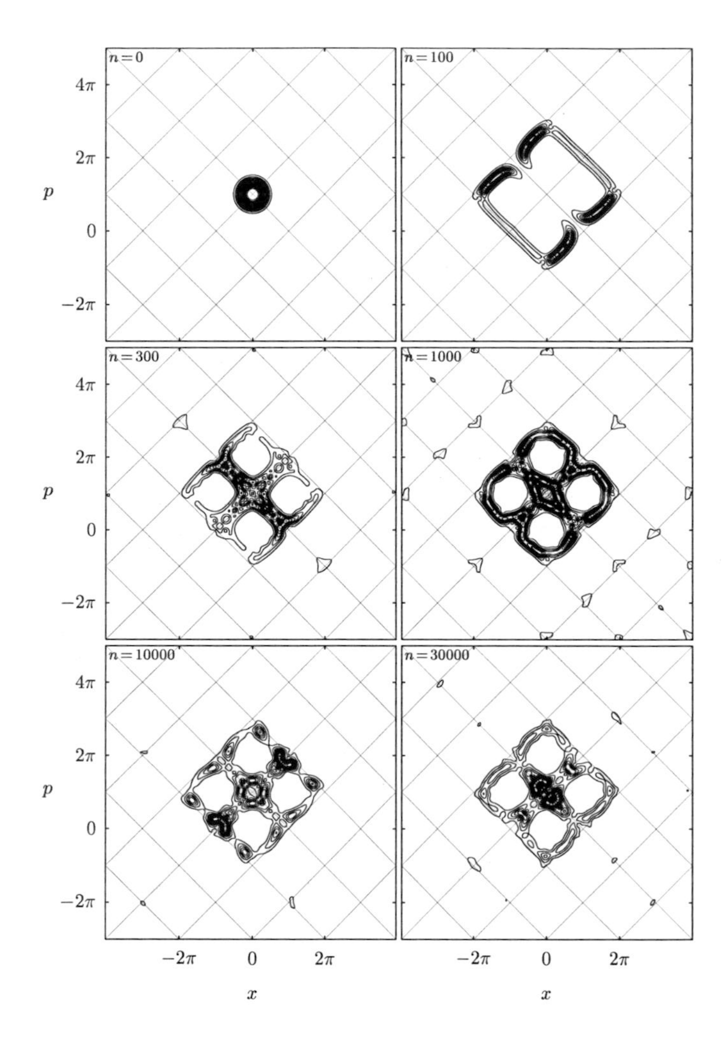

Figure C.12: Contour plots of HUSIMI distributions of $|\psi_n\rangle$.
Parameters: $T = \pi/2$, $\hbar = 0.1$, $V_0 = 1.0$. Initial state: $|\psi_0\rangle = \hat{D}(0, \pi)\,|5\rangle$.

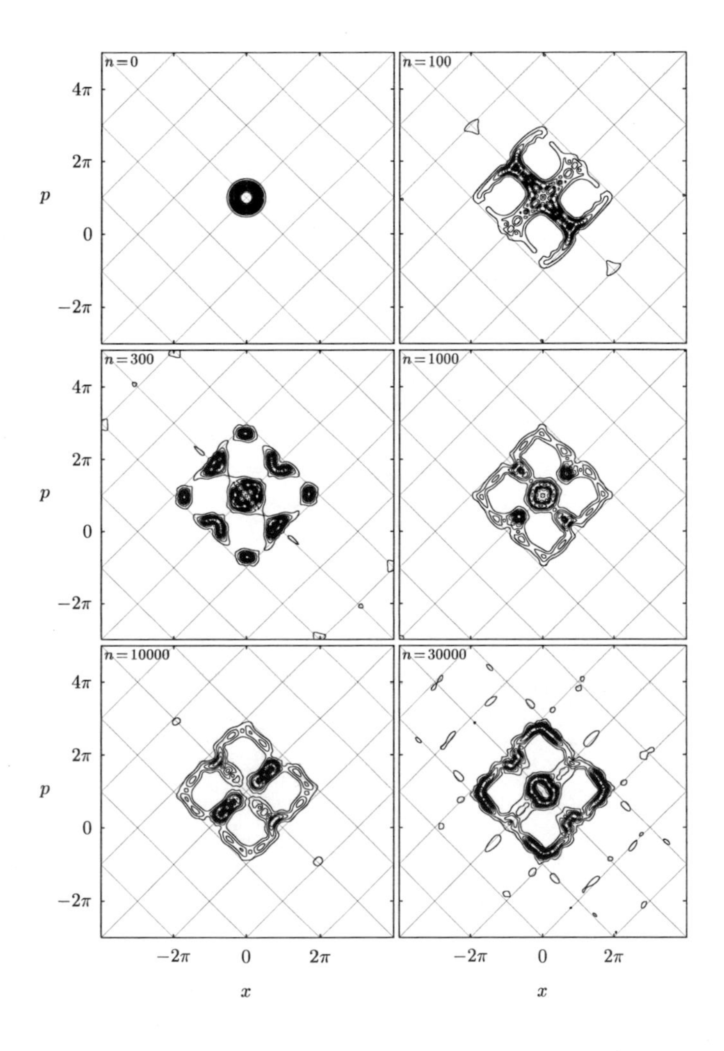

Figure C.13: Contour plots of HUSIMI distributions of $|\psi_n\rangle$.
Parameters: $T = \pi/2$, $\hbar = 0.1$, $V_0 = 3.0$. Initial state: $|\psi_0\rangle = \hat{D}\big(0, \pi\big)\,|5\rangle$.

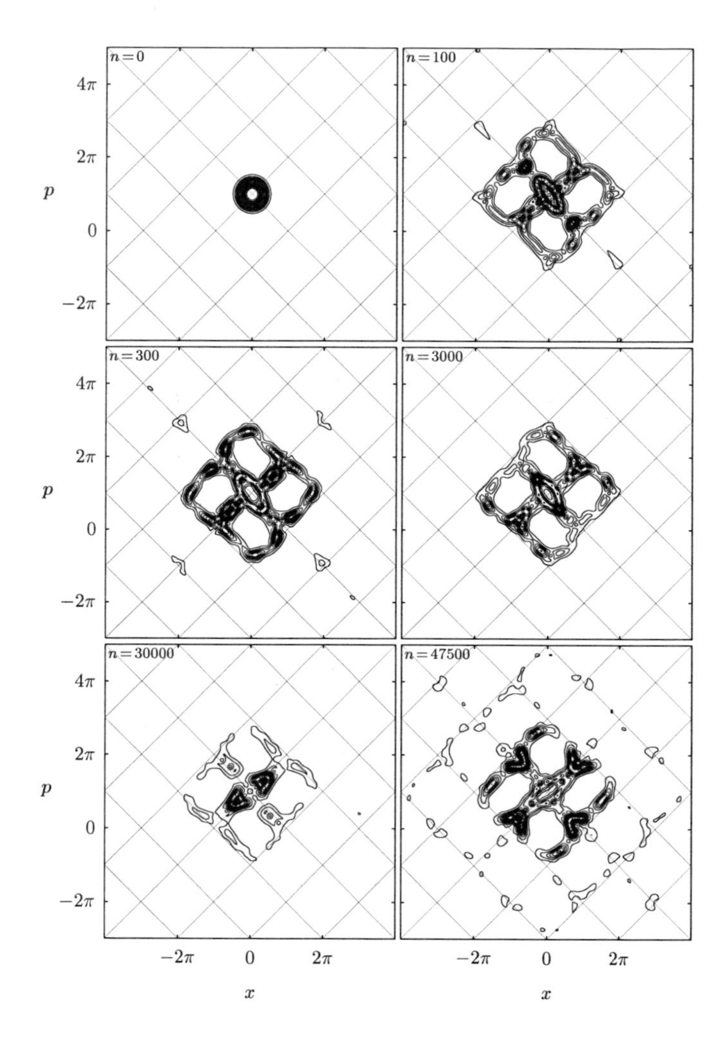

Figure C.14: Contour plots of Husimi distributions of $|\psi_n\rangle$.
Parameters: $T = \pi/2$, $\hbar = 0.1$, $V_0 = 5.0$. Initial state: $|\psi_0\rangle = \hat{D}(0, \pi)\,|5\rangle$.

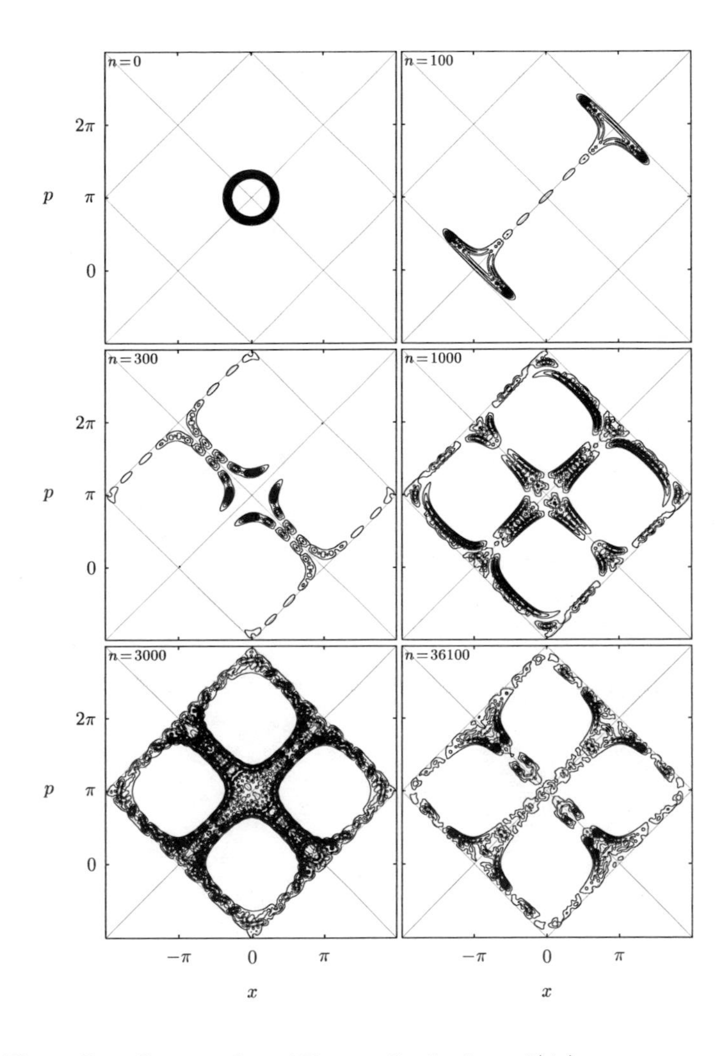

Figure C.15: Contour plots of HUSIMI distributions of $|\psi_n\rangle$.
Parameters: $T = \pi/2$, $\hbar = 0.01$, $V_0 = 5.0$. Initial state: $|\psi_0\rangle = \hat{D}(0, \pi)\,|50\rangle$.

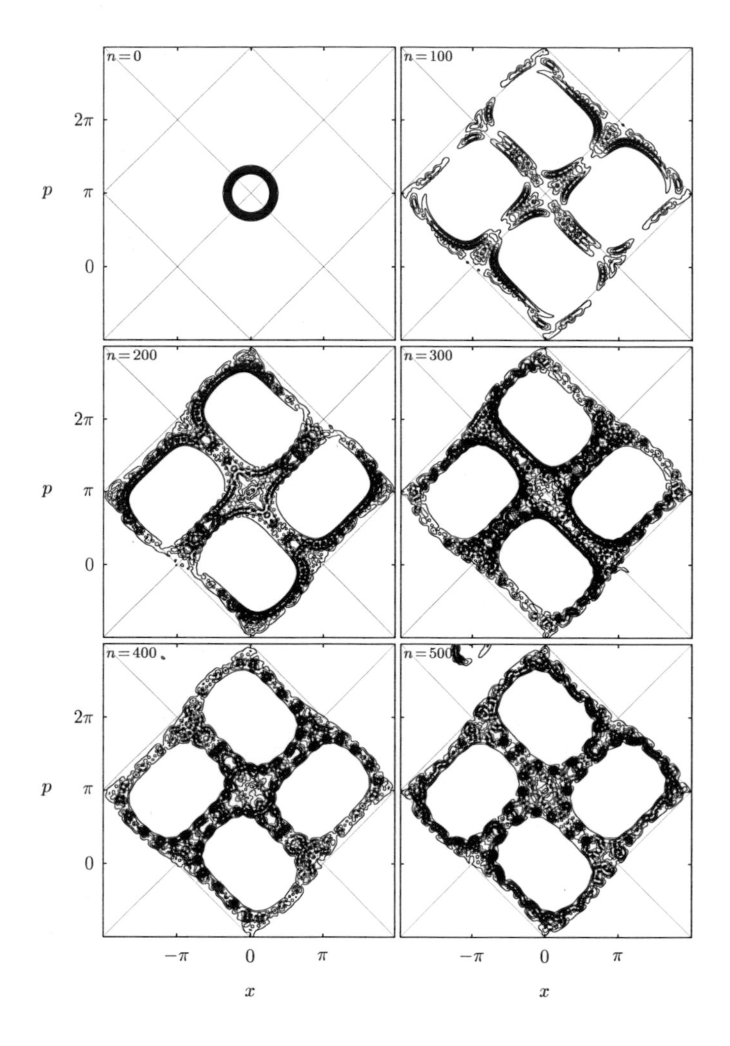

Figure C.16: Contour plots of HUSIMI distributions of $|\psi_n\rangle$.
Parameters: $T = \pi/2$, $\hbar = 0.01$, $V_0 = 50.0$. Initial state: $|\psi_0\rangle = \hat{D}(0, \pi)\,|50\rangle$.

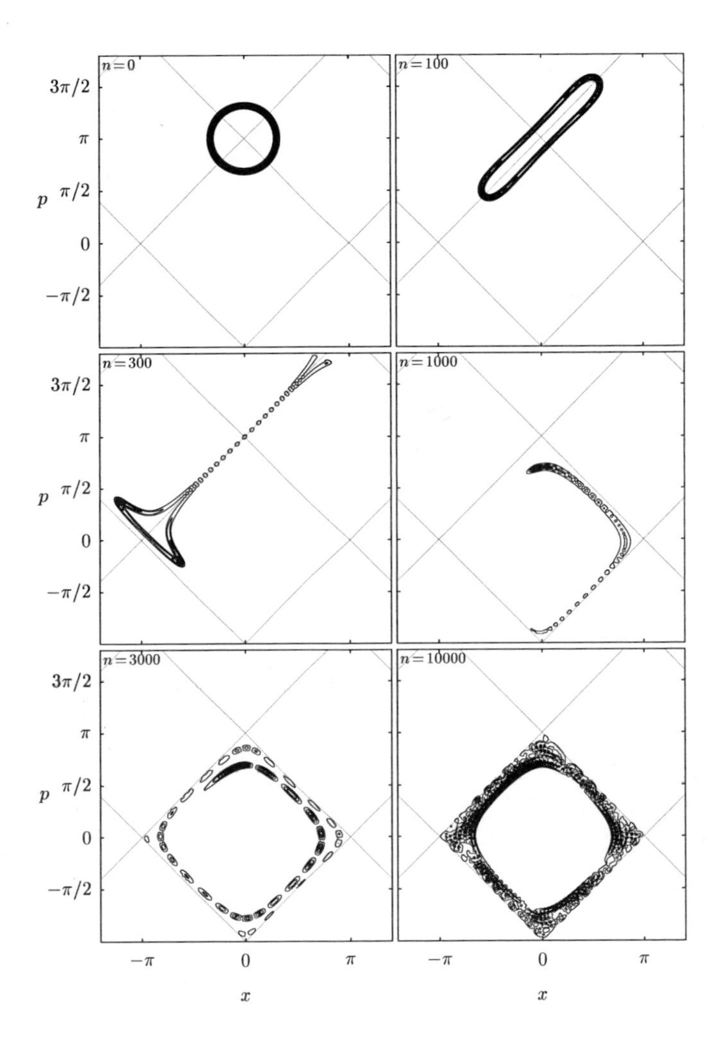

Figure C.17: Contour plots of HUSIMI distributions of $|\psi_n\rangle$.
Parameters: $T = \pi/2$, $\hbar = 0.003$, $V_0 = 5.0$. Initial state: $|\psi_0\rangle = \hat{D}(0, \pi)\,|166\rangle$.

C.2 Hexagonal Quantum Stochastic Webs

Similar phenomena as for $T = \pi/2$ can be observed for $T = 2\pi/3$ ($q = 3$). Here, the grid — again displayed via thin lines in the figures — underlying the skeleton of the classical webs is hexagonal, as can be concluded from figure 1.10a or by inspection of equation (1.47a). As in the previous section, depending on the values of $\hbar$ and V_0, the quantum states evolve into a phase space pattern that is more or less similar to their classical counterpart in figure 1.3a. (That figure is for $T = \pi/3$, actually, but $T = 2\pi/3$ generates a very similar picture).

Note that for $T = \pi/3$ ($q = 6$), qualitatively equivalent quantum dynamics is obtained, with HUSIMI distributions reproducing the classical webs in quite the same way as for $T = 2\pi/3$. See figures 4.4–4.6 for some typical examples.

C.2.1 $T = 2\pi/3$, $p_0 = 0.0$

Figures C.18–C.28.

While I do not discuss explicitly each of the figures of this section, attention should be called to figure C.20: there, for smaller values of n ($n \lesssim 300$), no regular structure of the HUSIMI distribution seems to develop, as a consequence of the quite large values of $V_0 = 20.0$ and $\hbar = 1.0$. But the sub-figures for larger n indicate that this absence of a regular pattern is a transitional feature of the dynamics: for $n \gtrsim 1000$, a relatively regular hexagonal pattern has been generated around the position of the initial state.

C.2.2 $T = 2\pi/3$, $p_0 = 2\pi/\sqrt{3}$

Figures C.29–C.37.

For larger n, the figures C.20 and C.31, using the same parameter values but $p_0 = 0.0$ for the former and $p_0 = 2\pi/\sqrt{3}$ for the later, exhibit very similar phase portraits, despite the different initial conditions. For $n = 13100$, figure C.31 shows a (numerical) state the norm of which has already decayed considerably (to less than 10^{-10}). Nevertheless, the most prominent feature of the phase portrait, its hexagonal structure, is still clearly visible.

The sequence of figures in the present subsection also nicely demonstrates some aspects of the transition from genuinely quantum behaviour (for larger values of $\hbar$) to more classical behaviour (for smaller $\hbar$): compare figure C.32 with figure C.35 and figure C.34 with figure C.36, for example. The smaller value of $\hbar$ leads to smaller structures in phase space, and to a closer resemblance with the classical hexagonal stochastic web as displayed in figure 1.3a.

What is more, the figures exhibit some fingerprints of the classical POINCARÉ map (1.21) in the quantum phase space dynamics generated by the quantum map (2.37). In particular the figures C.35 ($n = 300$) and C.36 ($n = 100/300$) apparently demonstrate the way in which the classical heteroclinic connections of the periodic points forming the skeleton of the web act as separatrices, separating an incident (part of the) quantum wave packet into two parts that are driven away from the corresponding fixed point along the unstable manifold of that fixed point.

The same observations, but for $T = \pi/2$ rather than $T = 2\pi/3$, can be made with respect to figures C.12–C.17 and 4.3. Especially the figures 4.3 ($n = 3000$) and C.12 ($n = 100/300$) show the effect quite clearly. See also figure 4.5 ($n = 100/300$) for the same pattern in the quantum $T = \pi/3$-dynamics.

It should be kept in mind, though, that these are just phenomenological observations; the transition from quantum to classical behaviour cannot be described in these terms in a mathematically sound way. This is reflected by the fact that for smaller $\hbar$ a (possibly much) larger basis is needed to describe quantum states in the same region of phase space — cf. table C.1.

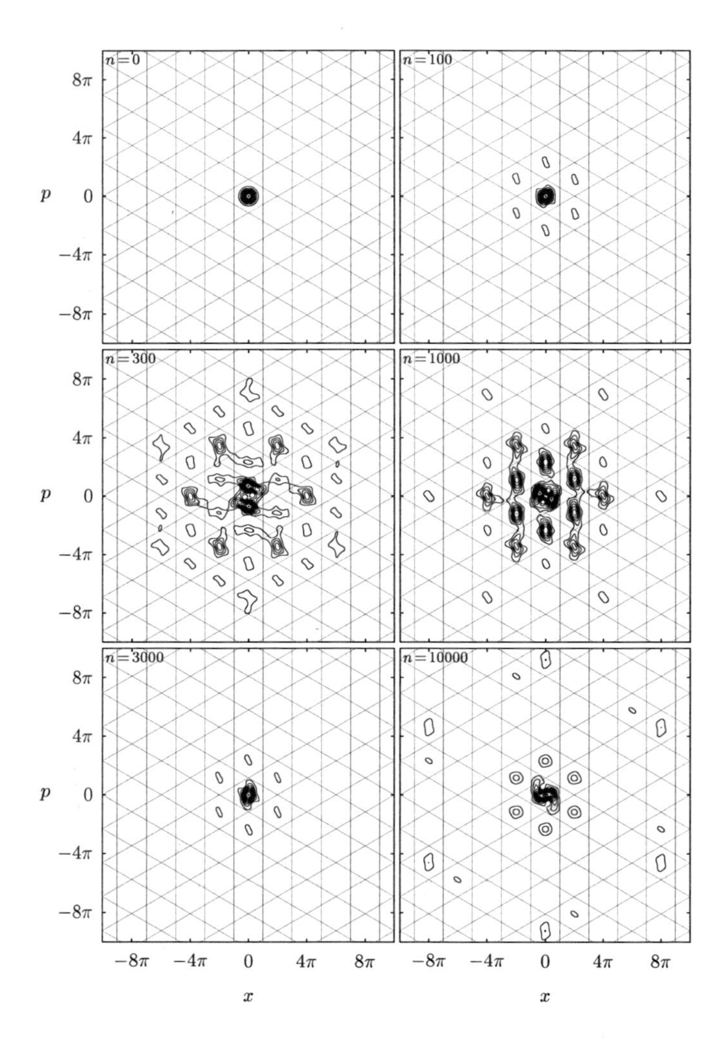

Figure C.18: Contour plots of HUSIMI distributions of $|\psi_n\rangle$.
Parameters: $T = 2\pi/3$, $\hbar = 1.0$, $V_0 = 1.0$. Initial state: $|\psi_0\rangle = |0\rangle$.

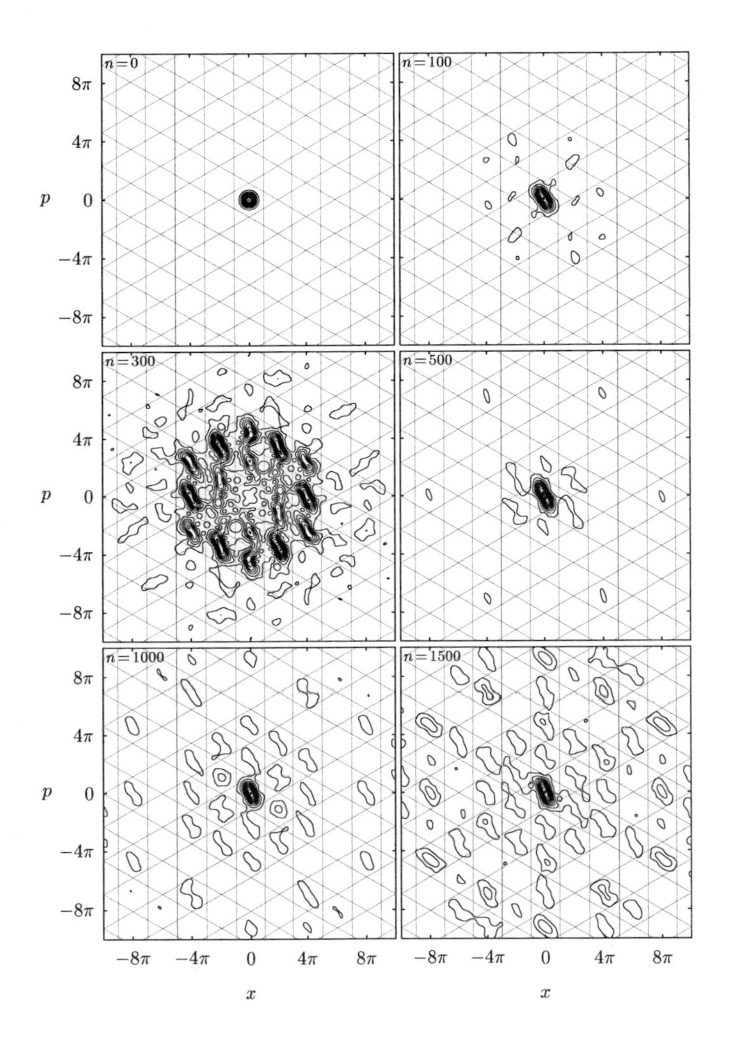

Figure C.19: Contour plots of HUSIMI distributions of $|\psi_n\rangle$.
Parameters: $T = 2\pi/3$, $\hbar = 1.0$, $V_0 = 2.0$. Initial state: $|\psi_0\rangle = |0\rangle$.

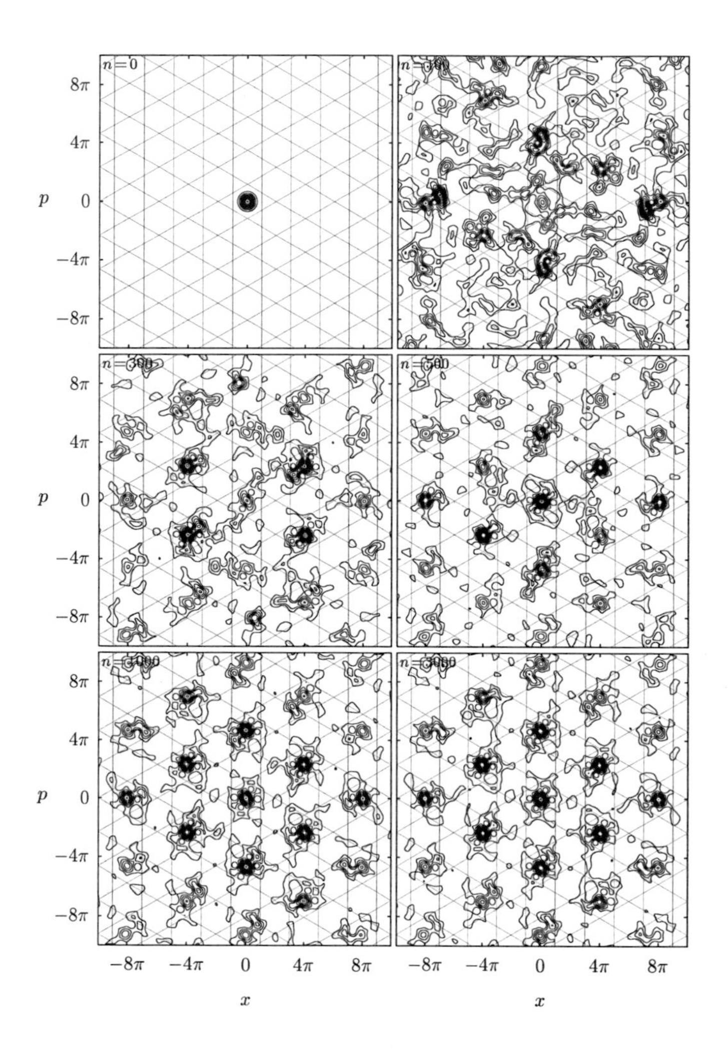

Figure C.20: Contour plots of HUSIMI distributions of $|\psi_n\rangle$.
Parameters: $T = 2\pi/3$, $\hbar = 1.0$, $V_0 = 20.0$. Initial state: $|\psi_0\rangle = |0\rangle$.

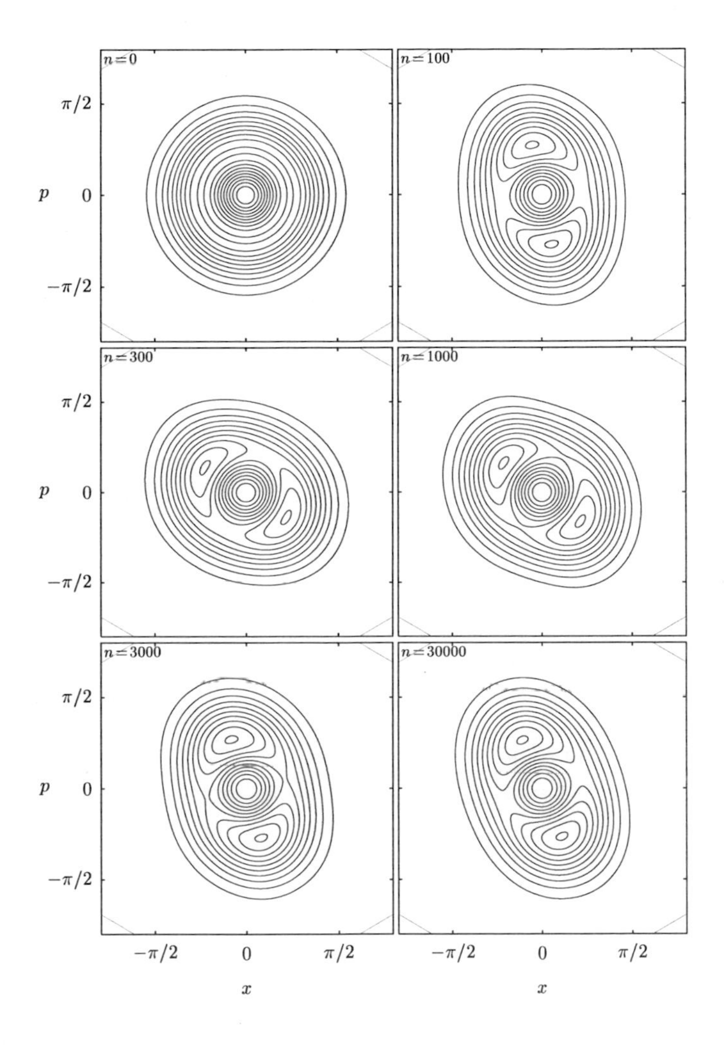

Figure C.21: Contour plots of HUSIMI distributions of $|\psi_n\rangle$.
Parameters: $T = 2\pi/3$, $\hbar = 0.3$, $V_0 = 1.0$. Initial state: $|\psi_0\rangle = |1\rangle$.

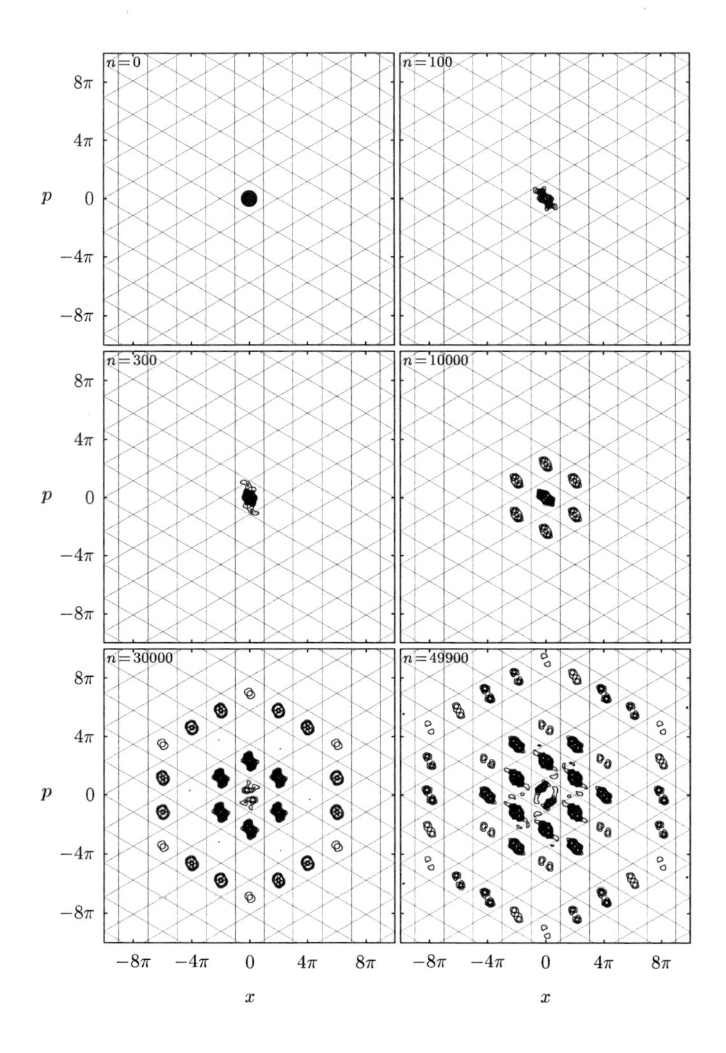

Figure C.22: Contour plots of HUSIMI distributions of $|\psi_n\rangle$.
Parameters: $T = 2\pi/3$, $\hbar = 0.3$, $V_0 = 3.0$. Initial state: $|\psi_0\rangle = |1\rangle$.

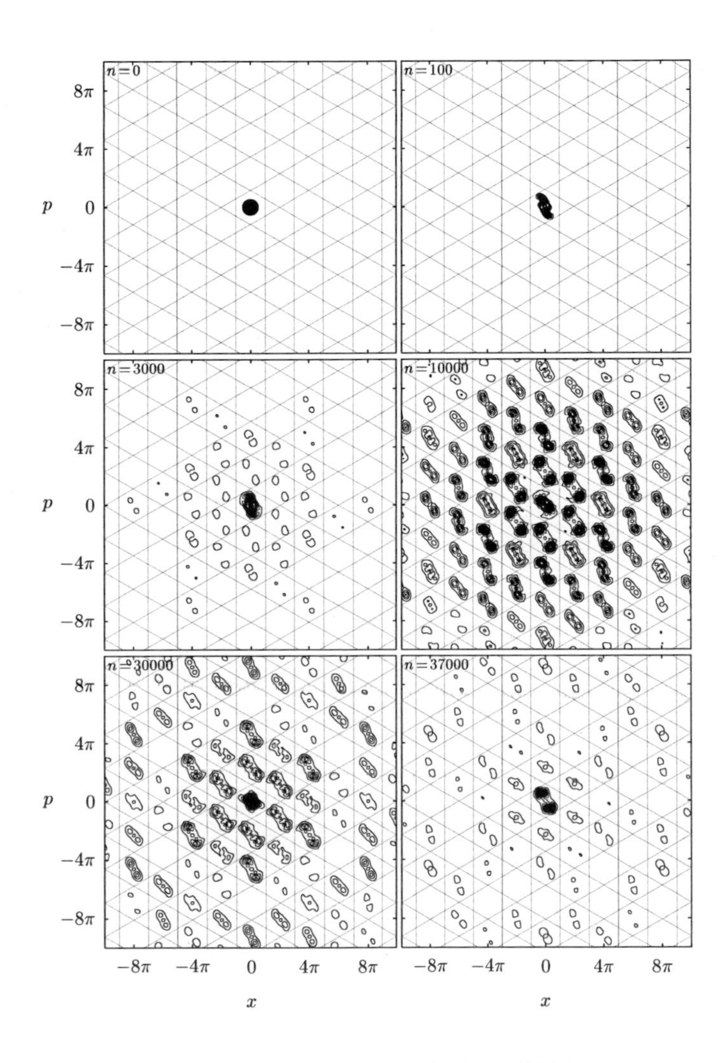

Figure C.23: Contour plots of HUSIMI distributions of $|\psi_n\rangle$.
Parameters: $T = 2\pi/3$, $\hbar = 0.3$, $V_0 = 5.0$. Initial state: $|\psi_0\rangle = |1\rangle$.

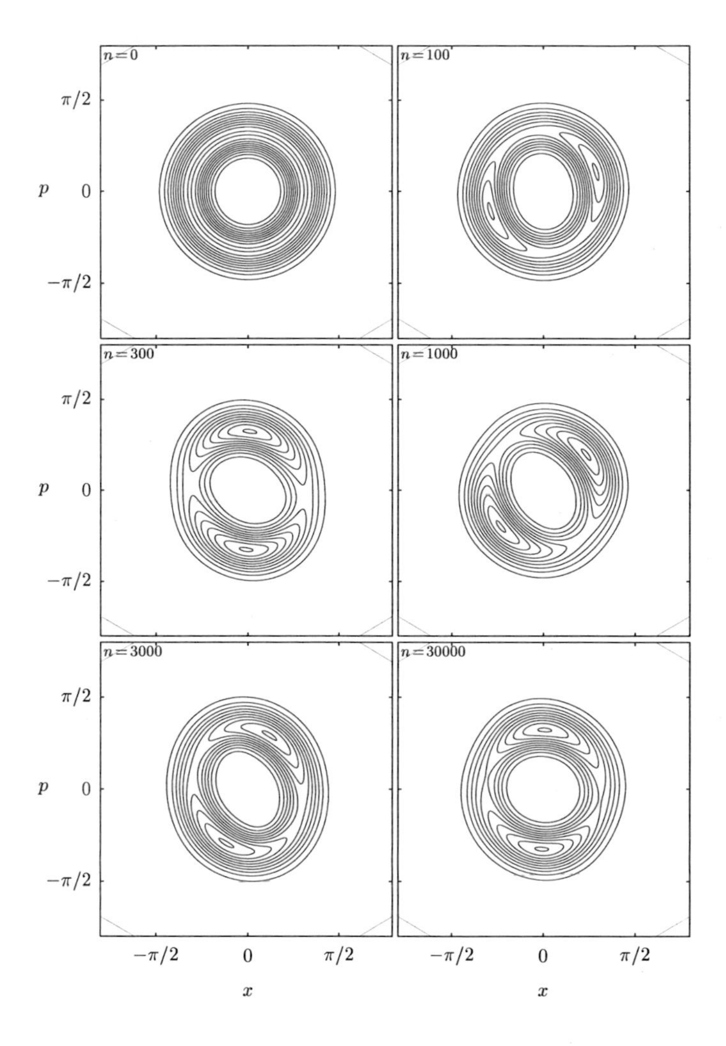

Figure C.24: Contour plots of HUSIMI distributions of $|\psi_n\rangle$.
Parameters: $T = 2\pi/3$, $\hbar = 0.1$, $V_0 = 1.0$. Initial state: $|\psi_0\rangle = |5\rangle$.

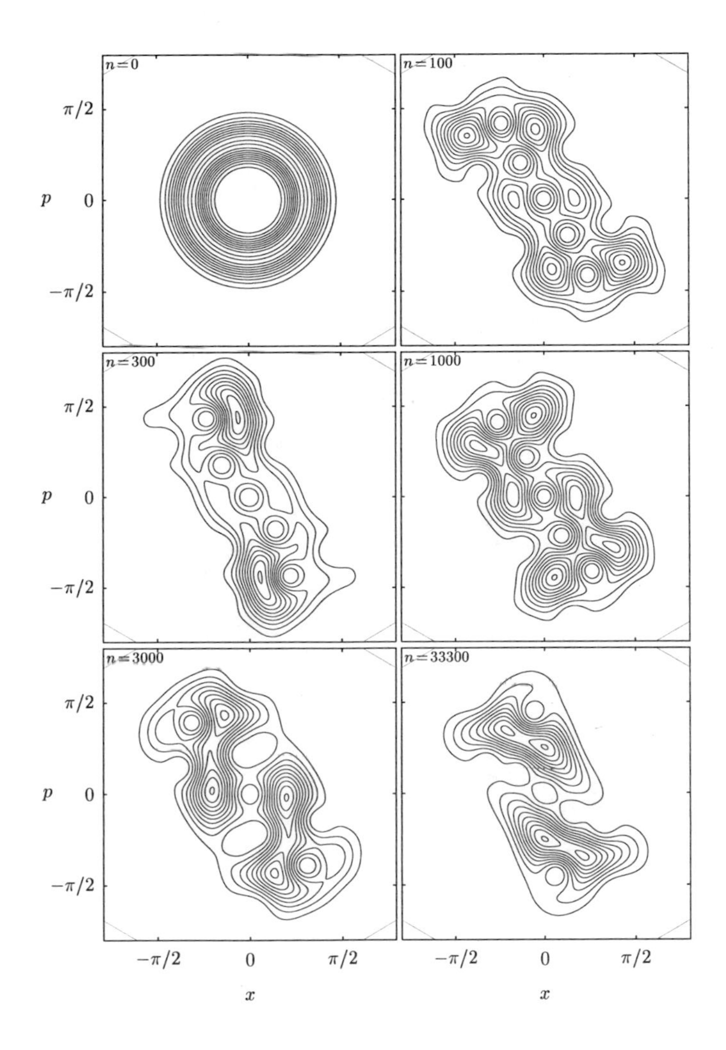

Figure C.25: Contour plots of HUSIMI distributions of $|\psi_n\rangle$.
Parameters: $T = 2\pi/3$, $\hbar = 0.1$, $V_0 = 10.0$. Initial state: $|\psi_0\rangle = |5\rangle$.

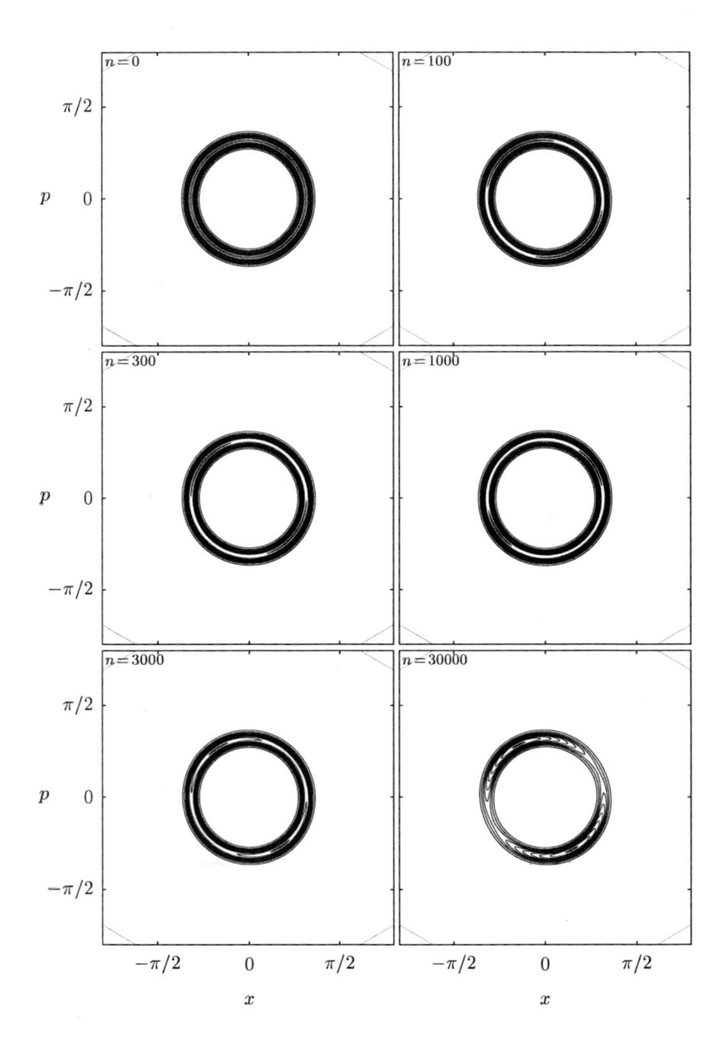

Figure C.26: Contour plots of HUSIMI distributions of $|\psi_n\rangle$.
Parameters: $T = 2\pi/3$, $\hbar = 0.01$, $V_0 = 1.0$. Initial state: $|\psi_0\rangle = |50\rangle$.

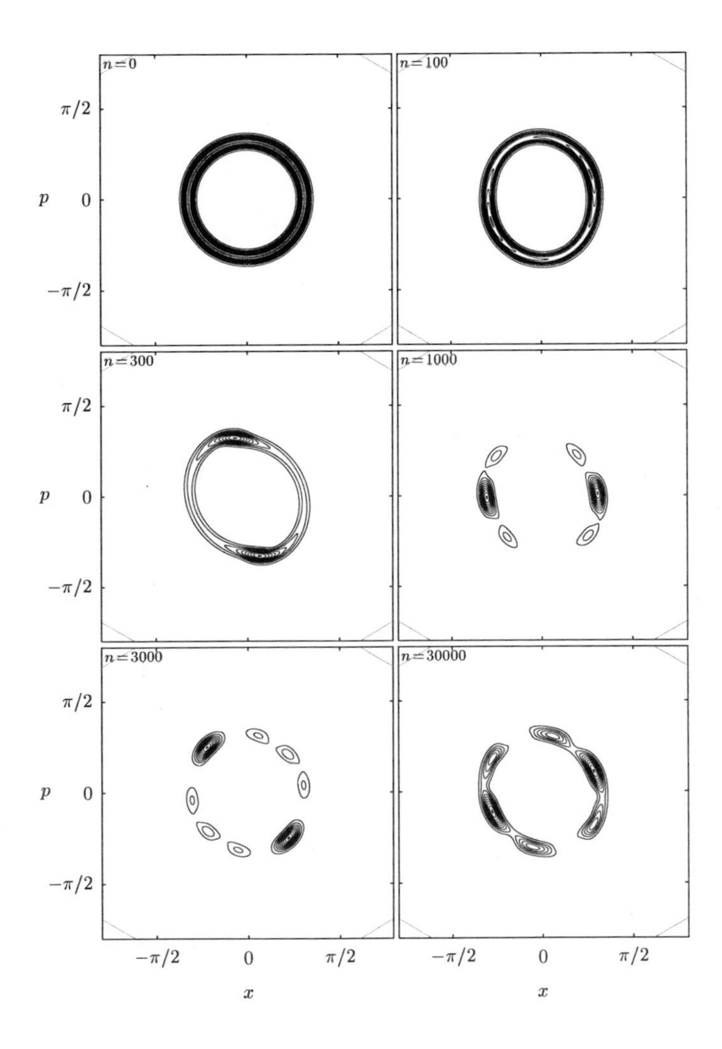

Figure C.27: Contour plots of HUSIMI distributions of $|\psi_n\rangle$.
Parameters: $T = 2\pi/3$, $\hbar = 0.01$, $V_0 = 10.0$. Initial state: $|\psi_0\rangle = |50\rangle$.

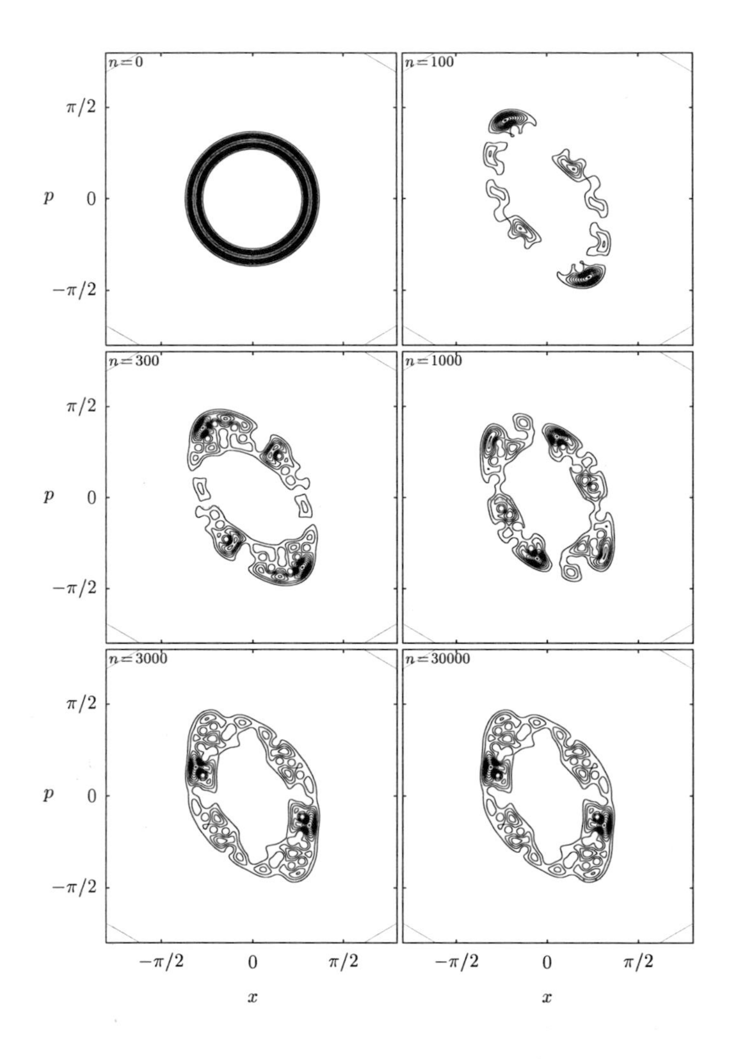

Figure C.28: Contour plots of HUSIMI distributions of $|\psi_n\rangle$.
Parameters: $T = 2\pi/3$, $\hbar = 0.01$, $V_0 = 50.0$. Initial state: $|\psi_0\rangle = |50\rangle$.

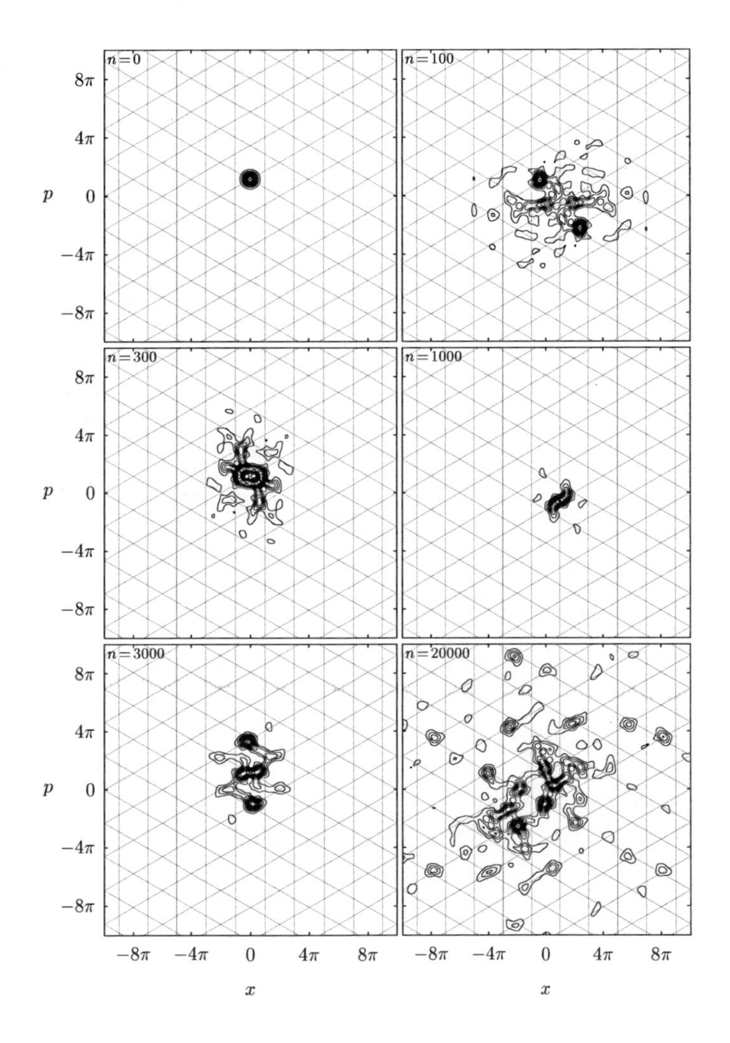

Figure C.29: Contour plots of HUSIMI distributions of $|\psi_n\rangle$.
Parameters: $T = 2\pi/3$, $\hbar = 1.0$, $V_0 = 1.0$. Initial state: $|\psi_0\rangle = \hat{D}(0, 2\pi/\sqrt{3})\,|0\rangle$.

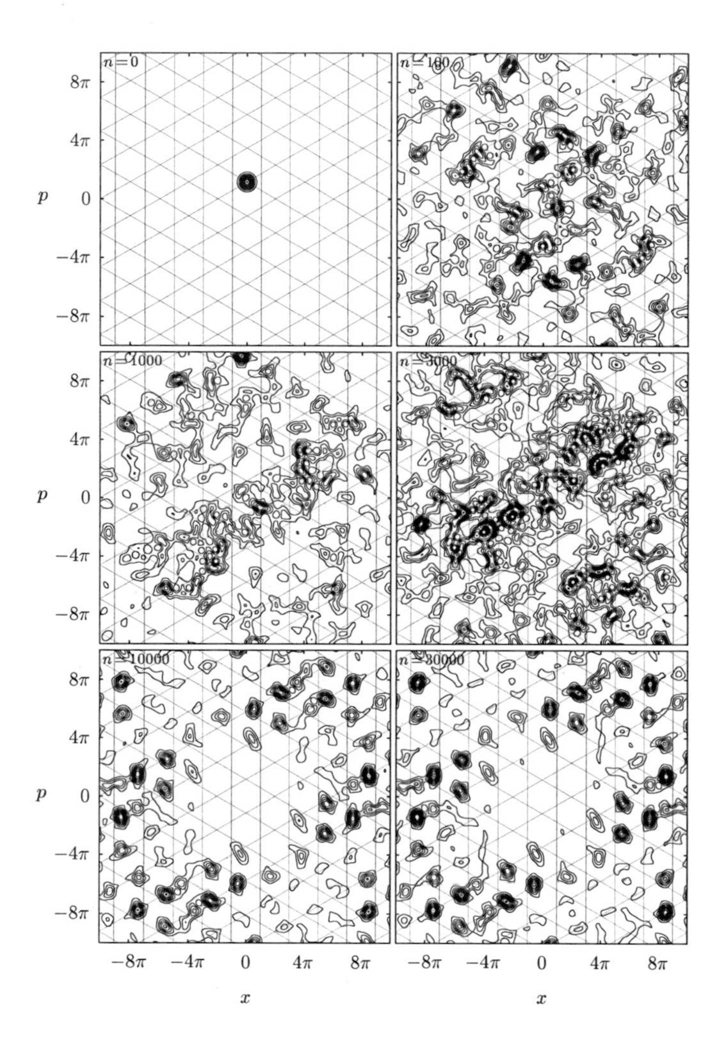

Figure C.30: Contour plots of HUSIMI distributions of $|\psi_n\rangle$.
Parameters: $T = 2\pi/3$, $\hbar = 1.0$, $V_0 = 10.0$. Initial state: $|\psi_0\rangle = \hat{D}\big(0, 2\pi/\sqrt{3}\big)\,|0\rangle$.

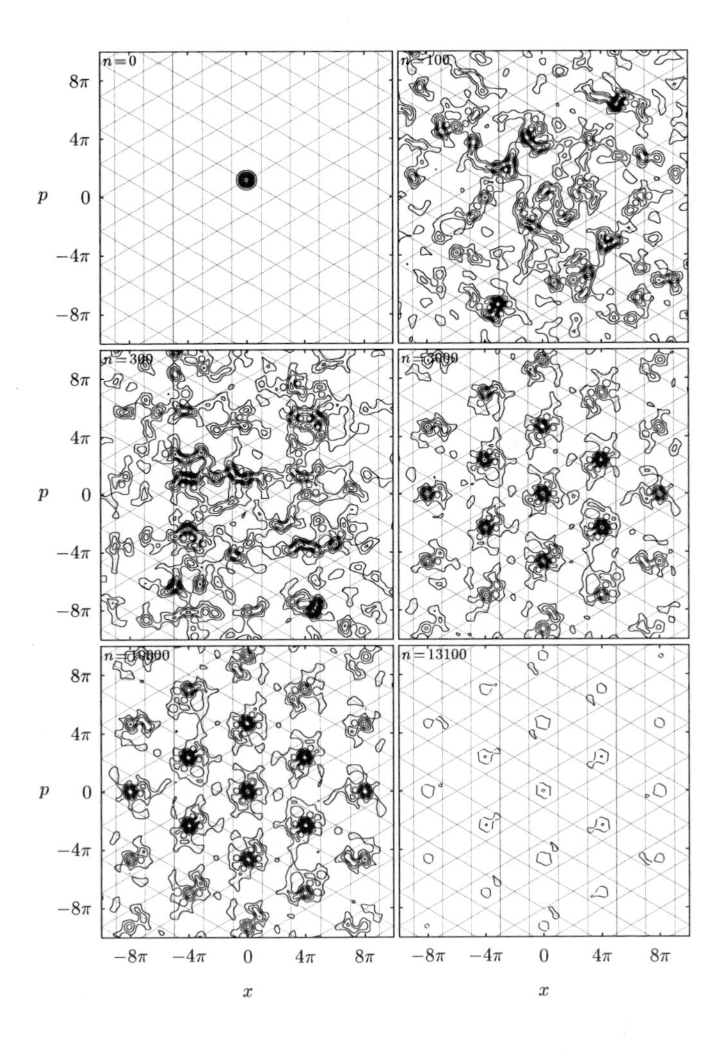

Figure C.31: Contour plots of HUSIMI distributions of $|\psi_n\rangle$.
Parameters: $T = 2\pi/3$, $\hbar = 1.0$, $V_0 = 20.0$. Initial state: $|\psi_0\rangle = \hat{D}\big(0, 2\pi/\sqrt{3}\big)\,|0\rangle$.

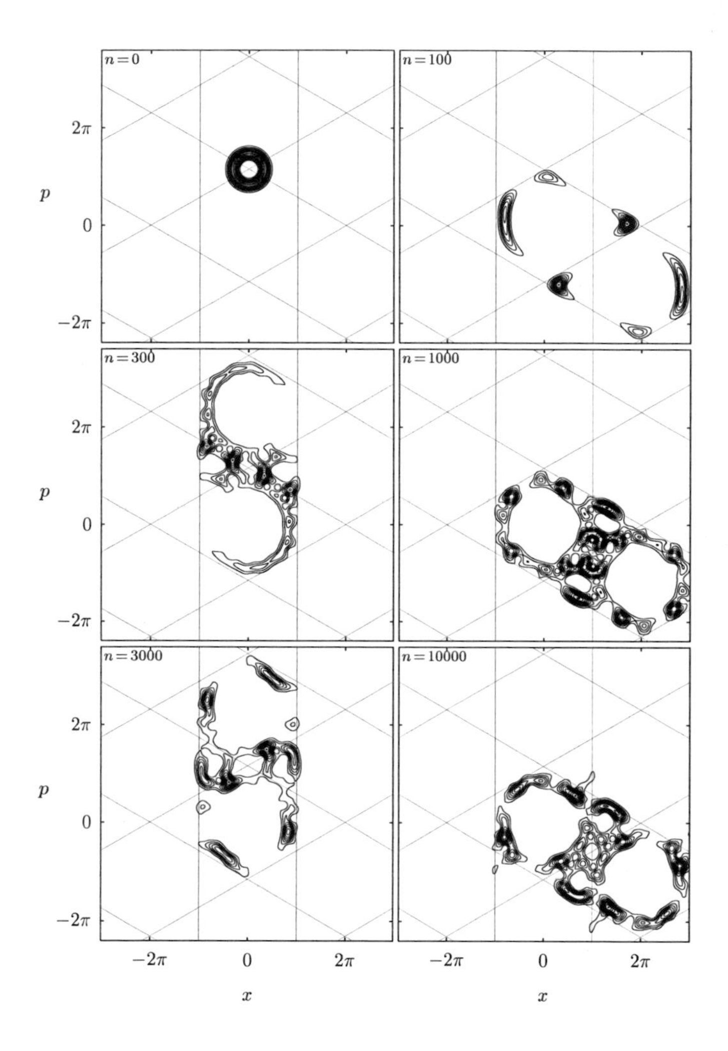

Figure C.32: Contour plots of HUSIMI distributions of $|\psi_n\rangle$.
Parameters: $T = 2\pi/3$, $\hbar = 0.1$, $V_0 = 1.0$. Initial state: $|\psi_0\rangle = \hat{D}\big(0, 2\pi/\sqrt{3}\big)\,|5\rangle$.

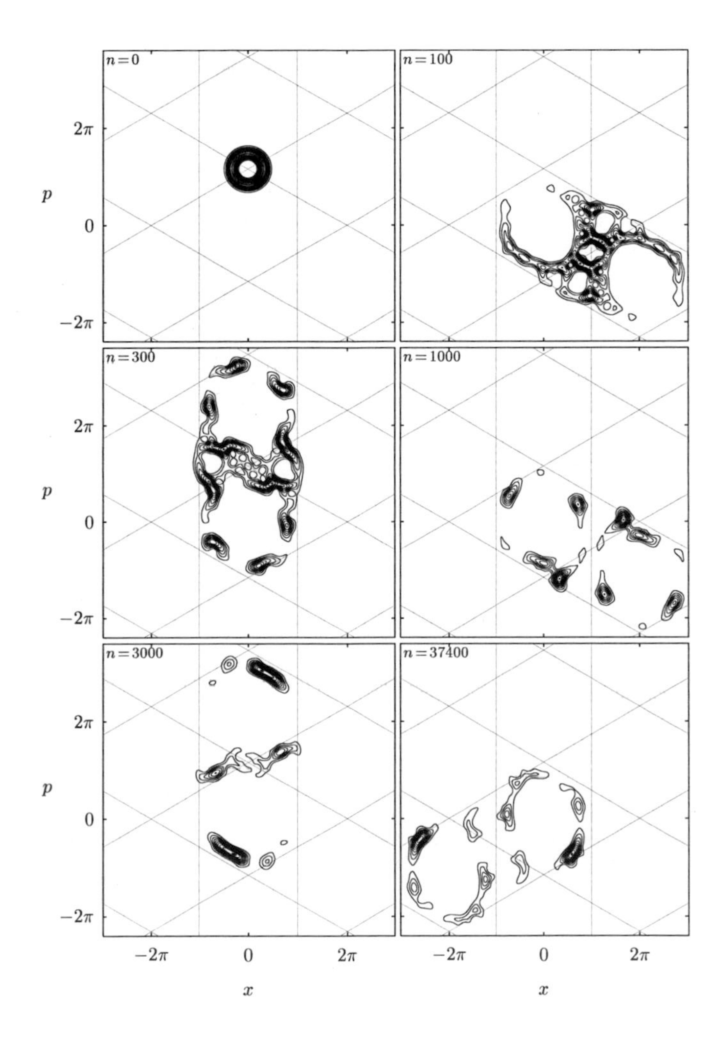

Figure C.33: Contour plots of Husimi distributions of $|\psi_n\rangle$.
Parameters: $T = 2\pi/3$, $\hbar = 0.1$, $V_0 = 3.0$. Initial state: $|\psi_0\rangle = \hat{D}(0, 2\pi/\sqrt{3})\,|5\rangle$.

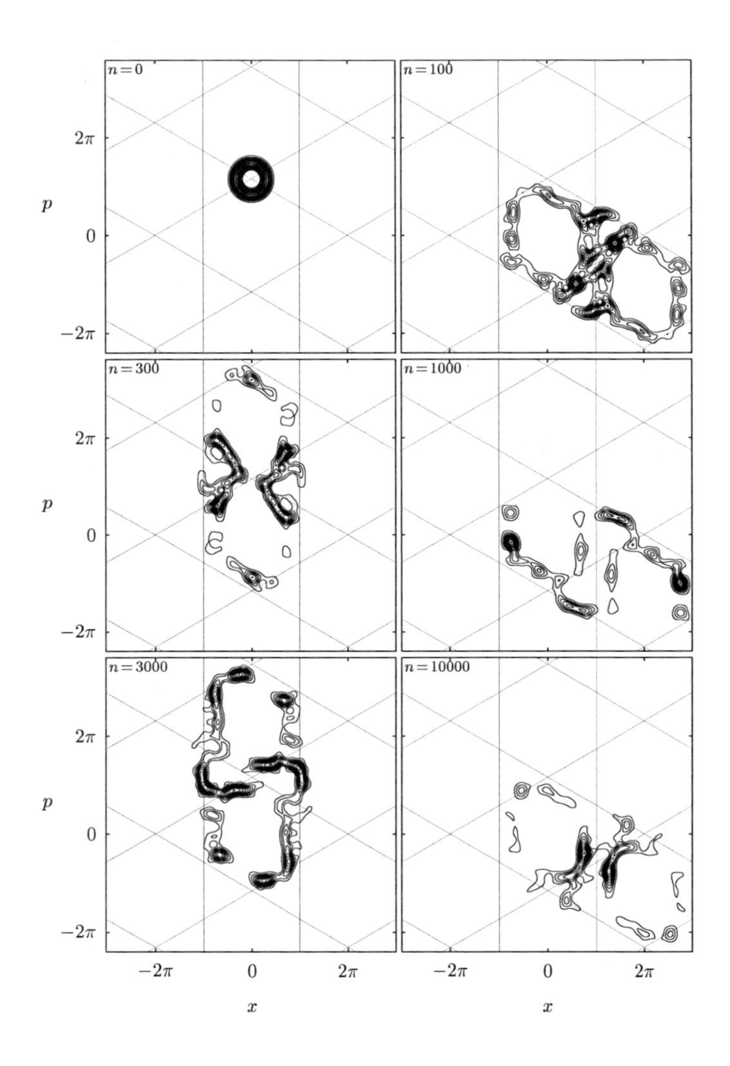

Figure C.34: Contour plots of HUSIMI distributions of $|\psi_n\rangle$.
Parameters: $T = 2\pi/3$, $\hbar = 0.1$, $V_0 = 5.0$. Initial state: $|\psi_0\rangle = \hat{D}(0, 2\pi/\sqrt{3})\,|5\rangle$.

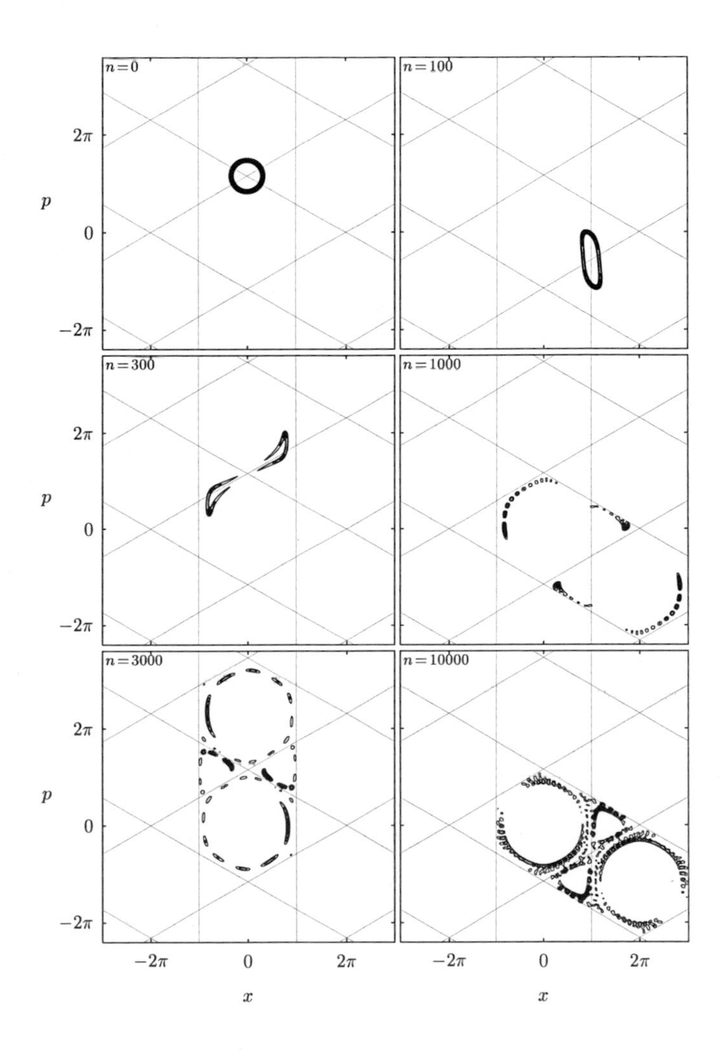

Figure C.35: Contour plots of HUSIMI distributions of $|\psi_n\rangle$.
Parameters: $T = 2\pi/3$, $\hbar = 0.01$, $V_0 = 1.0$. Initial state: $|\psi_0\rangle = \hat{D}\big(0, 2\pi/\sqrt{3}\big)\,|50\rangle$.

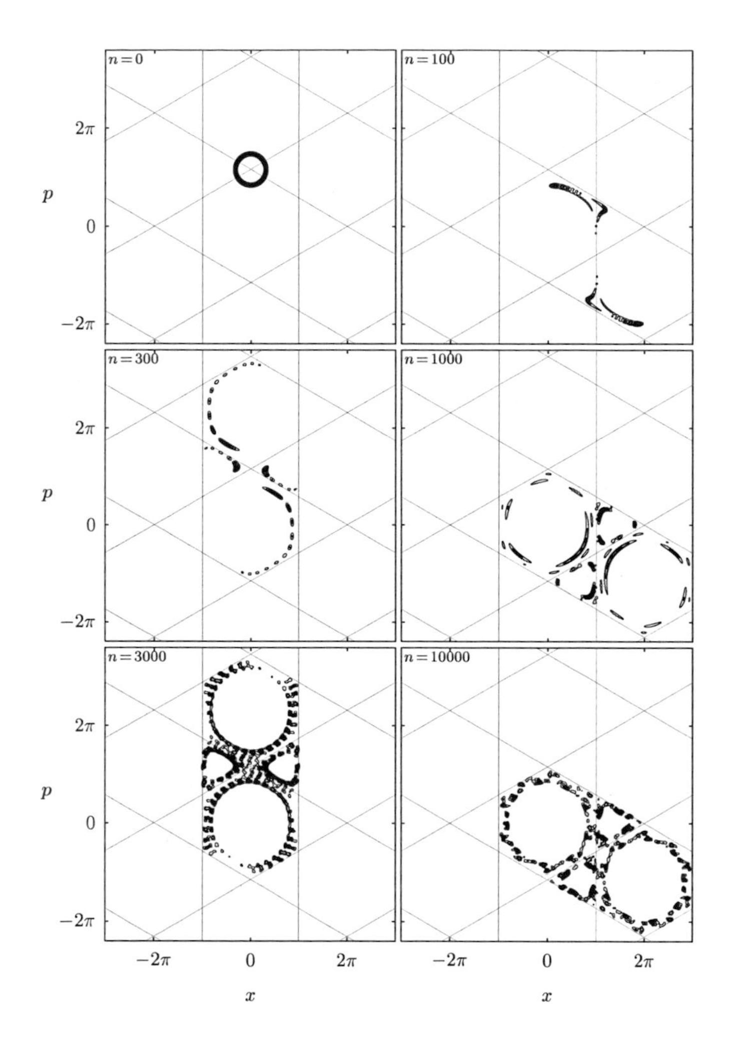

Figure C.36: Contour plots of HUSIMI distributions of $|\psi_n\rangle$.
Parameters: $T = 2\pi/3$, $\hbar = 0.01$, $V_0 = 5.0$. Initial state: $|\psi_0\rangle = \hat{D}\big(0, 2\pi/\sqrt{3}\big)\,|50\rangle$.

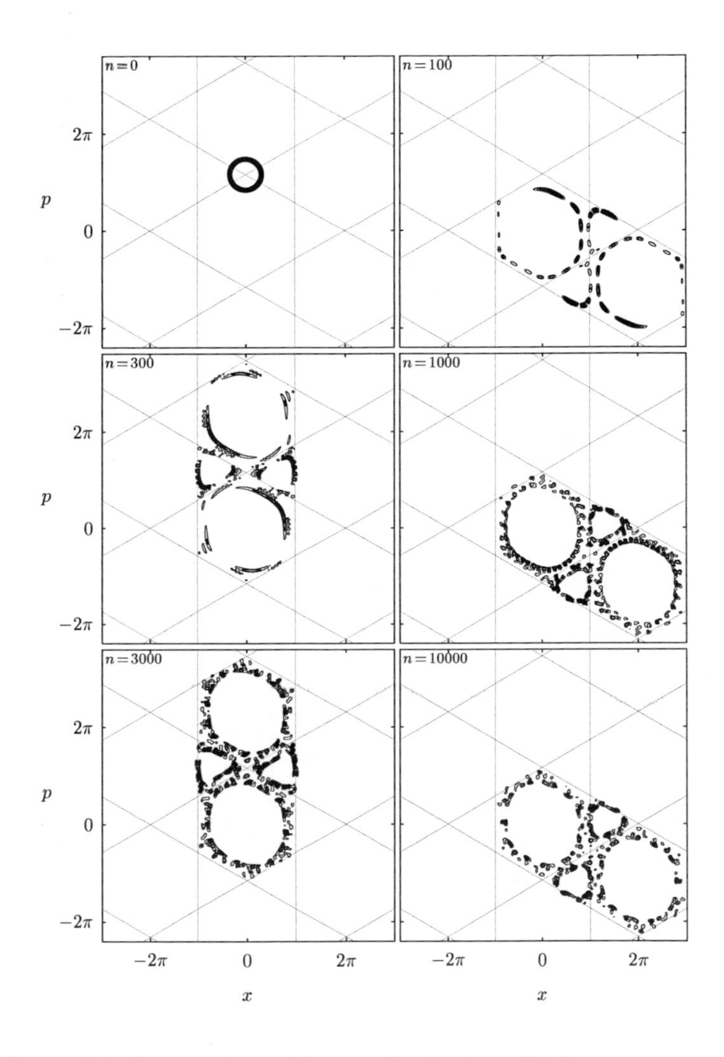

Figure C.37: Contour plots of HUSIMI distributions of $|\psi_n\rangle$.
Parameters: $T = 2\pi/3$, $\hbar = 0.01$, $V_0 = 20.0$. Initial state: $|\psi_0\rangle = \hat{D}\big(0, 2\pi/\sqrt{3}\big)\,|50\rangle$.

C.3 $T = 2\pi$, $p_0 = 0.0$

Figures C.38–C.40.

$T = 2\pi$ $(q = 1)$ classically generates a trivial stochastic web which is translation invariant with respect to translations by 2π in x-direction and arbitrary translations in p-direction — cf. figure 1.9. The solution (1.26) shows that the classical dynamics is confined to vertical lines in phase space and that p_n grows linearly with n, such that ballistic energy growth $E_n \sim n^2$ is obtained. The figures in this section display corresponding quantum pictures for several values of $\hbar$. Note the different scaling of the x- and p-axes in these pictures, as opposed to all other pictures in this appendix.

After just a few kicks, a strong stretching mechanism in p-direction becomes obvious, the magnitude of which increases with $\hbar$, thereby indicating a genuine quantum effect. In all three figures, after at most 100 kicks the quantum state has propagated near to the boundary of the numerically accessible region of phase space.

It is clear that for $p_0 \neq 0.0$ no different kind of dynamics is to be expected; therefore such pictures are omitted here.

Similar pictures are obtained for $T = \pi$ $(q = 2)$.

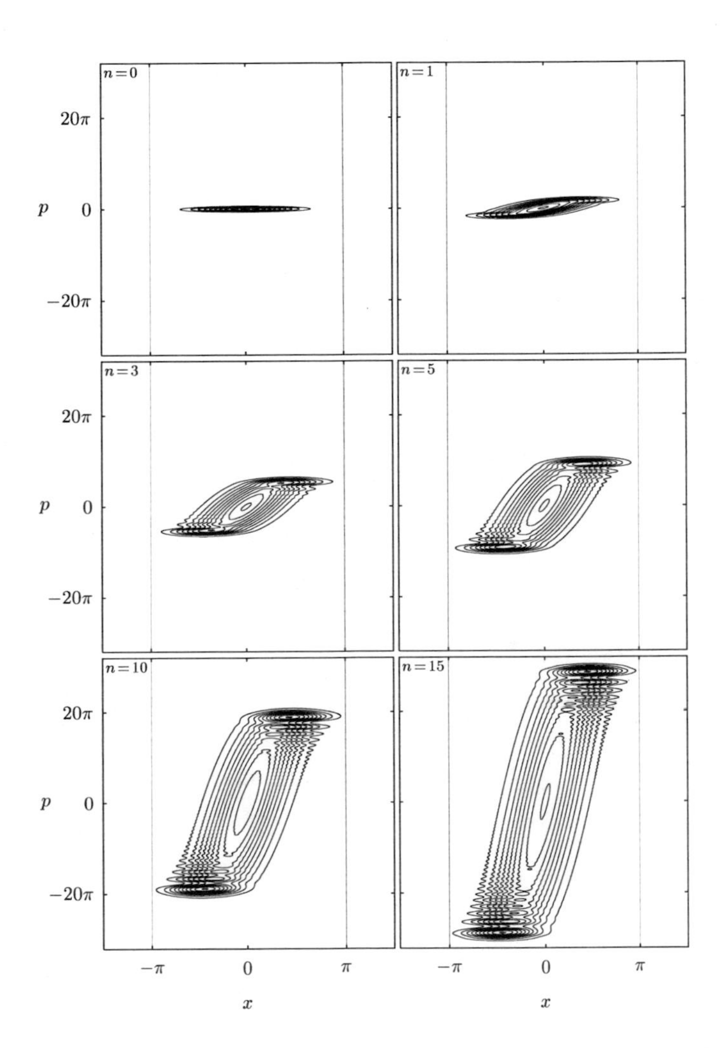

Figure C.38: Contour plots of HUSIMI distributions of $|\psi_n\rangle$.
Parameters: $T = 2\pi$, $\hbar = 1.0$, $V_0 = 1.0$. Initial state: $|\psi_0\rangle = |0\rangle$.

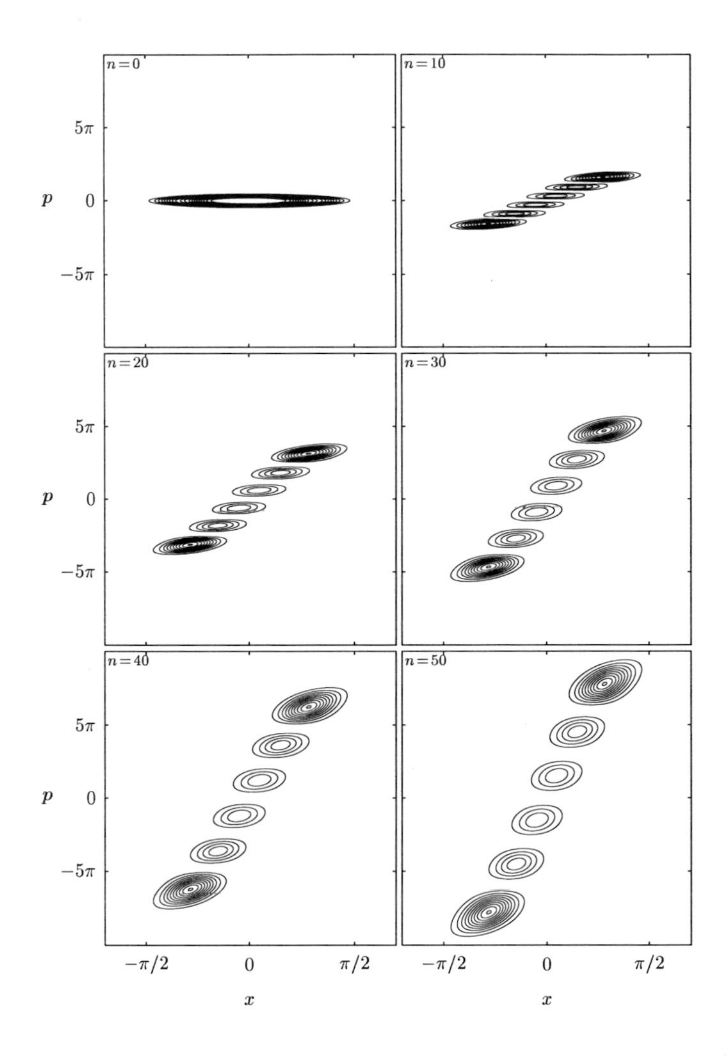

Figure C.39: Contour plots of HUSIMI distributions of $|\psi_n\rangle$.
Parameters: $T = 2\pi$, $\hbar = 0.1$, $V_0 = 1.0$. Initial state: $|\psi_0\rangle = |5\rangle$.

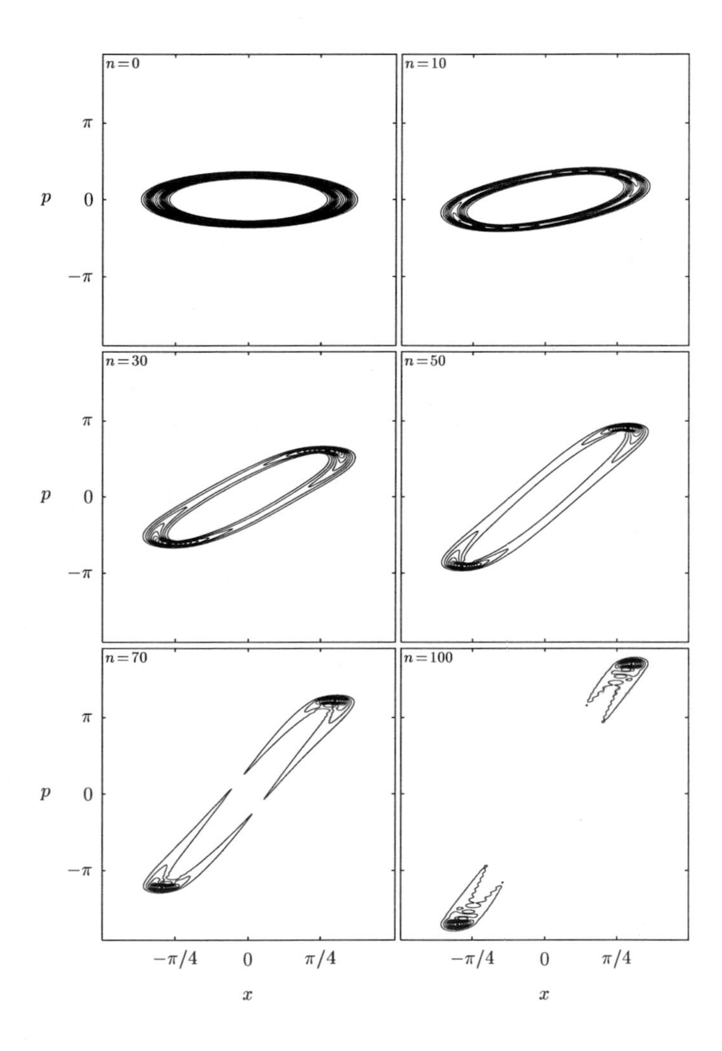

Figure C.40: Contour plots of HUSIMI distributions of $|\psi_n\rangle$.
Parameters: $T = 2\pi$, $\hbar = 0.01$, $V_0 = 1.0$. Initial state: $|\psi_0\rangle = |50\rangle$.

C.4 Examples of Quantum Localization

In the absence of resonances as specified by equation (1.33), the quantum dynamics typically is localized. In particular, no web-like phase space structures, as displayed in the previous sections, are generated. This is demonstrated here using the nonresonant example of $T = 1.0$.

C.4.1 $T = 1.0, \; p_0 = 0.0$

Figures 5.8–5.10 and C.41–C.46.

As discussed in section 5.2, the experience gained while numerically iterating the quantum map for many different combinations of parameters and initial states indicates that in the case of localization the dynamics is "robust" with respect to the initial state: having iterated just long enough, for a given parameter combination always the same type of quantum dynamics evolves, regardless of the exact choice of $|\psi_0\rangle$: rotation around the phase space origin, with the HUSIMI distribution essentially nonzero only in an annular area that includes the location of the initial $F^{\mathrm{H}}(x, p, -0; 1)$.

Note that there are indications of *quantum revivals* [BL57, YMS90, AP89] — see for example figure C.41, $n = 1000/10000$ and $n = 300/100000$; many more examples can be identified in the numerically collected data. I do not discuss this feature any further here.

C.4.2 $T = 1.0, \; p_0 = 12.5$

Figures C.47–C.52.

The classical figure 1.4 shows that, at least for $V_0 = 1.0$, there are invariant lines in the phase portrait encircling the origin. In order to exclude the possibility that these classical invariant lines account for the observed localized quantum dynamics, in the present section initial states are used that are shifted to $(0, 12.5)^t$, well beyond the outermost surviving invariant line. Essentially, the same localized rotation dynamics as in the previous subsection is obtained.

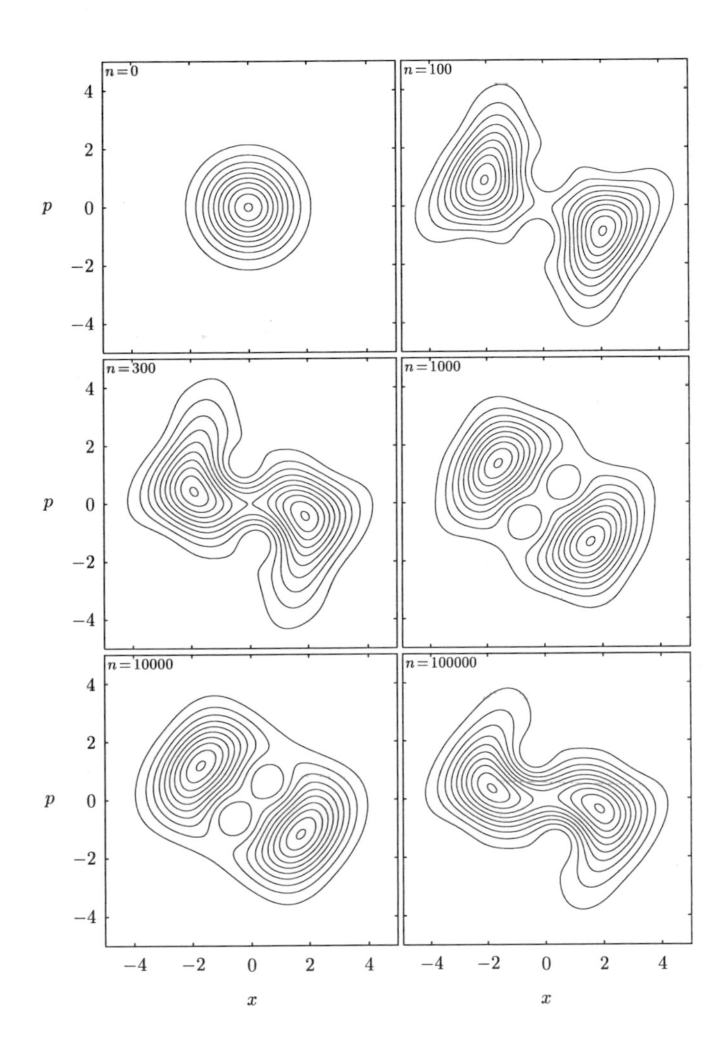

Figure C.41: Contour plots of HUSIMI distributions of $|\psi_n\rangle$.
Parameters: $T = 1.0$, $\hbar = 1.0$, $V_0 = 2.0$. Initial state: $|\psi_0\rangle = |0\rangle$.

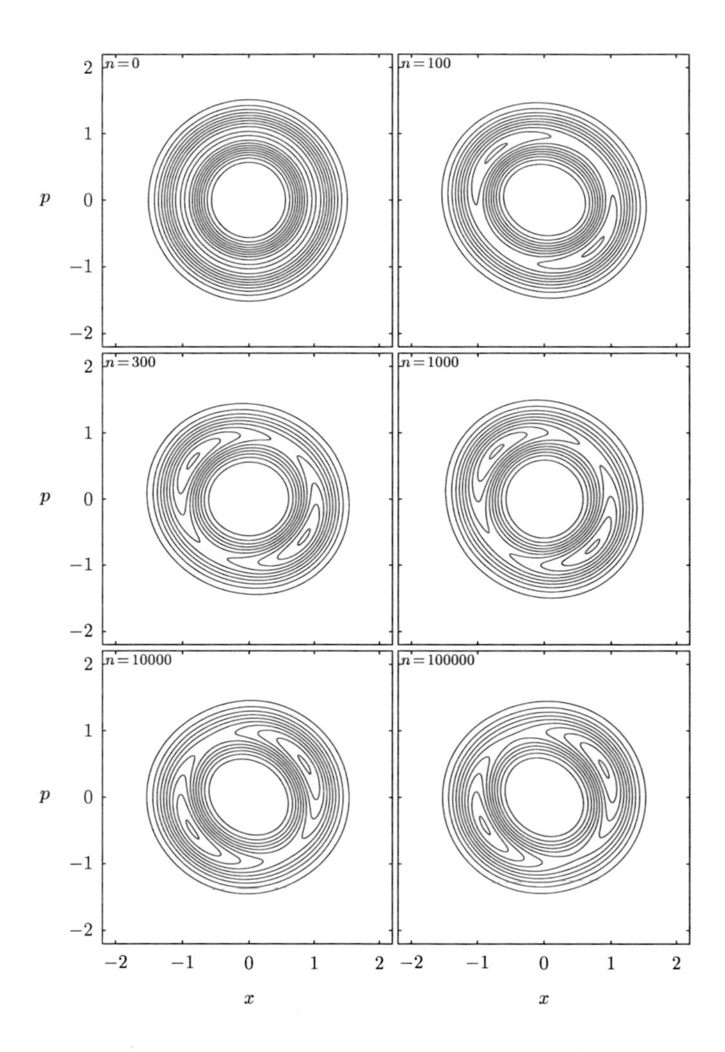

Figure C.42: Contour plots of HUSIMI distributions of $|\psi_n\rangle$.
Parameters: $T = 1.0$, $\hbar = 0.1$, $V_0 = 1.0$. Initial state: $|\psi_0\rangle = |5\rangle$.

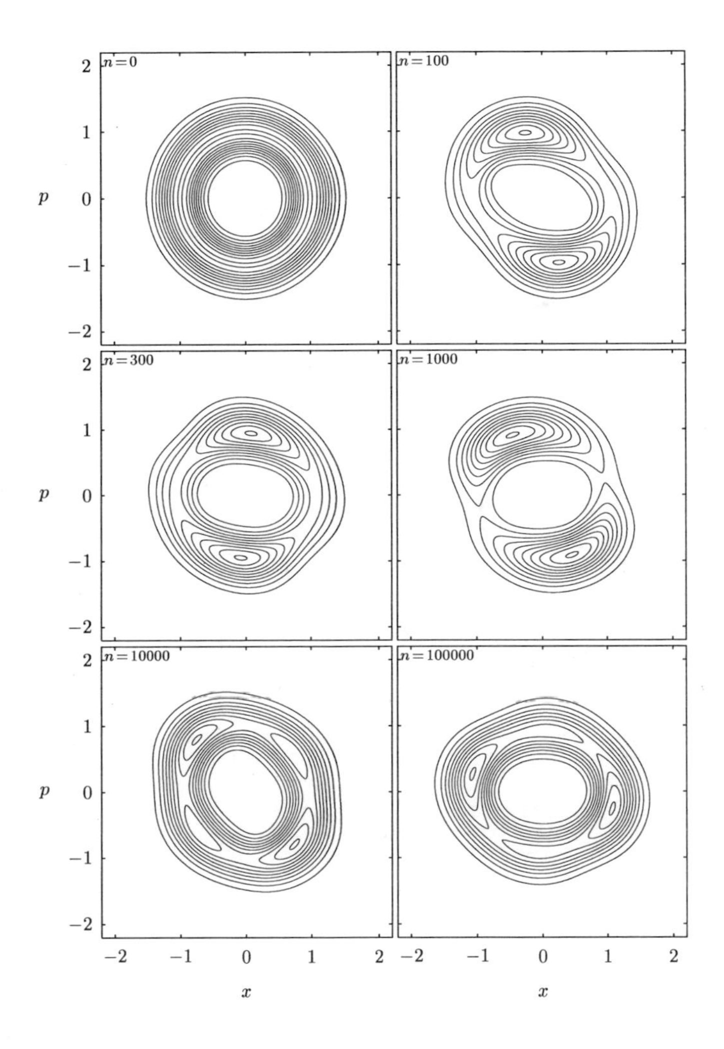

Figure C.43: Contour plots of HUSIMI distributions of $|\psi_n\rangle$.
Parameters: $T = 1.0$, $\hbar = 0.1$, $V_0 = 3.0$. Initial state: $|\psi_0\rangle = |5\rangle$.

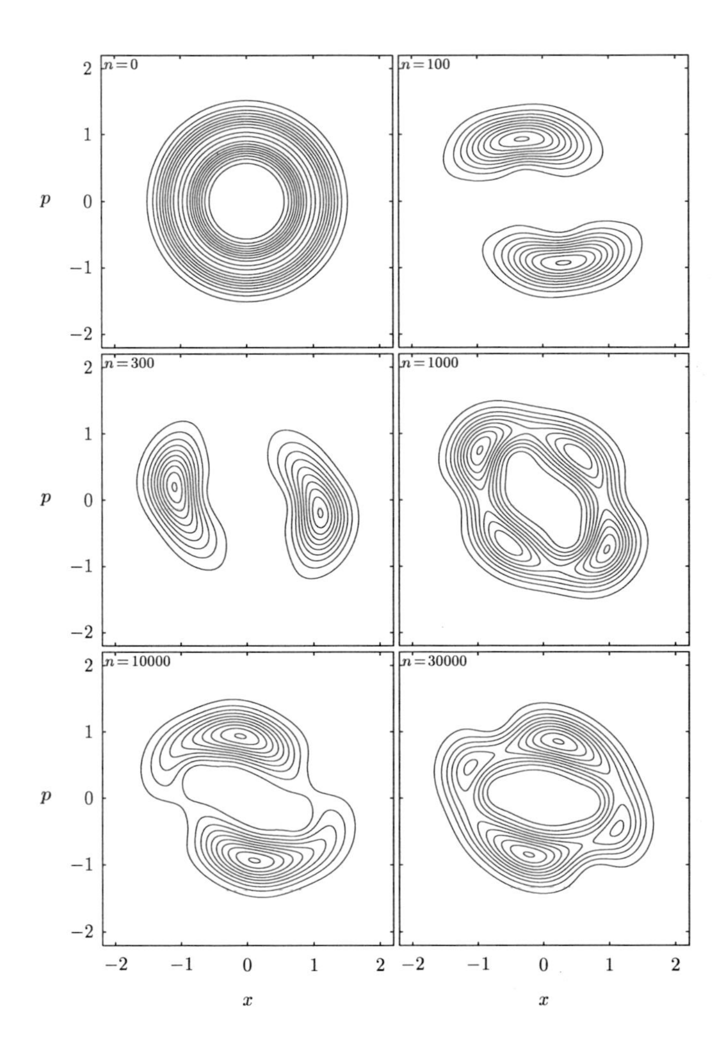

Figure C.44: Contour plots of HUSIMI distributions of $|\psi_n\rangle$.
Parameters: $T = 1.0$, $\hbar = 0.1$, $V_0 = 5.0$. Initial state: $|\psi_0\rangle = |5\rangle$.

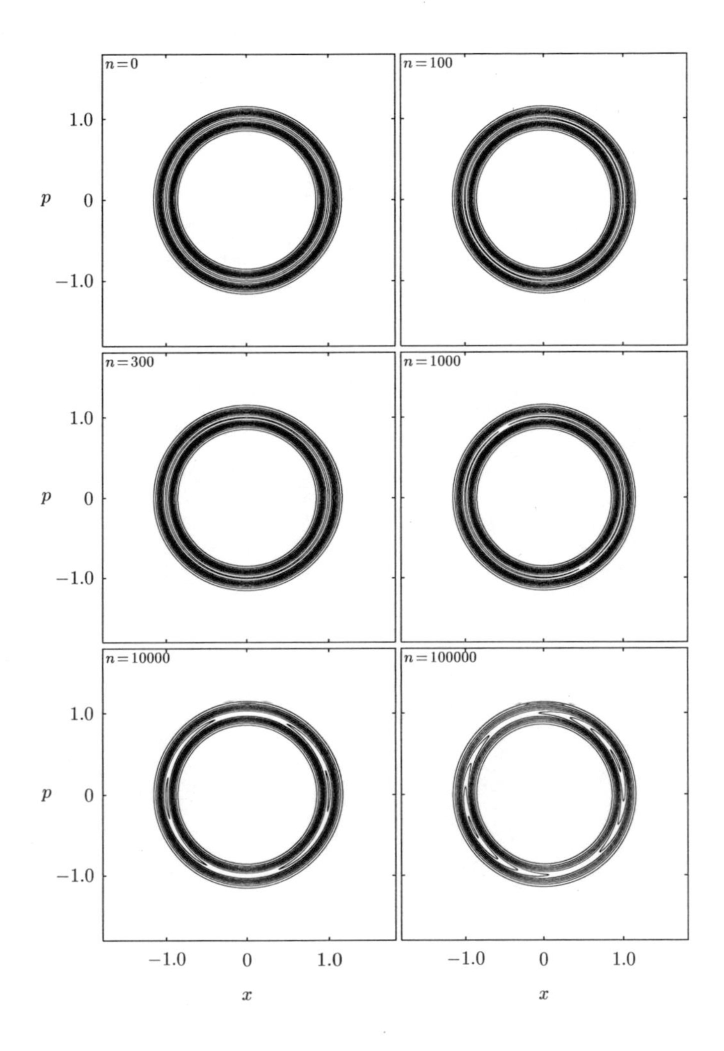

Figure C.45: Contour plots of HUSIMI distributions of $|\psi_n\rangle$.
Parameters: $T = 1.0$, $\hbar = 0.01$, $V_0 = 1.0$. Initial state: $|\psi_0\rangle = |50\rangle$.

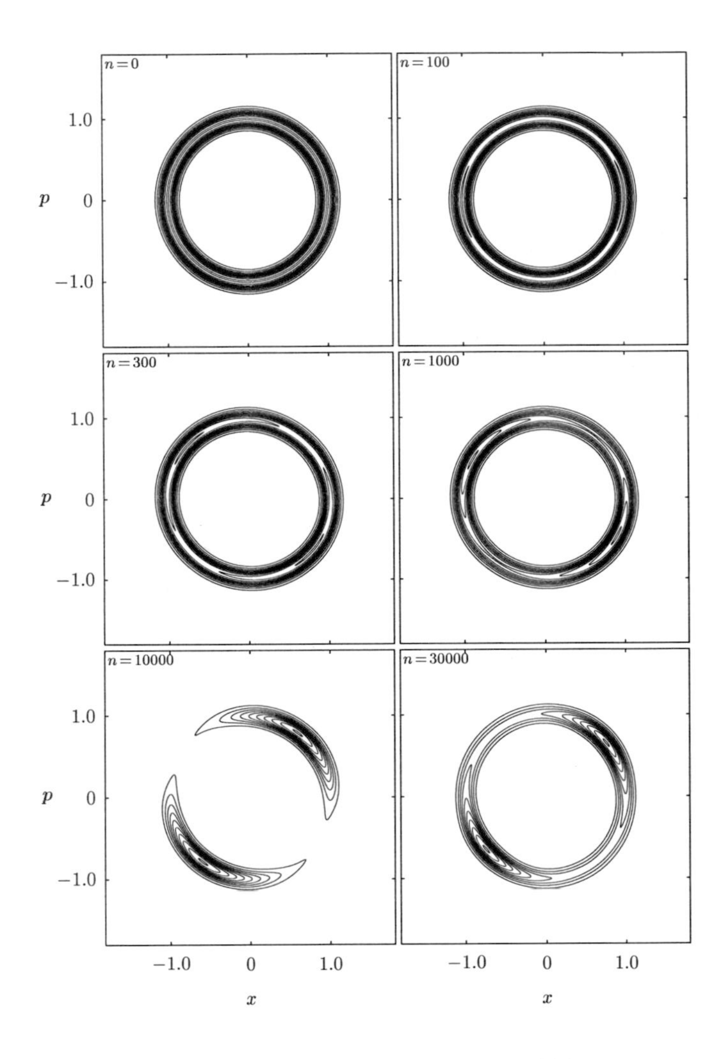

Figure C.46: Contour plots of HUSIMI distributions of $|\psi_n\rangle$.
Parameters: $T = 1.0$, $\hbar = 0.01$, $V_0 = 5.0$. Initial state: $|\psi_0\rangle = |50\rangle$.

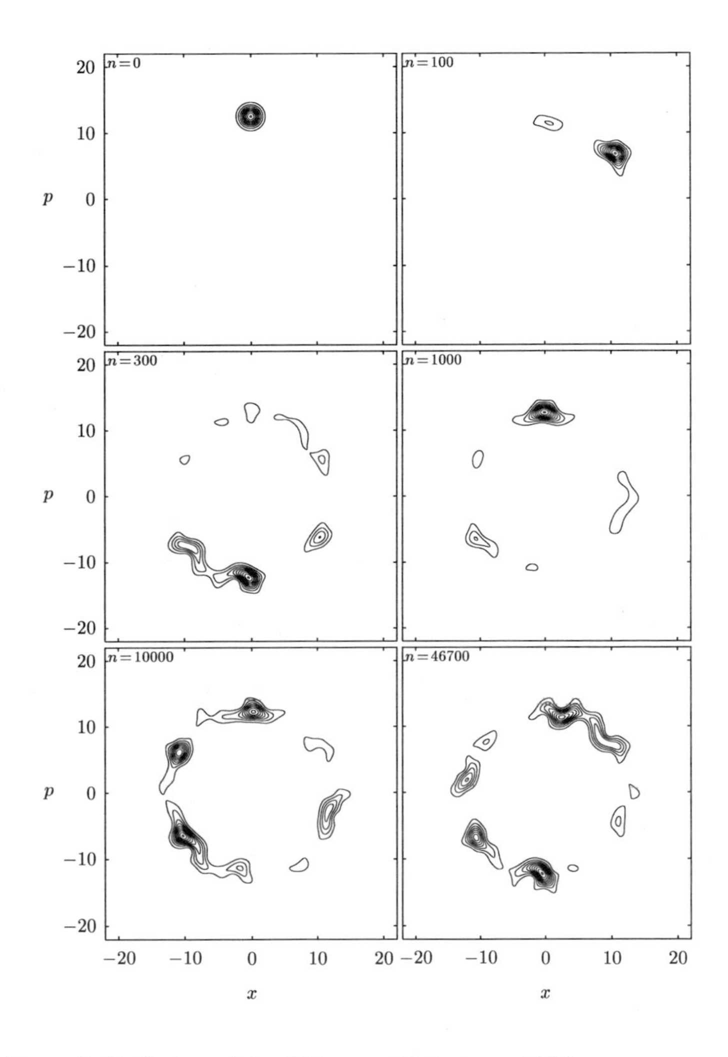

Figure C.47: Contour plots of HUSIMI distributions of $|\psi_n\rangle$.
Parameters: $T = 1.0$, $\hbar = 1.0$, $V_0 = 1.0$. Initial state: $|\psi_0\rangle = \hat{D}(0, 12.5)\,|0\rangle$.

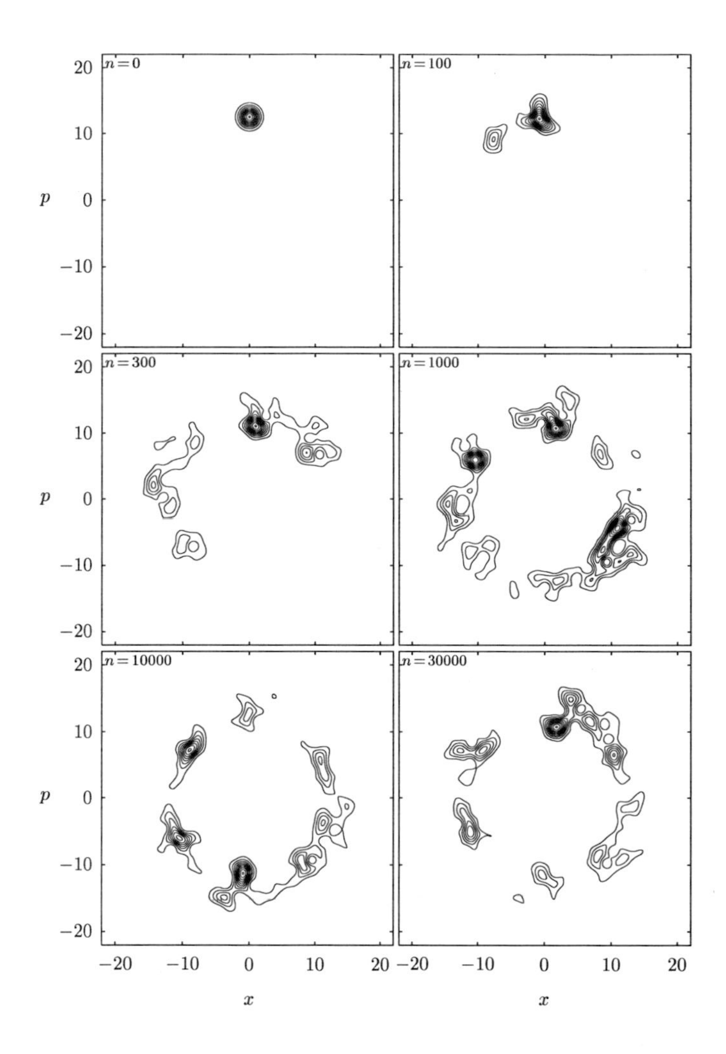

Figure C.48: Contour plots of HUSIMI distributions of $|\psi_n\rangle$.
Parameters: $T = 1.0$, $\hbar = 1.0$, $V_0 = 2.0$. Initial state: $|\psi_0\rangle = \hat{D}(0, 12.5)\,|0\rangle$.

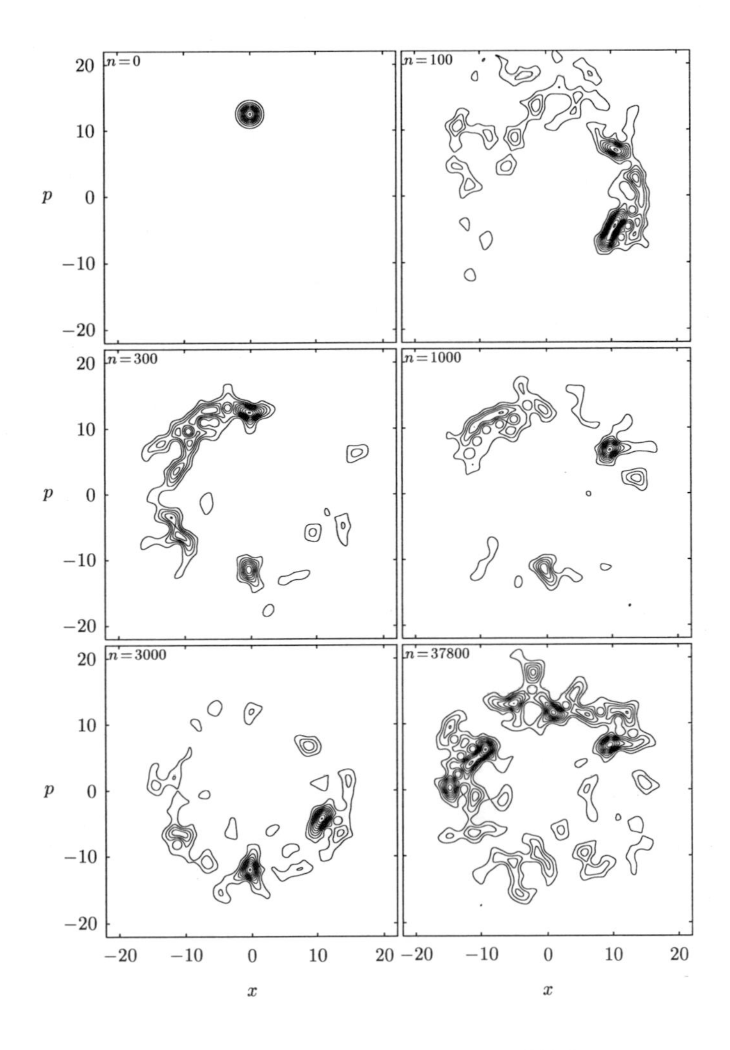

Figure C.49: Contour plots of HUSIMI distributions of $|\psi_n\rangle$.
Parameters: $T = 1.0$, $\hbar = 1.0$, $V_0 = 3.0$. Initial state: $|\psi_0\rangle = \hat{D}\big(0, 12.5\big)\,|0\rangle$.

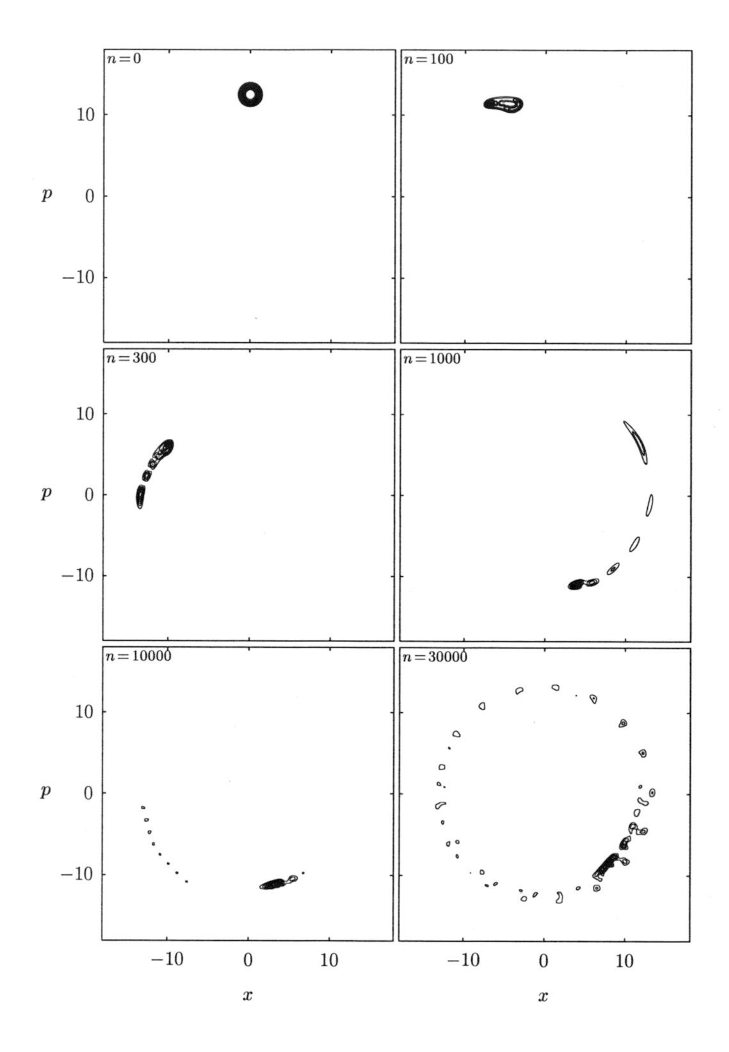

Figure C.50: Contour plots of HUSIMI distributions of $|\psi_n\rangle$.
Parameters: $T = 1.0$, $\hbar = 0.1$, $V_0 = 1.0$. Initial state: $|\psi_0\rangle = \hat{D}(0, 12.5)\,|5\rangle$.

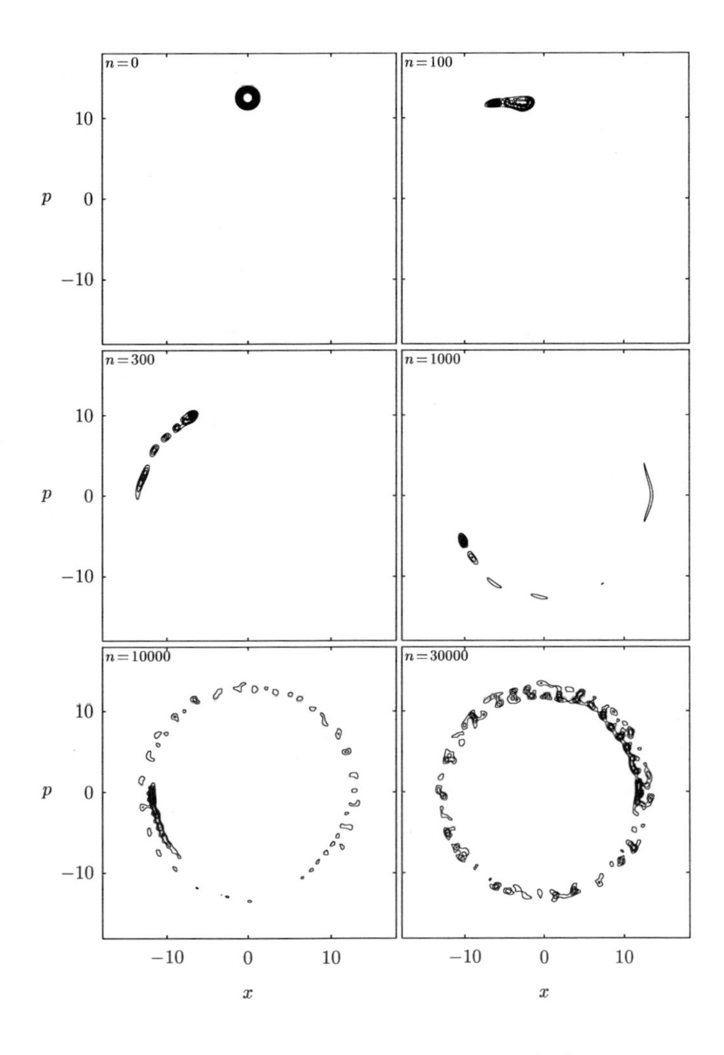

Figure C.51: Contour plots of HUSIMI distributions of $|\psi_n\rangle$.
Parameters: $T = 1.0$, $\hbar = 0.1$, $V_0 = 2.0$. Initial state: $|\psi_0\rangle = \hat{D}(0, 12.5)\,|5\rangle$.

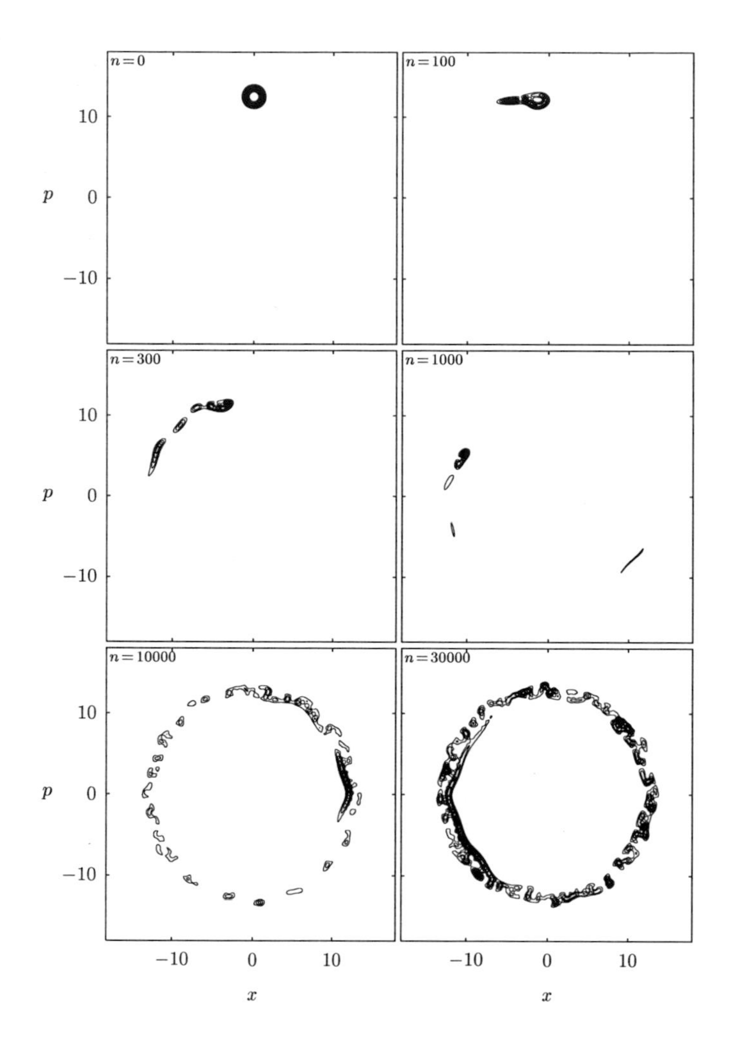

Figure C.52: Contour plots of HUSIMI distributions of $|\psi_n\rangle$.
Parameters: $T = 1.0$, $\hbar = 0.1$, $V_0 = 3.0$. Initial state: $|\psi_0\rangle = \hat{D}(0, 12.5)\,|5\rangle$.

List of Figures

List of Tables

Bibliography

Hier soll ich finden, was mir fehlt?
Soll ich vielleicht in tausend Büchern lesen?

Faust I
JOHANN WOLFGANG V. GOETHE

[ABB96] F. J. ARRANZ, F. BORONDO and R. M. BENITO, *Distribution of zeros of the Husimi function in a realistic Hamiltonian molecular system*, Phys. Rev. E **54** (1996) 2458–2464.

[AFH95] J. ARGYRIS, G. FAUST and M. HAASE, *Die Erforschung des Chaos. Studienbuch für Naturwissenschaftler und Ingenieure*, Vieweg, Braunschweig (1995).

[AH00] Y. ASHKENAZY and L. P. HORWITZ, *The effect of radiation on the stochastic web*, Discrete Dyn. Nat. Soc. **4** (2000) 283–292.

[AM77] G. P. AGRAWAL and C. L. MEHTA, *Ordering of the exponential of a quadratic in boson operators. II. Multimode case*, J. Math. Phys. **18** (1977) 408–412.

[And58] P. W. ANDERSON, *Absence of diffusion in certain random lattices*, Phys. Rev. **109** (1958) 1492–1505.

[And61] P. W. ANDERSON, *Localized magnetic states in metals*, Phys. Rev. **124** (1961) 41–53.

[And78] P. W. ANDERSON, *Local moments and localized states*, Rev. Mod. Phys. **50** (1978) 191–201.

[AP89] I. S. AVERBUKH and N. F. PERELMAN, *Fractional regenerations of wave packets in the course of long-term evolution of highly excited quantum systems*, Sov. Phys. JETP **69** (1989) 464–469.

[AS72] M. ABRAMOWITZ and I. A. STEGUN, *Handbook of Mathematical Functions*, Dover Publications, New York (1972).

[AS83] L. A. ARTSIMOWITSCH and R. S. SAGDEJEW, *Plasmaphysik für Physiker*, Teubner, Stuttgart (1983).

[ASZ91] V. V. AFANASIEV, R. Z. SAGDEEV and G. M. ZASLAVSKY, *Chaotic jets with multifractal space-time random walk*, Chaos **1** (1991) 143–159.

[AZ96] A. ALTLAND and M. R. ZIRNBAUER, *Field theory of the quantum kicked rotor*, Phys. Rev. Lett. **77** (1996) 4536–4539.

[AZ98] A. ALTLAND and M. R. ZIRNBAUER, Reply to [CIS98], Phys. Rev. Lett. **80** (1998) 641.

[BB97] M. BRACK and R. K. BHADURI, *Semiclassical Physics*, vol. 96 of *Frontiers in Physics*, Addison-Wesley Publishing Company, Reading (1997).

[Ber83] M. V. BERRY, *Semiclassical mechanics of regular and irregular motion*, in: G. IOOSS, R. H. G. HELLEMAN and R. STORA (eds.), *Chaotic Behaviour of Deterministic Systems*, Les Houches: École d'Été de Physique Théorique, Session XXXVI (1981), 171–271. North Holland Publishing Company, Amsterdam (1983).

[Ber89] M. V. BERRY, *Quantum chaology, not quantum chaos*, Physica Scripta **40** (1989) 335–336.

[Ber01] M. V. BERRY, *Chaos and the semiclassical limit of quantum mechanics (is the moon there when somebody looks?)*, in: R. J. RUSSELL, P. CLAYTON, K. WEGTER-MCNELLY and J. POLKINGHORNE (eds.), *Quantum Mechanics: Scientific Perspectives on Divine Action*, 41–54. Vatican Observatory Publications, Vatican City State, and Center for Theology and the Natural Sciences, Berkeley (2001).

[BGS84] O. BOHIGAS, M. J. GIANNONI and C. SCHMIT, *Characterization of chaotic spectra and universality of level fluctuation laws*, Phys. Rev. Lett. **52** (1984) 1–4.

[BJ84] N. L. BALAZS and B. K. JENNINGS, *Wigner's function and other distribution functions in mock phase spaces*, Phys. Rep. **104** (1984) 347–391.

[BK97] M. V. BERRY and S. KLEIN, *Transparent mirrors: rays, waves and localization*, Eur. J. Phys. **18** (1997) 222–228.

[BL57] P. BOCCHIERI and A. LOINGER, *Quantum recurrence theorem*, Phys. Rev. **107** (1957) 337–338.

[Bor63] R. E. BORLAND, *The nature of the electronic states in disordered one-dimensional systems*, Proc. Roy. Soc. London A **274** (1963) 529–545.

[BR95] R. BORGONOVI and L. REBUZZINI, *Translational invariance in the kicked harmonic oscillator*, Phys. Rev. E **52** (1995) 2302–2309.

[BR97] R. BLÜMEL and W. P. REINHARDT, *Chaos in Atomic Physics*, vol. 10 of *Cambridge Monographs on Atomic, Molecular and Chemical Physics*, Cambridge University Press, Cambridge (1997).

[Bru94] A. D. BRUNO, *The Restricted 3-Body Problem: Plane Periodic Orbits*, vol. 17 of *de Gruyter Expositions in Mathematics*, Walter de Gruyter, Berlin (1994).

[BRWK99] K. BANASZEK, C. RADZEWICZ, K. WÓDKIEWICZ and J. S. KRASIŃSKI, *Direct measurement of the Wigner function by photon counting*, Phys. Rev. A **60** (1999) 674–677.

[BRZ91] G. P. BERMAN, V. Y. RUBAEV and G. M. ZASLAVSKY, *The problem of quantum chaos in a kicked harmonic oscillator*, Nonlinearity **4** (1991) 543–566.

[BS93] C. BECK and F. SCHLÖGL, *Thermodynamics of Chaotic Systems*, vol. 4 of *Cambridge Nonlinear Science Series*, Cambridge University Press, Cambridge (1993).

[Bun95] G. W. BUND, *Classical distribution functions derived from Wigner distribution functions*, J. Phys. A: Math. Gen. **28** (1995) 3709–3717.

[BW96] K. BANASZEK and K. WÓDKIEWICZ, *Direct probing of quantum phase space by photon counting*, Phys. Rev. Lett. **76** (1996) 4344–4347.

[BW97] K. BANASZEK and K. WÓDKIEWICZ, *Accuracy of sampling quantum phase space in photon counting experiment*, J. Mod. Optics **44** (1997) 2441–2454.

[CAM+03] P. CVITANOVIĆ, R. ARTUSO, R. MAINIERI, G. TANNER and G. VATTAY, *Chaos: Classical and Quantum*, `http://www.nbi.dk/ChaosBook/`, Niels Bohr Institute, Copenhagen (2003).

[CC95] G. CASATI and B. V. CHIRIKOV (eds.), *Quantum Chaos. Between Order and Disorder*, Cambridge University Press, Cambridge (1995).

[CCIF79] G. CASATI, B. V. CHIRIKOV, F. M. IZRAELEV and J. FORD, *Stochastic behavior of a quantum pendulum under a periodic perturbation*, in: G. CASATI and J. FORD (eds.), *Stochastic Behaviour in Classical and Quantum Hamiltonian Systems*, vol. 93 of *Lecture Notes in Physics*, 334–352. Springer, Berlin (1979).

[CGS93] G. CASATI, I. GUARNERI and U. SMILANSKY (eds.), *Quantum Chaos*, Proceedings of the International School of Physics "Enrico Fermi", Course CXIX (1991). North-Holland, Amsterdam (1993).

[Che89] J.-Q. CHEN, *Group Representation Theory for Physicists*, World Scientific, Singapore (1989).

[Chi79] B. V. CHIRIKOV, *A universal instability of many-dimensional oscillator systems*, Phys. Rep. **52** (1979) 263–379.

[CIS98] G. CASATI, F. M. IZRAILEV and V. V. SOKOLOV, *Dynamical theory of quantum chaos or a hidden random matrix ensemble?* Comment on [AZ96], Phys. Rev. Lett. **80** (1998) 640.

[CKM87] R. CARMONA, A. KLEIN and F. MARTINELLI, *Anderson localization for Bernoulli and other singular potentials*, Commun. Math. Phys. **108** (1987) 41–66.

[CM81] J. R. CARY and J. D. MEISS, *Rigorously diffusive deterministic map*, Phys. Rev. A **24** (1981) 2664–2668.

[Coh66] L. COHEN, *Generalized phase-space distribution function*, J. Math. Phys. **7** (1966) 781–786.

[Coh94] D. COHEN, *Noise, dissipation and the classical limit in the quantum kicked-rotator problem*, J. Phys. A: Math. Gen. **27** (1994) 4805–4829.

[Con02] G. CONTOPOULOS, *Order and Chaos in Dynamical Astronomy*, Springer, Berlin (2002).

[CR62] C. W. CURTIS and I. REINER, *Representation Theory of Finite Groups and Associate Algebras*, vol. XI of *Pure and Applied Mathematics*, Wiley Interscience Publishers, New York (1962).

[CSU+87] A. A. CHERNIKOV, R. Z. SAGDEEV, D. A. USIKOV, M. Y. ZAKHAROV and G. M. ZASLAVSKY, *Minimal chaos and stochastic webs*, Nature **326** (1987) 559–563.

[CSUZ87] A. A. CHERNIKOV, R. Z. SAGDEEV, D. A. USIKOV and G. M. ZASLAVSKY, *The Hamiltonian method for quasicrystal symmetry*, Phys. Lett. A **125** (1987) 101–106.

[CSUZ89] A. A. CHERNIKOV, R. Z. SAGDEEV, D. A. USIKOV and G. M. ZASLAVSKY, *Weak chaos and structures*, Sov. Sci. Rev. C. Math. Phys. **8** (1989) 83–170.

[CT65] J. W. COOLEY and J. W. TUKEY, *An algorithm for the machine calculation of complex Fourier series*, Math. Comput. **19** (1965) 297–301.

[DA95] I. DANA and M. AMIT, *General approach to diffusion of periodically kicked charges in a magnetic field*, Phys. Rev. E **51** (1995) R2731–R2734.

[Dan95] I. DANA, *Kicked Harper models and kicked charge in a magnetic field*, Phys. Lett. A **197** (1995) 413–416.

[DEGN95] M. DINEYKHAN, G. EFIMOV, G. GANBOLD and S. N. NEDLEKO, *Oscillator Representation in Quantum Physics*, vol. m26 of *Monographs*, Springer, Berlin (1995).

[DGS72] S. R. DE GROOT and L. G. SUTTORP, *Foundations of Electrodynamics*, North-Holland, Amsterdam (1972).

[DH95] M. V. DALY and D. M. HEFFERNAN, *Chaos in a resonantly kicked oscillator*, J. Phys. A: Math. Gen. **28** (1995) 2515–2528.

[DH97] M. V. DALY and D. M. HEFFERNAN, *Dynamically enhanced chaotic transport*, Chaos, Sol. & Fract. **8** (1997) 933–939.

[DK96] I. DANA and T. KALISKY, *Symbolic dynamics for strong chaos on stochastic webs: General quasisymmetry*, Phys. Rev. E **53** (1996) R2025–R2028.

[DLS85a] F. DELYON, Y.-E. LÉVY and B. SOUILLARD, *Anderson localization for one- and quasi-one-dimensional systems*, J. Stat. Phys. **41** (1985) 375–388.

[DLS85b] F. DELYON, Y.-E. LÉVY and B. SOUILLARD, *Approach* à la *Borland to multidimensional localization*, Phys. Rev. Lett. **55** (1985) 618–621.

[DM98] F. C. DELGADO and B. MIELNIK, *Are there Floquet quanta?*, Phys. Lett. A **249** (1998) 369–375.

[DMFV96a] R. L. DE MATHOS FILHO and W. VOGEL, *Nonlinear coherent states*, Phys. Rev. A **54** (1996) 4560–4563.

[DMFV96b] R. L. DE MATHOS FILHO and W. VOGEL, *Even and odd coherent states of the motion of a trapped ion*, Phys. Rev. Lett. **76** (1996) 608–611.

[DR87] H. De Raedt, *Product formula algorithms for solving the time dependent Schrödinger equation*, Comp. Phys. Rep. **7** (1987) 1–72.

[DR96] H. De Raedt, *Quantum dynamics in nanoscale devices*, in: K. H. Hoffmann and M. Schreiber (eds.), *Computational Physics — Selected Methods, Simple Exercises, Serious Applications*, 209–224. Springer, Berlin (1996).

[Dro95] A. N. Drozdov, *Power series expansion for the time evolution operator with a harmonic-oscillator reference system*, Phys. Rev. Lett. **75** (1995) 4342–4345.

[DV73] P. Du Val, *Elliptic Functions and Elliptic Curves*, vol. 9 of *London Mathematical Society Lecture Note Series*, Cambridge University Press, Cambridge (1973).

[Ein06] A. Einstein, *Zur Theorie der Brownschen Bewegung*, Ann. d. Phys. **4** (1906) 371–381.

[EMR93] G. Engeln-Müllges and F. Reutter, *Numerik-Algorithmen mit ANSI-C-Programmen*, BI-Wissenschaftsverlag, Mannheim (1993).

[Eng93] U. M. Engel, *Normalformen und Quasiintegrale für Magnetische Flaschen*, Diploma thesis, Westfälische Wilhelms-Universität Münster (1993).

[Esc94] M. C. Escher, *Zon en maan (1948)*, in: J. L. Locher (ed.), *Leben und Werk M. C. Escher*, 295. RVG Interbook Verlagsgesellschaft, Remseck (1994). More examples can be viewed online at `http://ftp.sunet.se/pub/pictures/art/M.C.Escher/`.

[ESE95] U. M. Engel, B. Stegemerten and P. Eckelt, *Normal forms and quasi-integrals for the Hamiltonians of magnetic bottles*, J. Phys. A: Math. Gen. **28** (1995) 1425–1448.

[FGKP95] R. Fleischmann, T. Geisel, R. Ketzmerick and G. Petschel, *Quantum diffusion, fractal spectra, and chaos in semiconductor microstructures*, Physica D **86** (1995) 171–181.

[FGP82] S. Fishman, D. R. Grempel and R. E. Prange, *Chaos, quantum recurrences, and Anderson localization*, Phys. Rev. Lett. **49** (1982) 509–512.

[FH65] R. P. Feynman and A. R. Hibbs, *Quantum Mechanics and Path Integrals*, McGraw-Hill Book Company, New York (1965).

[Fis93] S. FISHMAN, *Quantum localization*, in: G. CASATI, I. GUARNERI and U. SMILANSKY (eds.), *Quantum Chaos*, Proceedings of the International School of Physics "Enrico Fermi", Course CXIX (1991), 187–219. North-Holland, Amsterdam (1993).

[Fis96] W. FISCHER, *Zur elektronischen Lokalisierung durch gaußsche zufällige Potentiale*, PhD thesis, Friedrich-Alexander-Universität Erlangen-Nürnberg (1996).

[Flo83] M. G. FLOQUET, *Sur les équations différentielles linéaires à coéfficients périodiques*, Ann. Scient. de l'Ecole Norm. Sup. **12** (1883) 47–88.

[FMSS85] J. FRÖHLICH, F. MARTINELLI, E. SCOPPOLA and T. SPENCER, *Constructive proof of localization in the Anderson tight binding model*, Commun. Math. Phys. **101** (1985) 21–46.

[For88] J. FORD, *Quantum Chaos — is there any?*, in: B.-L. HAO (ed.), *Directions in Chaos Vol. 2*, vol. 4 of *World Scientific Series on Directions in Condensed Matter Physics*, 128–147. World Scientific, Singapore (1988).

[Fur63] H. FURSTENBERG, *Noncommuting random products*, Trans. Am. Math. Soc. **108** (1963) 377–428.

[GB93] I. GUARNERI and F. BORGONOVI, *Generic properties of a class of translation invariant quantum maps*, J. Phys. A: Math. Gen. **26** (1993) 119–132.

[Ger92] H. A. GERSCH, *Time evolution of minimum uncertainty states of a harmonic oscillator*, Am. J. Phys. **60** (1992) 1024–1030.

[GFP82] D. R. GREMPEL, S. FISHMAN and R. E. PRANGE, *Localization in an incommensurate potential: An exactly solvable model*, Phys. Rev. Lett. **49** (1982) 833–836.

[GH83] J. GUCKENHEIMER and P. HOLMES, *Nonlinear Oscillations, Dynamical Systems, and Bifurcations of Vector Fields*, vol. 42 of *Applied Mathematical Sciences*, Springer-Verlag, New York (1983).

[Gla63a] R. J. GLAUBER, *The quantum theory of optical coherence*, Phys. Rev. **130** (1963) 2529–2539.

[Gla63b] R. J. GLAUBER, *Coherent and incoherent states of the radiation field*, Phys. Rev. **131** (1963) 2766–2788.

[Gla63c] R. J. GLAUBER, *Photon correlations*, Phys. Rev. Lett. **10** (1963) 84–86.

[Gla65] R. J. GLAUBER, *Optical coherence and photon statistics*, in: C. DE WITT, A. BLANDIN and C. COHEN-TANNOUDJI (eds.), *Quantum Optics and Electronics*, Les Houches: École d'Été de Physique Théorique, Session XIV (1964), 63–185. Gordon and Breach, Science Publishers, New York (1965).

[Gla66] R. J. GLAUBER, *Classical behavior of systems of quantum oscillators*, Phys. Lett. **21** (1966) 650–652.

[Gol80] H. GOLDSTEIN, *Classical Mechanics*, Addison-Wesley Series in Physics. Addison-Wesley, Reading, 2nd edition (1980).

[GR00] I. S. GRADSHTEYN and I. M. RYZHIK, *Table of Integrals, Series, and Products*, Academic Press, San Diego, 6th edition (2000).

[Gra89] G. GRAWERT, *Quantenmechanik*, Aula-Verlag, Wiesbaden, 5th edition (1989).

[GS87] B. GRÜNBAUM and G. C. SHEPHARD, *Tilings and Patterns*, W. H. Freeman and Company, New York (1987).

[Gus66] F. G. GUSTAVSON, *On constructing formal integrals of a Hamiltonian system near an equilibrium point*, Astron. J. **71** (1966) 670–686.

[Gut90] M. C. GUTZWILLER, *Chaos in Classical and Quantum Mechanics*, vol. 1 of *Interdisciplinary Applied Mathematics*, Springer-Verlag, New York (1990).

[Gut91] M. C. GUTZWILLER, *The semi-classical quantization of chaotic Hamiltonian systems*, in: M.-J. GIANNONI, A. VOROS and J. ZINN-JUSTIN (eds.), *Chaos and Quantum Physics*, Les Houches: École d'Été de Physique Théorique, Session LII (1989), 201–250. North-Holland, Amsterdam (1991).

[GVZJ91] M.-J. GIANNONI, A. VOROS and J. ZINN-JUSTIN (eds.), *Chaos and Quantum Physics*, Les Houches: École d'Été de Physique Théorique, Session LII (1989). North-Holland, Amsterdam (1991).

[HA99] L. P. HORWITZ and Y. ASHKENAZY, *Chaos and maps in relativistic dynamical systems*, e-Print: chao-dyn/9901016 (1999).

[Haa01] F. HAAKE, *Quantum Signatures of Chaos*, vol. 54 of *Springer Series in Synergetics*, Springer-Verlag, Berlin, 2nd edition (2001).

[Har55] P. G. HARPER, *Single band motion of conduction electrons in a uniform magnetic field*, Proc. Phys. Soc. London A **68** (1955) 874–878.

[Hau97] M. HAUKE, *Geladenes Teilchen im elektromagnetischen Kick-Feld*, Diploma thesis, Westfälische Wilhelms-Universität Münster (1997).

[Hau00] M. HAUKE, *Semiklassische Streuung an einem offenen dispersiven Billard*, PhD thesis, Westfälische Wilhelms-Universität Münster (2000).

[HB95] B. S. HELMKAMP and D. A. BROWNE, *Inhibition of mixing in chaotic quantum dynamics*, Phys. Rev. E **51** (1995) 1849–1857.

[HB96] B. S. HELMKAMP and D. A. BROWNE, *The role of the environment in chaotic quantum dynamics*, Phys. Rev. Lett. **76** (1996) 3691–3694.

[Hea92] J. F. HEAGY, *A physical interpretation of the Hénon map*, Physica D **57** (1992) 436–446.

[Hei92] W. D. HEISS (ed.), *Chaos and Quantum Chaos*, vol. 411 of *Lecture Notes in Physics*, Springer, Berlin (1992).

[Hel83] R. H. G. HELLEMAN, *One mechanism for the onsets of large-scale chaos in conservative and dissipative systems*, in: C. W. HORTON, L. E. REICHL and V. G. SZEBEHELY (eds.), *Long-Time Prediction in Dynamics*, 95–126. Wiley, New York (1983).

[Hén76] M. HÉNON, *A two-dimensional map with a strange attractor*, Commun. Math. Phys. **50** (1976) 69–77.

[Hén83] M. HÉNON, *Numerical exploration of Hamiltonian systems*, in: G. IOOSS, R. H. G. HELLEMAN and R. STORA (eds.), *Chaotic Behaviour of Deterministic Systems*, Les Houches: École d'Été de Physique Théorique, Session XXXVI (1981), 53–170. North Holland Publishing Company, Amsterdam (1983).

[HG88] R. W. HENRY and S. C. GLOTZER, *A squeezed-state primer*, Am. J. Phys. **56** (1988) 318–328.

[HG94] F. HIPPERT and D. GRATIAS (eds.), *Lectures on Quasicrystals*, Les Editions de Physique, Les Ulis (1994).

[Hip94] C. HIPPEL, *Stochastische Netze: Struktur und diffusives Wachstum*, Diploma thesis, Westfälische Wilhelms-Universität Münster (1994).

[Hip97] C. HIPPEL, *Chaotische reaktive Dreiteilchenstreuung in der Ebene*, PhD thesis, Westfälische Wilhelms-Universität Münster (1997).

[HM87] R. HEATHER and H. METIU, *An efficient procedure for calculating the evolution of the wave function by fast Fourier transform methods for systems with spatially extended wave function and localized potential*, J. Chem. Phys. **86** (1987) 5009–5017.

[Hor93] R. HORSTMANN, *Quantenchaos im Sinai-Billard*, PhD thesis, Westfälische Wilhelms-Universität Münster (1993).

[HOSW84] E. HILLERY, R. F. O'CONNELL, M. O. SCULLY and E. P. WIGNER, *Distribution functions in physics: Fundamentals*, Phys. Rep. **106** (1984) 121–167.

[Hov92] I. HOVEIJN, *Symplectic Reversible Maps, Tiles and Chaos*, Chaos, Solitons & Fractals **2** (1992) 81–90.

[HS96] K. H. HOFFMANN and M. SCHREIBER (eds.), *Computational Physics — Selected Methods, Simple Exercises, Serious Applications*, Springer–Verlag, Berlin (1996).

[Hus40] 伏見康治 (K. HUSIMI), *Some formal properties of the density matrix*, Proc. Phys. Math. Soc. Japan **22** (1940) 264–314.

[HvMW79] A. HERMANN, K. v. MEYENN and V. F. WEISSKOPF (eds.), *Wolfgang Pauli. Wissenschaftlicher Briefwechsel mit Bohr, Einstein, Heisenberg u. a., Band I: 1919–1929*, Springer-Verlag, New York (1979).

[JJ72] S. H. JEFFREYS and B. S. JEFFREYS, *Methods of Mathematical Physics*, Cambridge University Press, Cambridge, 3rd edition (1972).

[Jor97] S. JORDA, *Ein quantenmechanischer Steckbrief*, Phys. Bl. **53** (1997) 510–511.

[JS98] J. V. JOSÉ and E. J. SALETAN, *Classical Dynamics. A Contemporary Approach*, Cambridge University Press, Cambridge (1998).

[Jun95] B. JUNGLAS, *Geladenes Teilchen im elektromagnetischen Kick-Feld: Untersuchung einer vierdimensionalen Abbildung*, Diploma thesis, Westfälische Wilhelms-Universität Münster (1995).

[Jun97] B. JUNGLAS, *Quantenchaos in einem dispersiven Billard*, PhD thesis, Westfälische Wilhelms-Universität Münster (1997).

[Ken27] E. H. KENNARD, *Zur Quantenmechanik einfacher Bewegungstypen*, Zeit. Phys. **44** (1927) 326–352.

[Kep11] J. KEPLER, *Strena, seu de Nive sexangula*, ad Tampach, Francofurt ad Moemum (1611).

[Kep19] J. KEPLER, *Harmonices Mundi. Liber II: De Congruentia Figurarum Harmonicarum*, Godofredus Tampachius, Francofurt ad Moemum (1619).

[Ket92] R. KETZMERICK, *Chaos, fraktale Spektren und Quantendynamik in Halbleiter-Mikrostrukturen*, vol. 9 of *Reihe Physik*, Verlag Harri Deutsch, Thun (1992).

[KH95a] A. KATOK and B. HASSELBLATT, *Introduction to the modern theory of dynamical systems*, vol. 54 of *Encylopedia of mathematics and its applications*, Cambridge University Press, Cambridge (1995).

[KH95b] S. Y. KILIN and D. B. HOROSHKO, *Fock state generation by the methods of nonlinear optics*, Phys. Rev. Lett. **74** (1995) 5206–5207.

[Kir33] J. G. KIRKWOOD, *Quantum statistics of almost classical assemblies*, Phys. Rev. **44** (1933) 31–37.

[Kle97] A. KLEIN, *Localization in the Anderson model with long range hopping*, Braz. J. of Phys. **23** (1997) 363–371.

[Klu97] T. KLUTH, *Zum Einfluß des Massenverhältnisses auf die Dynamik des kollinearen symmetrischen Dreikörperproblems*, Diploma thesis, Westfälische Wilhelms-Universität Münster (1997).

[KM90] H.-T. KOOK and J. D. MEISS, *Diffusion in symplectic maps*, Phys. Rev. A **41** (1990) 4143–4150.

[Koo86] S. E. KOONIN, *Computational Physics*, The Benjamin/Cummings Publishing Company, Menlo Park (1986).

[KS85] J. R. KLAUDER and B.-S. SKAGERSTAM, *Coherent States*, World Scientific, Singapore (1985).

[KS95] W. KUHN and J. STRNAD, *Quantenfeldtheorie. Photonen und ihre Deutung*, Vieweg, Braunschweig (1995).

[KSD92] J. C. KIMBALL, V. A. SINGH and M. D'SOUZA, *Quantum approximation to regular and chaotic classical motion: An electron in two periodic potentials*, Phys. Rev. A **45** (1992) 7065–7072.

[KW96] H. J. KORSCH and H. WIESCHER, *Quantum chaos*, in: K. H. HOFFMANN and M. SCHREIBER (eds.), *Computational Physics — Selected Methods, Simple Exercises, Serious Applications*, 225–244. Springer–Verlag, Berlin (1996).

[KWZ94] L.-M. KUANG, F.-B. WANG and Y.-G. ZHOU, *Coherent states of a harmonic oscillator in a finite-dimensional Hilbert space and their squeezing properties*, J. Mod. Opt. **41** (1994) 1307–1318.

[Lam93] J. S. W. LAMB, *Crystallographic symmetries of stochastic webs*, J. Phys. A: Math. Gen. **26** (1993) 2921–2933.

[Lan94] B. L. LAN, *Wave-packet initial motion, spreading, and energy in the periodically kicked pendulum*, Phys. Rev. E **50** (1994) 764–769.

[Lax74] M. LAX, *Symmetry Principles in Solid State and Molecular Physics*, John Wiley & Sons, New York (1974).

[Lee95] H.-W. LEE, *Theory and application of the quantum phase-space distribution functions*, Phys. Rep. **259** (1995) 147–211.

[Leo95] U. LEONHARDT, *Quantum-state tomography and discrete Wigner function*, Phys. Rev. Lett. **74** (1995) 4101–4105.

[Leo96] U. LEONHARDT, *Discrete Wigner function and quantum-state tomography*, Phys. Rev. A **53** (1996) 2998–3013.

[LL73] L. D. LANDAU and E. M. LIFSCHITZ, *Mechanik*, vol. 1 of *Lehrbuch der theoretischen Physik*, Akademie-Verlag, Berlin, 8th edition (1973).

[LL92] A. J. LICHTENBERG and M. A. LIEBERMAN, *Regular and Chaotic Dynamics*, vol. 38 of *Applied Mathematical Sciences*, Springer-Verlag, New York, 2nd edition (1992).

[Lou64] W. H. LOUISELL, *Radiation and Noise in Quantum Electronics*, McGraw-Hill Physical and Quantum Electronic Series. McGraw-Hill, New York (1964).

[Lou73] W. H. LOUISELL, *Quantum Statistical Properties of Radiation*, Wiley Series in Pure and Applied Optics. John Wiley & Sons, New York (1973).

[Low91] J. H. LOWENSTEIN, *Parameter dependence of stochastic layers in a quasicrystalline web*, Chaos **1** (1991) 473–481.

[Low92] J. H. LOWENSTEIN, *Interpolating Hamiltonians for a stochastic-web map with quasicrystalline symmetry*, Chaos **2** (1992) 413–422.

[Low96] J. H. LOWENSTEIN, *Equal abundance of positive- and negative-residue fixed points for resonantly kicked harmonic oscillators*, Nonlinearity **9** (1996) 1071–1088.

[LQ94] J. S. W. LAMB and G. R. W. QUISPEL, *Reversing k-symmetries in dynamical systems*, Physica D **73** (1994) 277–304.

[LR85] P. A. LEE and T. V. RAMAKRISHNAN, *Disordered electronic systems*, Rev. Mod. Phys. **57** (1985) 287–337.

[LS87] D. W. LONGCOPE and R. N. SUDAN, *Arnol'd diffusion in $1\frac{1}{2}$ dimensions*, Phys. Rev. Lett. **59** (1987) 1500–1503.

[LW89] A. J. LICHTENBERG and B. P. WOOD, *Diffusion through a stochastic web*, Phys. Rev. A **39** (1989) 2153–2159.

[Mad78] O. MADELUNG, *Introduction to Solid-State-Theory*, vol. 2 of *Springer Series in Solid-State Sciences*, Springer-Verlag, Berlin (1978).

[Mar97] P. Marian, *Second-order squeezed states*, Phys. Rev. A **55** (1997) 3051–3058.

[MCK93] H. Moya-Cessa and P. L. Knight, *Series representation of quantum-field quasiprobabilities*, Phys. Rev. A **48** (1993) 2479–2481.

[Meh77] C. L. Mehta, *Ordering of the exponential of a quadratic in boson operators. I. Single mode case*, J. Math. Phys. **18** (1977) 404–407.

[Mes91] A. Messiah, *Quantenmechanik I*, de Gruyter, Berlin, 2nd edition (1991).

[MM71] A. Malkin and V. I. Man'ko, *Coherent states and Green's function of a charged particle in variable electric and magnetic fields*, Sov. Phys. JETP **32** (1971) 949–953.

[MMT97] S. Mancini, V. I. Man'ko and P. Tombesi, *Beyond the standard "marginalizations" of Wigner function*, e-Print: quant-ph/9707018 (1997).

[MRB+95] F. L. Moore, J. C. Robinson, C. F. Bharucha, B. Sudaram and M. G. Raizen, *Atom optics realization of the quantum δ-kicked rotor*, Phys. Rev. Lett. **75** (1995) 4598–4601.

[MS99] P. A. Miller and S. Sarkar, *Entropy production, dynamical localization and criteria for quantum chaos in the open quantum kicked rotor*, Nonlinearity **12** (1999) 419–442.

[Nie97a] M. M. Nieto, *The discovery of squeezed states — in 1927*, e-Print: quant-ph/9708012 (1997).

[Nie97b] M. M. Nieto, *Towards even and odd squeezed number states*, e-Print: quant-ph/9711015 (1997).

[Nol02] W. Nolting, *Quantenmechanik. Methoden und Anwendungen*, vol. 5/2 of *Grundkurs Theoretische Physik*, Springer-Verlag, Berlin, 4th edition (2002).

[Olv74] F. W. J. Olver, *Asymptotics and Special Functions*, Computer Science and Applied Mathematics. Academic Press, New York (1974).

[Ote91] J. A. Oteo, *The Baker-Campbell-Hausdorff formula and nested commutator identities*, J. Math. Phys. **32** (1991) 419–424.

[Pen94] J. B. Pendry, *Symmetry and transport of waves in one-dimensional disordered systems*, Adv. Phys. **43** (1994) 461–542.

[Per93] A. PERES, *Quantum Theory: Concepts and Methods*, vol. 57 of *Fundamental Theories of Physics*, Kluwer Academic Publishers, Dordrecht (1993).

[PGF83] R. E. PRANGE, D. R. GREMPEL and S. FISHMAN, *Wave functions at a mobility edge: An example of a singular continuous spectrum*, Phys. Rev. B **28** (1983) 7370–7372.

[PGF85] R. E. PRANGE, D. R. GREMPEL and S. FISHMAN, *Quantum chaos and Anderson localization*, in: G. CASATI (ed.), *Chaotic Behavior in Quantum Systems. Theory and Applications*, vol. 120 of *NATO ASI Series B: Physics*, 205–216. Plenum Press, New York (1985).

[PTVF94] W. H. PRESS, S. A. TEUKOLSKY, W. T. VETTERLING and B. P. FLANNERY, *Numerical Recipes in C. The Art of Scientific Computing*, Cambridge University Press, Cambridge, 2nd edition (1994).

[Rei98] C. REICH, *Analyse einer zweidimensionalen Kick-Abbildung*, Diploma thesis, Westfälische Wilhelms-Universität Münster (1998).

[Rih68] A. W. RIHACZEK, *Signal energy distribution in time and frequency*, IEEE Trans. Inf. Theory **14** (1968) 369–374.

[Sal74] W. R. SALZMAN, *Quantum mechanics of systems periodic in time*, Phys. Rev. A **10** (1974) 461–465.

[Sam73] H. SAMBE, *Steady states and quasienergies of a quantum-mechanical system in an oscillating field*, Phys. Rev. A **7** (1973) 2203–2213.

[SB00] J. STOER and R. BULIRSCH, *Numerische Mathematik 2*, Springer-Verlag, Berlin, 4th edition (2000).

[Sch26] E. SCHRÖDINGER, *Der stetige Übergang von der Mikro- zur Makromechanik*, Naturwissenschaften **14** (1926) 664–666.

[Sch81] L. S. SCHULMAN, *Techniques and Applications of Path Integration*, Wiley-Interscience, New York (1981).

[Sch89] H. G. SCHUSTER, *Deterministic Chaos*, VCH Verlagsgesellschaft, Weinheim, 2nd edition (1989).

[Sch93] T. SCHRIDDE, *Chaos und Ordnung in einem zentralsymmetrischen Kickpotential*, Diploma thesis, Westfälische Wilhelms-Universität Münster (1993).

[Sch02] F. SCHWABL, *Quantenmechanik*, Springer-Verlag, Berlin, 6th edition (2002).

[See95] B. SEEGERS, *Quantenchaos in eindimensionalen Stoßkomplexen*, PhD thesis, Westfälische Wilhelms-Universität Münster (1995).

[SHM00] A. J. SCOTT, C. A. HOLMES and G. J. MILBURN, *Quantum and classical chaos for a single trapped ion*, Phys. Rev. A **61** (2000) 013401.1–013401.7.

[SJM92] G. SCHMERA, P. JUNG and F. MOSS, *Diffusion on the chaotic web of a Hamiltonian oscillator with incommensurate forcing*, Phys. Rev. A **45** (1992) 5462–5468.

[SK91] M. SCHWÄGERL and J. KRUG, *Subdiffusive transport in stochastic webs*, Physica D **52** (1991) 143–156.

[Smi85] G. D. SMITH, *Numerical Solution of Partial Differential Equations: Finite Difference Methods*, Oxford Applied Mathematics and Computing Science Series. Oxford University Press, Oxford, 3rd edition (1985).

[SPM99] R. SALA, J. P. PALAO and J. G. MUGA, *Phase space formalisms of quantum mechanics with singular kernel*, e-Print: quant-ph/9901042 (1999).

[SS92] D. SHEPELYANSKY and C. SIRE, *Quantum evolution in a dynamical quasi-crystal*, Europhys. Lett. **20** (1992) 95–100.

[Stö99] H.-J. STÖCKMANN, *Quantum Chaos. An Introduction*, Cambridge University Press, Cambridge (1999).

[Sto99] J. STOER, *Numerische Mathematik 1*, Springer-Verlag, Berlin, 8th edition (1999).

[Str01a] K. STRAMM, *private communication* (2001).

[Str01b] N. STRAUMANN, *Schrödingers Entdeckung der Wellenmechanik*, e-Print: quant-ph/0110097 (2001).

[Sud63] E. C. G. SUDARSHAN, *Equivalence of semiclassical and quantum mechanical descriptions of statistical light beams*, Phys. Rev. Lett. **10** (1963) 277–279.

[SUZ88] R. Z. SAGDEEV, D. A. USIKOV and G. M. ZASLAVSKY, *Nonlinear Physics. From the Pendulum to Turbulence and Chaos*, vol. V of *Contemporary Concepts in Physics*, Harwood Academic Publishers, Chur (1988).

[SZ94] W. SŁOMCZYŃSKI and K. ŻYCZKOWSKI, *Quantum chaos: An entropy approach*, J. Math. Phys. **35** (1994) 5674–5700.

[Ten83] J. L. TENNYSON, *Resonance streaming in electron-positron colliding beam systems*, in: C. W. HORTON, L. E. REICHL and V. G. SZEBEHELY (eds.), *Long-Time Prediction in Dynamics*, 427–451. Wiley, New York (1983).

[TS85] K. Takahashi and Saitô, *Chaos and Husimi distribution function in quantum mechanics*, Phys. Rev. Lett. **55** (1985) 645–648.

[TV93a] G. Torres-Vega, *Chebyshev scheme for the propagation of quantum wave functions in phase space*, J. Chem. Phys. **99** (1993) 1824–1827.

[TV93b] G. Torres-Vega, *Lanczos method for the numerical propagation of quantum densities in phase space with an application to the kicked harmonic oscillator*, J. Chem. Phys. **98** (1993) 7040–7045.

[TVF93] G. Torres-Vega and J. H. Frederick, *A quantum mechanical representation in phase space*, J. Chem. Phys. **98** (1993) 3103–3120.

[TVZ+96] G. Torres-Vega, A. Zúñiga-Segundo and J. D. Morales-Guzmán, *Special functions and quantum mechanics in phase space: Airy functions*, Phys. Rev. A **53** (1996) 3792–3797.

[Vec95] V. V. Vecheslavov, *Instability of weakly nonlinear chaotic structures*, Phys. Rev. E **51** (1995) 5106–5108.

[Vou94] A. Vourdas, *Coherent states on the m-sheeted complex plane*, J. Math. Phys. **35** (1994) 2687–2697.

[VT99] D. Vitali and P. Tombesi, *Using parity kicks for decoherence control*, Phys. Rev. A **59** (1999) 4178–4186.

[Wey31] H. Weyl, *The Theory of Groups and Quantum Mechanics*, Dover, New York (1931).

[Wey82] H. Weyl, *Symmetry*, Princeton University Press, Princeton (1982).

[Wig32] E. P. Wigner, *On the quantum correction for thermodynamic equilibrium*, Phys. Rev. **40** (1932) 749–759.

[Wil67] R. M. Wilcox, *Exponential operators and parameter differentiation in quantum physics*, J. Math. Phys. **8** (1967) 962–982.

[Wir99] A. Wirzba, *Quantum Mechanics and Semiclassics of Hyperbolic n-Disk Scattering Systems*, Phys. Rep. **309** (1999) 1–116.

[WK93] F.-B. Wang and L.-M. Kuang, *Even and odd q-coherent states and their optical statistics properties*, J. Phys. A: Math. Gen. **26** (1993) 293–300.

[YMS90] J. A. Yeazell, M. Mallalieu and C. R. Stroud Jr., *Observation of the collapse and revival of a Rydberg electronic wave packet*, Phys. Rev. Lett. **64** (1990) 2007–2010.

[YP92] L. Y. YU and R. H. PARMENTER, *Some new systems that generate a uniform stochastic web*, Chaos **2** (1992) 581–588.

[YS75] V. A. YAKUBOVICH and V. M. STARZHINSKII, *Linear Differential Equations with Periodic Coefficients*, vol. 1, John Wiley & Sons, New York (1975).

[Yue76] H. P. YUEN, *Two-photon coherent states of the radiation field*, Phys. Rev. A **13** (1976) 2226–2243.

[Zas85] G. M. ZASLAVSKY, *Chaos in Dynamic Systems*, Harwood Academic Publishers, Chur (1985).

[Zas91] G. M. ZASLAVSKY, *Stochastic webs and their applications*, Chaos **1** (1991) 1–12.

[Zel67] Y. B. ZEL'DOVICH, *The quasienergy of a quantum-mechanical system subjected to a periodic action*, Sov. Phys. JETP **24** (1967) 1006–1008.

[Zim79] J. M. ZIMAN, *Models of Disorder. The Theoretical Physics of Homogeneously Disordered Systems*, Cambridge University Press, Cambridge (1979).

[ZK94] J.-Y. ZHU and L.-M. KUANG, *Even and odd coherent states of a harmonic oscillator in a finite-dimensional Hilbert space and their squeezing properties*, Phys. Lett. A **193** (1994) 227–234.

[ZSUC88] G. M. ZASLAVSKY, R. Z. SAGDEEV, D. A. USIKOV and A. A. CHERNIKOV, *Minimal chaos, stochastic webs, and structures of quasicrystal symmetry*, Sov. Phys. Usp. **31** (1988) 887–915.

[ZSUC91] G. M. ZASLAVSKY, R. Z. SAGDEEV, D. A. USIKOV and A. A. CHERNIKOV, *Weak Chaos and Quasi-Regular Patterns*, vol. 1 of *Cambridge Nonlinear Science Series*, Cambridge University Press, Cambridge (1991).

[Zum97] T. ZUMKLEY, *Diffusion in Aluminium-(Si,Ge) Mischkristallen sowie in ikosaedrischen AlPdMn-Quasikristallen*, PhD thesis, Westfälische Wilhelms-Universität Münster (1997).

[ZZN+89] G. M. ZASLAVSKY, M. Y. ZAKHAROV, A. I. NEISHTADT, R. Z. SAGDEEV and D. A. USIKOV, *Multidimensional Hamiltonian chaos*, Sov. Phys. JETP **69** (1989) 885–897.

Danksagung

Zum Schluß möchte ich mich bei allen bedanken, die mich bei der Anfertigung dieser Arbeit unterstützt haben.

Mein besonderer Dank gilt Prof. Dr. Peter Eckelt. Er gab nicht nur die Anregung zu dieser Arbeit. Er unterstützte mich auch während ihrer Fertigstellung in vielfacher Weise und gewährleistete eine hervorragende, kritische Betreuung.

Meiner Arbeitsgruppe und vor allem meinen Zimmerkollegen im Institut für Theoretische Physik bin ich für einiges dankbar: Prof. Dr. Berthold Stegemerten und ganz besonders Dr. Arne Beeker halfen beim Korrekturlesen des Typoskriptes. Beide sorgten überdies für eine anregende, humorvolle und angenehme Atmosphäre in Zimmer 421. Das gilt gleichermaßen für Dr. Constanze Hippel, der ich auch für viele gute Gespräche danke, die oft weit über physikalische Dinge hinausgingen. Sie trug wesentlich zur Überwindung mancher verfahrenen Situation bei. Dr. Björn Seegers danke ich für seine Anmerkungen als HUSIMI-Spezialist, Dr. Markus Hauke und Dr. Bodo Junglas für etliche Ratschläge in Computer-Fragen und Achim Schulte für viele Unterhaltungen über fachliche und weniger fachliche Dinge.

Dr. Klaus Stramm danke ich für diverse nützliche mathematische Hinweise, Paula Stramm für interessante Diskussionen über klassische Erhaltungsgrößen.

Maria Shinoto danke ich für 伏見康治 und einiges mehr, ebenso wie Dr. Claudia Schwalfenberg.

Dem Institut für Geometrie und Praktische Mathematik der RWTH Aachen, insbesondere Frank Knoben, danke ich dafür, daß ich einen großen Teil der umfangreichen numerischen Berechnungen auf dem dortigen Linux-Cluster durchführen konnte. Ebenso danke ich Gero Bornefeld und Tim Pähler für die freundliche Bereitstellung weiterer Rechenkapazität.

Schließlich danke ich meinen Eltern, die mir mein Studium ermöglicht haben, ferner meiner Familie und meinen Freunden für all die Unterstützung, die ich in diesen Jahren von ihnen erfahren habe, und für die Aufmunterung in schwierigen Phasen dieser Arbeit.